Fundamentals of
Automobile Body Structure Design

Other SAE books of interest

The Passenger Car Body: Design, Deformation Behavior, Accident Repairs
by Dieter Anselm
(Product Code: R-307)

Advanced Vehicle Technology, Second Edition
by Heinz Heisler
(Product Code: R-337)

Handbook of Automotive Engineering
Edited by Herman Braess and Ulrich Seiffert
(Product Code: R-312)

For more information or to order a book, contact SAE International at
400 Commonwealth Drive, Warrendale, PA 15096-0001 USA;
phone 877-606-7323 (U.S. and Canada only) or 724-776-4970
(outside U.S. and Canada);
fax 724-776-0790;
email CustomerService@sae.org;
website: http://books.sae.org

Fundamentals of
Automobile Body Structure Design

By Donald E. Malen

SAE International®
Warrendale, Pennsylvania
USA

400 Commonwealth Drive
Warrendale, PA 15096-0001 USA

E-mail: CustomerService@sae.org
Phone: 877-606-7323 (inside USA and Canada)
724-776-4970 (outside USA)
Fax: 724-776-1615

ISBN 978-0-7680-2169-1
SAE Order No. R-394

Library of Congress Cataloging-in-Publication Data

Malen, Donald E.
Fundamentals of automobile body structure design / by Donald E. Malen.
p. cm.
"SAE order no. R-394."
ISBN 978–0–7680–2169–1
1. Automobiles—Bodies—Design and construction. I. Title.
TL255.M255 2011
629.2´6—dc22   2010041059

Information contained in this work has been obtained by SAE International from sources believed to be reliable. However, neither SAE International nor its authors guarantee the accuracy or completeness of any information published herein and neither SAE International nor its authors shall be responsible for any errors, omissions, or damages arising out of use of this information. This work is published with the understanding that SAE International and its authors are supplying information, but are not attempting to render engineering or other professional services. If such services are required, the assistance of an appropriate professional should be sought.

To purchase bulk quantities, please contact:
SAE Customer Service
E-mail: CustomerService@sae.org
Phone: 877-606-7323 *(inside USA and Canada)*
724-776-4970 *(outside USA)*
Fax: 724-776-1615

Visit the SAE Bookstore at
http://store.sae.org

Table of Contents

Table of Contents cont.

Dedication

This book is dedicated to my wife Char, who shared in the long hours required to bring this book to fruition.

Acknowledgments

The author is indebted to several individuals who significantly influenced the trajectory of his professional life. First, a very special thanks to Professor Noboru Kikuchi of the University of Michigan, Department of Mechanical Engineering, for his guidance on my new career as an adjunct faculty member, and his collaboration in developing the course *ME513-Fundamentals of Auto Body Structure*, on which this book is based. His contributions and suggestions were invaluable. Mr. William Elliott encouraged me to pursue the discipline of automobile body engineering as a new engineer at General Motors. His wise and steady mentoring at the onset of my career was greatly appreciated. Mr. Klaus Winkelmann of GM took a great risk by sending an engineer at mid-career back to school, and I am grateful for that opportunity. Finally, deep gratitude is extended to my colleagues in the Advance Product Engineering Structures Group at General Motors. It was an honor to work with this competent and creative group.

Thanks to Professor Raghu Echempati of Kettering University, who offered many valuable comments on early drafts of this manuscript, and to Martha Swiss of SAE International, who patiently guided me through the editing process.

Appreciation is also extended to several organizations who have given permission to use their materials in this text. These include the American Iron and Steel Institute, General Motors Corporation, the Motor Industry Research Association, and the SAE International.

Portions of materials contained herein have been reprinted with the permission of General Motors Corporation, Service and Parts Operations, License Agreement #0910987.

Preface

The objective of this book is to give the reader an understanding of the behavior of auto body structural elements, and to provide insight into how changing design parameters will influence this behavior. We seek to answer questions such as: What is the best way to shape a structural element to achieve a desired function? What is the behavior of the structure relative to required performance? Why does it behave in this way? How can the performance be improved? If we have the ability to answer these questions without resorting to complex mathematical models, we have *structural intuition*; this is particularly relevant today when bringing new products with unique configurations to market quickly is so important.

This book emphasizes simple models of a structure's physical behavior—first-order models. Students are often surprised that the first-order models they have encountered earlier in courses in Mechanics, Strength of Materials, and Vibration may be applied so successfully to a complex structure like the auto body. The benefit of these first-order models lays in the insight the student develops.

In addition to physical testing, finite element analysis (FEA) is today a reliable and precise means to determine structural performance. Both these tools are important during design, but each has limitations. Both testing and FEA require a complete definition of structure shape. Often this complete geometrical definition does not exist during preliminary design. Both testing and FEA require considerable time for construction of the model and for evaluation. Often decisions on design alternatives must be made very quickly, and testing or FEA just takes too long. Finally, both testing and FEA can tell us very precisely <u>how</u> a structure will behave, but not <u>why</u> it behaves as it does or how we may improve it. We need to understand the *whys* in order to create the structure in the first place.

The intent of this book is to supplement these more precise structural design tools with first-order models and structural intuition. First-order models enable *instant analysis* for the what-if questions so important during early design. First-order models can evaluate structural concepts with incomplete geometrical definition. First-order models provide a sense check on the results of a physical test or FEA. First-order models help the designer identify potential problems in a design, which then requires a more precise analysis. In this, we are following the advice of Robert Cook and Warren Young given in their book, *Advanced Mechanics of Materials*:

> *Structural design seldom requires sophisticated mathematics or a powerful computer. Rather, it depends on the ability of the designer to gain clear insight into the phenomena, to identify simplifying assumptions, and to apply straightforward methods to the simplified problem.*

In this book we intend to *develop a sense of the relevant:* What are the most relevant requirements; the most relevant structural behavior; the most relevant first-order models to describe this behavior; and the most relevant interactions with other vehicle subsystems. In doing so, we hope to contribute to the development of skilled structural designers.

Disclaimer: The first-order models in this book are intended for preliminary layout and sizing of automotive structure. Actual structural performance should always be validated by more detailed analysis and testing.

Chapter 1
The Automobile Body

The automobile body, Figure 1.1, is an important vehicle subsystem that performs many functions [1]. These range from the very basic function of being the armature holding the parts of the vehicle together, to the function of noise and vibration refinement which differentiates a luxury vehicle from an economy vehicle. In this chapter, we briefly describe the contemporary body structure and terminology.

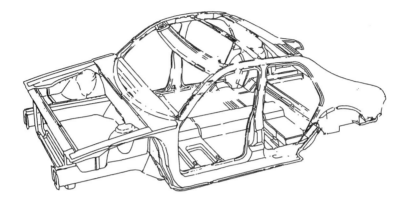

Figure 1.1 Typical body-in-white. (Courtesy of General Motors Corporation)

A typical body is an assembly of metal stampings, Figure 1.2, usually of steel but also of aluminum. Several material grades or alloys are used to meet the structural requirements at the formability needed to achieve the part shape, Figure 1.3 [2]. The stampings are assembled to form thin-walled structural elements, Figure 1.4. The general arrangement of these structural elements leads to several different body types.

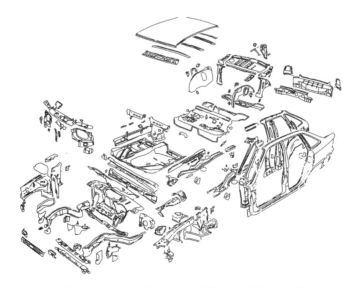

Figure 1.2 Body stampings. (Courtesy of General Motors Corporation)

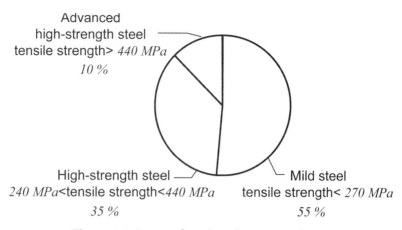

Figure 1.3 Range of steel grades in typical car.

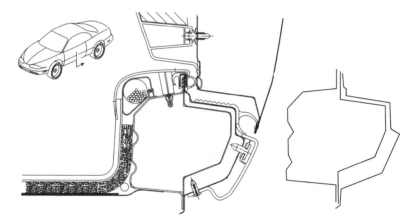

Figure 1.4 Typical section at rocker. (Courtesy of General Motors Corporation)

1.1 Description of the Automobile Body Types

There are several types of body structure configurations today. The predominant types are space frame, central frame, body-on-frame, and monocoque (integral body-frame), Figure 1.5.

The space frame configuration, Figure 1.5a, is characterized by a three-dimensional framework of beams connected at nodes. The framework provides the structural integrity with the exterior panels being unstressed. The space frame type can be fabricated using lower-cost tooling such as roll forming or hydroforming, and is usually targeted for lower-volume vehicles.

The central frame configuration, Figure 1.5b, is characterized by a large, closed structural member down the center of the vehicle. This member provides the structural integrity for this type. Because of the intrusion into the cabin, this arrangement is limited to two- or four-seat interior configurations.

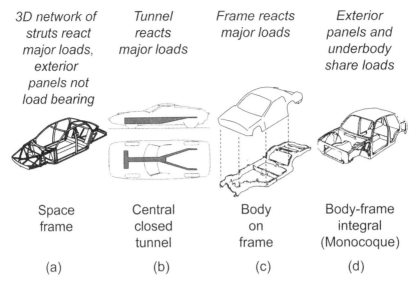

3D network of struts react major loads, exterior panels not load bearing	Tunnel reacts major loads	Frame reacts major loads	Exterior panels and underbody share loads
Space frame	Central closed tunnel	Body on frame	Body-frame integral (Monocoque)
(a)	(b)	(c)	(d)

Figure 1.5 Body configurations. (Courtesy of General Motors Corporation)

The body-on-frame configuration, Figure 1.5c, was the predominant passenger car type until the 1980s, and remains the predominant type for light trucks. This type is characterized by a ladder frame to which the suspensions and powertrain are attached, and a body shell which is connected to the frame by flexible body mounts.

The monocoque configuration, Figure 1.5d, is characterized by an integral structure which forms a shell, including exterior panels. The integral structure reacts to all major loads. This is currently the predominant type for passenger cars, and is considered the most mass-efficient configuration. Because of its predominance, this book will focus on the monocoque configuration; however, the principles we will develop are also applicable to the other configurations as well.

Within the monocoque configuration, there are topology variants. *Topology* is the arrangement of structural elements—beams and panels—to meet requirements in the most efficient manner. Besides structural concerns, an effective body topology also satisfies the additional constraints which the package, styling, and manufacturing place on the positioning and size of structural elements. Figure 1.6 shows three of the most common monocoque topologies in use today [3]. In later chapters, we will discuss the fundamentals of determining a body topology.

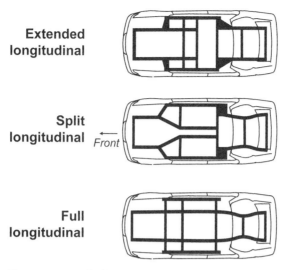

Extended longitudinal

Split longitudinal *Front*

Full longitudinal

Figure 1.6 Body-frame integral typical topologies.

1.2 Body Nomenclature

While a uniform naming convention for body structure elements does not exist, there are some more common terms in use. Figures 1.7 through 1.10 show common names for the major structural elements with respect to the overall body. In Figures 1.11 through 1.16, part names are shown in a useful hierarchical format. This hierarchy follows a typical manufacturing partition of the body structure.

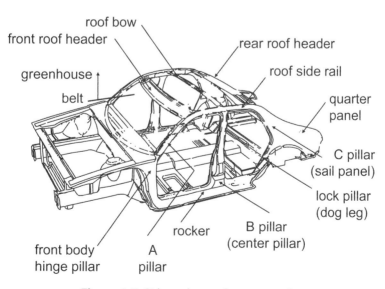

Figure 1.7 Side and greenhouse members.

5

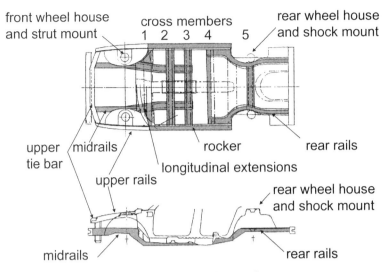

Figure 1.8 Underbody members.

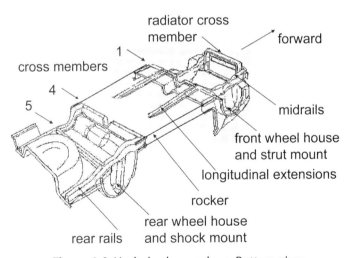

Figure 1.9 Underbody members: Bottom view.

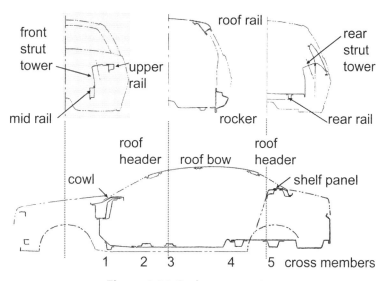

Figure 1.10 Body cross sections.

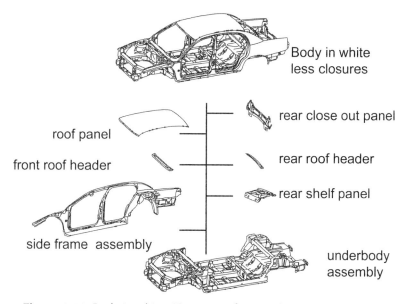

Figure 1.11 Body in white. (Courtesy of General Motors Corporation)

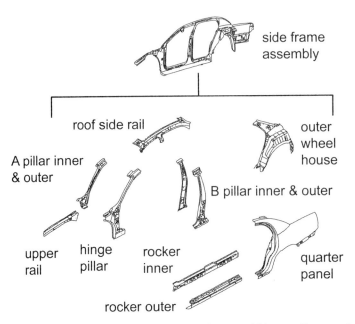

Figure 1.12 Side frame. (Courtesy of General Motors Corporation)

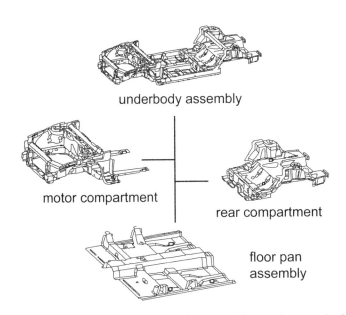

Figure 1.13 Underbody. (Courtesy of General Motors Corporation)

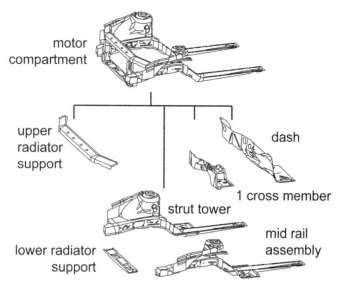

Figure 1.14 Motor compartment. (Courtesy of General Motors Corporation)

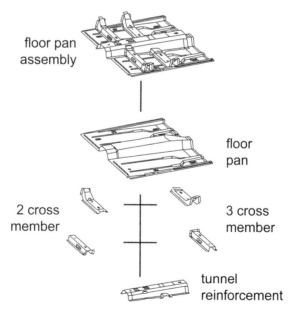

Figure 1.15 Floor pan assembly. (Courtesy of General Motors Corporation)

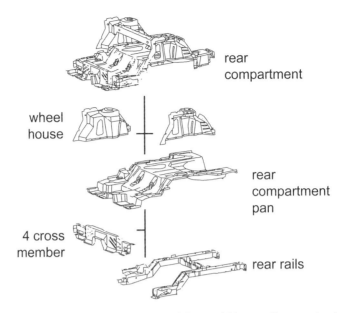

Figure 1.16 Rear compartment. (Courtesy of General Motors Corporation)

1.3 Body Mass Benchmarking

The body structure mass is a significant portion of the vehicle mass, and because of this, the body influences mass-sensitive vehicle functions such as fuel economy, acceleration performance, and handling. Looking at the body shell mass (no trim, glass, closures, or bolt-on panels) for several sedans, Figure 1.17, the average mass is approximately 325 kg (715 lb) [4]. A more useful way to look at body mass is in comparison with the other vehicle subsystems. Figure 1.18 shows this breakdown for a typical mid-size vehicle having integral body and front wheel drive.

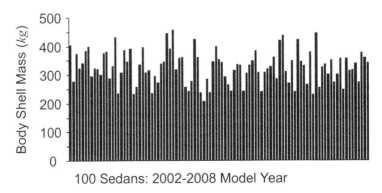

Figure 1.17 Body shell mass.

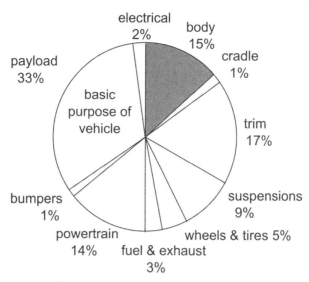

Figure 1.18 Vehicle mass breakdown.

As we are often interested in structural efficiency, it is useful to look at the body structure mass normalized by the gross vehicle mass [5]. This tells us how much structure mass is required to provide support for a unit mass of the vehicle. Figure 1.19 shows this useful ratio for several vehicles as well as other man-made and natural systems.

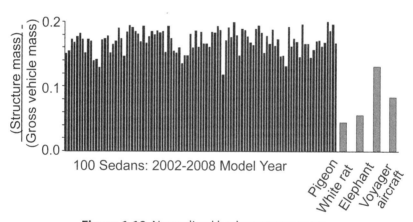

Figure 1.19 Normalized body structure mass.

1.4 The Body Structure as a System

Throughout this book, we will use the systems engineering approach to treat automobile body design. A system may be described in two ways: The first is by a breakdown of its physical parts or subsystems. The second is by examining the functions the system must provide. For example, in Section 1.2, we described the body using a breakdown by parts. In the next chapter on requirements, we will describe the body system by looking at the functions it must perform within the larger vehicle system (the systems engineering approach).

1.5 Note on Design Philosophy

The concepts and methods discussed in this book are directed at the *preliminary design stage* for automobile body structure. During preliminary design, the vehicle mission is translated into a vehicle concept including the selection of subsystems, layout, and styling to meet the mission. In the absence of a preexisting design, the body structure is synthesized from "a blank sheet of paper" during preliminary design. Frequently, design data during this phase are fuzzy and changing. Also, the vehicle design is rapidly evolving, and design decisions must be made rapidly, often within a turn-around time of hours. This lack of precise design data, and need for very rapid decision making, renders finite element analysis difficult to apply during preliminary design. Therefore, the structure designer relies heavily on *physical intuition* and *first-order analysis models* in defining load paths, sizing structural elements, and making design trade-offs. Having created a solid preliminary design, the number and severity of problems passed along to the detail design phase is reduced, allowing the detail design phase to proceed more rapidly. During the later detail design phase, application of physical intuition is again valuable in interpreting and sense-checking finite element analysis and test results.

The goal of this book is to help the reader develop this physical intuition for body structure layout, particularly developing the abilities to:

- Identify the small set of topology-defining structural requirements
- Gain an intuitive feel for thin-walled structure behavior
- Develop simple analytical models—first-order models—to approximate structure sizing
- Gain an appreciation for the vehicle and manufacturing context of body design and the common trade-off issues which must be balanced

References

1. Kikuchi, N. and Malen, D., Course Notes for ME513: Fundamentals of Auto Body Structures, University of Michigan, Ann Arbor, MI, 2007.

2. Hall, J., "50 Year Perspective of Automotive Engineering Body Materials and an Analysis of the Future," Great Designs in Steel 2008 Conference, American Iron and Steel Institute, Southfield, MI, 2008.

3. Fenton, J., *Vehicle Body Layout and Analysis*, Mechanical Engineering Publications, Ltd, London, 1980, p. 12.

4. A2Mac1.com Automotive Benchmarking, Auto Reverse Database, www.a2mac1.com, Ypsilanti, MI, 2009.

5. Malen, D., "Preliminary Vehicle Mass Estimation Using Empirical Subsystem Influence Coefficients," American Iron and Steel Institute, Southfield, MI, April 2007.

Chapter 2
Body Structural Requirements

Throughout this book we will use a systems engineering approach, Figure 2.1, to design body structure [1, 2, 3]. That is, we will first consider the function of a particular structure element. From the function we will identify structural requirements which the structure element must meet to fulfill its purpose. Then, based on the requirements, a specific structure concept will be synthesized. It is the intent that the analyses and examples provided in this book will give the reader considerable physical insight in efficiently synthesizing these preliminary design concepts. Finally, because our insight is not complete, we must evaluate the performance of our design concepts by test or by analysis, and compare the performance to the original requirement. Often, a gap between performance and requirement will be the case. When this occurs, corrections and refinements to the design concept are needed, and the process becomes iterative.

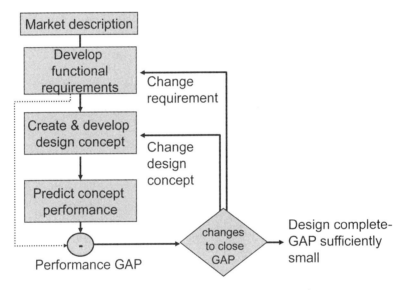

Figure 2.1 Systems engineering approach.

Requirements are then important as they are used to form the initial design concept, and then are used to guide the development and refinement of the concept. In this chapter we will consider a general approach to develop automotive body structural requirements.

2.1 Categories of Structural Requirements

Before looking at the specific case of automotive body structure, we will first consider the function of any general structure. At a basic level, the purpose of any structure is to:

1. React *loads* applied to it during use
2. React those loads with an amount of *deformation* which enhances (or does not restrict) the function of the system which the structure is part of.

Therefore, in considering structural requirements, we are dealing with *loads* and *deformations*. A helpful way to look at the reaction of a structure to loading and the resulting deformation is with the tensile test, Figure 2.2. In a typical test, the structure rests on the stationary bed and is loaded by a moving platen. As the platen imposes a displacement on the structure, the resulting reaction load is measured at the bed. Moving the platen through a displacement range and measuring the load reaction results in a load/deflection curve for the structure.

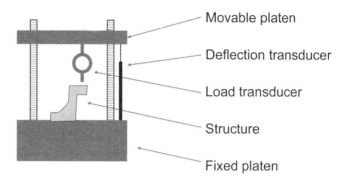

Figure 2.2 Tensile test machine.

A typical load-deflection curve, Figure 2.3, shows several regions; initially there is a linear region which is elastic—when the load is removed the deformation returns to zero. With increased load, some structures exhibit a nonlinear but elastic region. At some load, the structure will become inelastic; that is, when the load is removed there will remain some small permanent deformation. Continuing to increase deflection, the load will reach a maximum or ultimate load. Finally, at some maximum deformation, the structure will experience a catastrophic failure which can include a fracture (load goes to zero) or a bottoming out (load becomes very large).

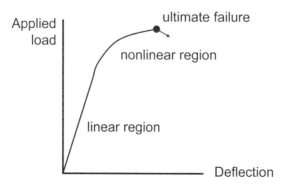

Figure 2.3 Typical load-deflection curve.

Earlier we said the purpose of a structure is to react to loads with an allowable deformation. It is useful to categorize structural requirements based on allowable deformation. What constitutes the particular allowable deformation depends on the function being satisfied by the structural element; the element's load-deflection curve helps categorization [4].

In the *stiffness category*, the deformations are small and elastic, and are characterized by stiffness—the ratio of applied load per unit of deformation, Figure 2.4. An example of an automotive structural requirement in the stiffness category is body stiffness at a suspension attachment point. The handling properties of the vehicle depend on the small elastic deformations relative to the vehicle center of mass. Another example is the perception of solidness of a panel, which is related to the panel's normal stiffness under a point load. The stiffness category is closely related to vibration performance of structural elements, as will be discussed in the chapter on vibration.

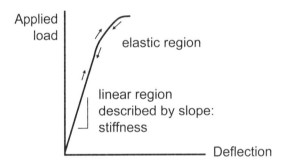

Figure 2.4 Stiffness behavior.

The *strength category* is characterized by the onset of a small permanent deformation, Figure 2.5. Here the requirement is stated as the lowest load at which a permanent deformation first appears. Physically this involves loading and unloading the structure with increasing levels until the structure does not return to its original shape. The specific degree of permanent deformation allowed depends on the loss of some functionality. For example, consider the body strength at a suspension attachment point. During impact with an unusually severe pot hole, a small permanent deformation may occur. Because of the severity, the user may expect a loss of some function such that the suspension must be realigned. The user would not expect a deformation so large the suspension becomes totally nonfunctional. Thus in this case, the strength requirement is the load at which a permanent deformation of a few millimeters occurs.

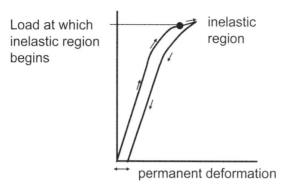

Strength describes the maximum load in extreme applications where some permanent deformation is expected

Load at which inelastic region begins

inelastic region

permanent deformation

Figure 2.5 Strength behavior.

The *energy absorption category* is characterized by very large permanent deformations where the amount of energy absorbed by the structure during deformation is of interest, Figure 2.6. The requirements in this category can be stated in terms of the area under the load-deformation curve, or as an equivalent square wave load having the same contained area. An example of an energy absorption requirement is the average crush force generated by the motor compartment midrail during a front collision. While all structural elements will have stiffness and strength requirements, only a subset will have energy absorption requirements.

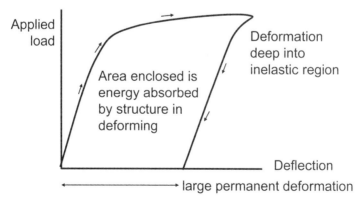

Energy absorbed by the structure describes load and deformation performance in severe impacts

Applied load

Deformation deep into inelastic region

Area enclosed is energy absorbed by structure in deforming

Deflection

large permanent deformation

Figure 2.6 Energy absorption behavior.

The task of the automobile body designer is to create a structure with these three qualities of strength, stiffness, and energy absorption at the required levels. In the sections following, we begin to look at the sources of those structural requirements.

2.2 The Locate and Retain Function

We can imagine all the pieces that comprise the vehicle initially in a pile. In order to function as a vehicle, all these pieces must be located in a specific orientation and retained in that orientation to form a usable vehicle. This task becomes the most basic function of the body: to *locate* each vehicle subsystem and to *retain* it in a way that allows each subsystem to function.

By examining each of the interfaces, Figure 2.7a, between the body structure and the subsystems of the vehicle, we can determine what loads are being applied and what deformations are allowed for proper functioning of each subsystem. These loads and deformations define structural requirements. As we are designing a fleet of vehicles, we must consider the loading of the population of vehicles and set the required capacity not at that of a typical vehicle but for that small fraction of vehicles which will experience an extreme or severe level of loading, Figure 2.7b. Precisely where to set the load requirement between reasonable but extreme loading and abusive loading is subjective.

A four-step process is used to identify requirements for the *locate and retain* function. After selecting a specific subsystem for analysis,

1. Choose a mode-of-use for the subsystem

2. Identify the loads being applied to the subsystem during this mode

3. Identify how much deformation of the structure is allowable without a detrimental effect on functioning for this mode

4. Using free-body analysis, determine the loads applied at the structure interfaces.

The implementation of this process is best understood using some examples.

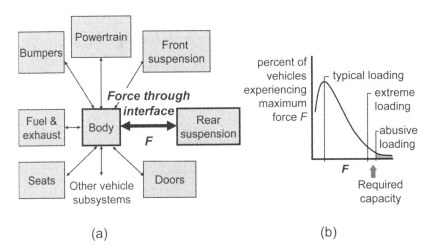

(a)

(b)

Figure 2.7 Body – subsystem interfaces.

Example: Steering column mount structure

The steering column mount structure locates and retains the steering column, Figure 2.8. It performs this function in several modes of use including *entry/exit* and *column vibration*. Below we analyze each of these modes to determine structural requirements.

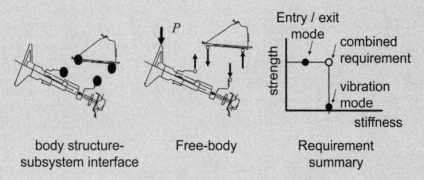

body structure-   Free-body   Requirement
subsystem interface   summary

Figure 2.8 Steering column mount.

1. *Choose a mode-of-use for the subsystem*: To aid entering and exiting the vehicle, the driver often grasps the steering wheel and applies a vertical load.

2. *Identify the loads being applied to the subsystem during this mode*: For a large person, this load is approximately *1500 N*.

3. *Identify how much deformation of the structure is allowable without a detrimental effect on functioning for this mode*: The driver would not be distressed if the column deflects elastically up to *10 mm* under this load (a required stiffness of *150 N/mm*) but would be distressed if a noticeable permanent deformation larger than *1 mm* remained (a required strength of *1500 N*).

4. *Using free-body analysis, determine the loads applied at the structure interfaces*: Figure 2.8 shows a free-body for this case.

So in this mode, we determine a requirement for strength of *1500 N* and a requirement for stiffness of *150 N/mm*. Now consider another mode of use: column vertical vibration. The column mount structure and column form a vibratory system, with the structural mount providing stiffness and the column providing inertia.

1. *Mode of use*: Column vertical vibration.

2. *Load being applied*: The applied loads are very small and are significant only when at the resonance frequency of the mounted column.

3. *Allowable deformation*: It is desirable for this system to have a resonance of at least *35 Hz* to avoid coupling with powertrain excitation at engine idle. Based on the relationship $k=m\omega_n^2$, and with the steering column effective mass of *5 kg*, a mount stiffness requirement of *240 N/mm* results. (Structure vibration will be more fully discussed in a later chapter.)

4. *Loads at structure interfaces*: We can find the required stiffnesses at the structure interfaces using a free body as shown in Figure 2.8.

For these two modes, the *entry/exit mode* provides the dominant strength requirement (*1500 N*), and the *column vibration mode* provides the dominant stiffness requirement (*240 N/mm*), Figure 2.8. To complete development of structural requirements for the steering column mount, other modes would be analyzed including *driver toggle arms, crash impact loading*, etc.

Example: Powertrain mounting structure

The powertrain mounting structure in a front-wheel-drive vehicle locates and retains the engine and transmission, usually at three to four mounting locations on the body structure. Example modes which provide dominant structural requirements include *stall torque* and *vibration isolation*, Figure 2.9.

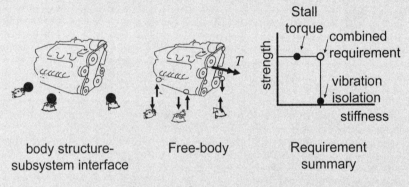

body structure-
subsystem interface   Free-body   Requirement
summary

Figure 2.9 Powertrain mount.

1. *Mode of use:* stall torque.

2. *Load being applied*: During stall torque, the maximum torque the powertrain can generate is applied at the drive shafts and is reacted at the mounting points.

3. *Allowable deformation*: In this strength requirement, a small amount of permanent deformation (approximately *2 mm*) is acceptable.

4. *Loads at structure interfaces*: For a particular mount geometry, the strength requirement at each mount point may be found by static equilibrium of the powertrain, Figure 2.9.

For the vibration isolation mode, we are interested in minimizing the transmission of vibratory forces from the engine.

1. *Mode of use*: vibration isolation.

2. *Load being applied*: The applied loads are very small.

3. *Allowable deformation*: Ensure the ability to tune the mount rate to minimize transmitted engine vibration. This is done when the deflection of the structure is less than *0.1* to *0.14* times the deflection across the mount. This translates into a required stiffness which is *7-10* times that of the elastomeric engine mount rate.

4. *Loads at structure interfaces*: We can find the required stiffnesses at the structure interfaces using a free body as shown in Figure 2.9.

For these two modes, the *stall torque mode* provides the dominant strength requirement and the *vibration isolation mode* provides the dominant stiffness requirement, Figure 2.9. To complete development of structural requirements, other modes would be analyzed such as *engine retention during low-speed impac*t.

2.3 Locate and Retain for Front Suspension Attachment Structure

The front suspension attachment structure, Figure 2.10, locates and retains the suspension at several points: the shock absorber, spring, and control arm. Example modes which provide dominant structural requirements include *braking, cornering, rollover, vertical bump, vibration isolation,* and *handling*. The first four modes define strength requirements, and the last two, stiffness requirements.

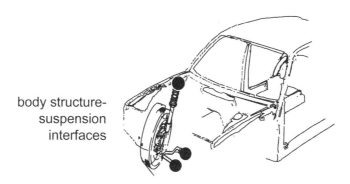

body structure-suspension interfaces

Figure 2.10 Suspension attachment.

1. *Modes of use*: braking, cornering, rollover, and vertical bump.
2. *Loads being applied:* In this analysis, we first look at the maximum loads at the tire patch during these modes, and then using static equilibrium of the suspension, determine the maximum loads (strength requirements) at the body attachment points.

The vehicle at rest is shown in Figure 2.11. Using static equilibrium, the front and rear tire patch loads, R_F and R_R, may be determined as

$$R_F = \frac{b}{a+b}W \quad \text{and} \quad R_R = \frac{a}{a+b}W \tag{2.1}$$

where:

R_F = Vertical load at a front tire patch

R_R = Vertical load at a rear tire patch

W = Vehicle weight

a, b = Dimensions shown in Figure 2.11

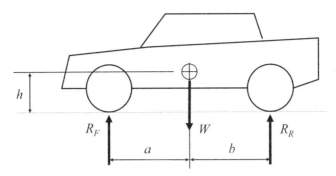

Figure 2.11 Vehicle at rest: Front tire patch load.

Now consider a steady-state braking deceleration of n times the acceleration due to gravity, Figure 2.12. Again using static equilibrium, the fore-aft load at the tire patch is

$$F_F = \frac{b+nh}{a+b} \, \mu \, \frac{W}{2}$$

(2.2)

where:

F_F = Fore-aft load at a front tire patch

n = Braking acceleration in $g's$

μ = Coefficient of friction between tire and road

h = Height of the CG above ground

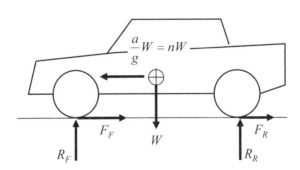

Figure 2.12 Vehicle braking: Front tire patch load.

For the same vehicle we now look at cornering loads at the tire patch, Figure 2.13. The steady-state lateral cornering acceleration is given by n in $g's$. Considering the front axle free body, the lateral load on the outer tire is:

$$L_O = \mu \left(\frac{1}{2} + \frac{hn}{t} \right) \left(\frac{b}{a+b} \frac{W}{2} \right)$$

(2.3)

where:

L_O = Lateral load on the outside front tire patch

W = Vehicle weight

t = Track width

Other variables are listed under Equations 2.1 and 2.2.

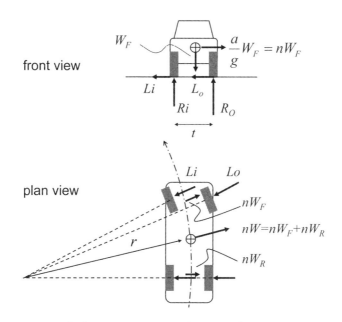

Figure 2.13 Vehicle cornering: Outside corner front tire patch load.

When rollover of the vehicle is incipient during cornering, the steady-state resultant acceleration at the center of gravity (CG) begins to go outside the base of the vehicle track, Figure 2.14. This occurs when the lateral acceleration in $g's$ is $n=t/(2h)$, then the lateral tire patch load at the front axle is

$$L_O = \frac{2h}{t}\left(\frac{b}{a+b}\frac{W}{2}\right)$$

(2.4)

where:

L_O = Lateral load at the tire patch

W = Vehicle weight

h = Height of CG above ground, shown in Figure 2.14

t = Vehicle track

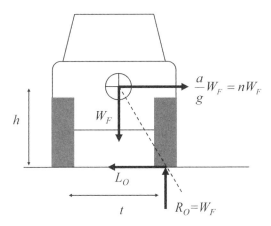

Figure 2.14 Vehicle incipient rollover: Outside corner front tire patch load.

Because the loads we have determined above are for steady-state conditions, a design load factor, r, is often applied to account for dynamic effects [5, 6]. Figure 2.15 summarizes the suggested design load factors for various modes including the vertical bump mode.

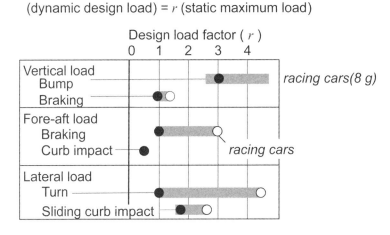

Figure 2.15 Dynamic load factors.

The maximum front tire patch loads for these modes are summarized in Figure 2.16. The steady-state maximum for each of these modes is taken to occur when the braking and cornering acceleration is $n=1$, and the coefficient of friction between tire and road is $\mu=1$. Again looking at Figure 2.16, the maximum lateral, fore-aft, and vertical loads are the greatest considering all modes. For lateral loads this is the rollover mode, for fore-aft it is the braking mode, and for vertical it is the 3-g bump mode.

Mode	Lateral	Fore-aft	Vertical
static	——	——	$R_F = \left(\dfrac{b}{a+b}\right)\dfrac{W}{2}$
braking	——	$F_F = \mu\left(\dfrac{b+nh}{a+b}\right)\dfrac{W}{2}$	$R_F = \left(\dfrac{b+nh}{a+b}\right)\dfrac{W}{2}$
cornering	$L_O = \mu\left(\dfrac{1}{2}+\dfrac{nh}{t}\right)W_F$	——	$R_O = \left(\dfrac{1}{2}+\dfrac{nh}{t}\right)W_F$
incipient rollover	$L_O = \left(\dfrac{2h}{t}\right)W_F$	——	$R_O = W_F$
max acceleration coefficient of friction dynamic load factor	$n =2h/t$ (rollover) $\mu =1$ $r =1$	$n =1$ (1g braking) $\mu=1$ $r =1$	$n =1$ $\mu =1$ $r =3$ (3g bump)
design load per tire (max of all the above)	$L_O = \left(\dfrac{2h}{t}\right)W_F$	$F_F = \left(\dfrac{b+h}{a+b}\right)\dfrac{W}{2}$	$R_F = 3\dfrac{W_F}{2}$

For each design load, allowable deflection is on-set of permanent deformation

Figure 2.16 Summary: Front tire patch loads.

3. *Allowable deformation*: For all modes, the allowable deformation of the structure is that which could be compensated for by a suspension realignment (about *1–2 mm* of permanent deformation).

4. *Loads at structure interfaces*: With these maximum loads identified at the tire patch, we can now use static equilibrium of the suspension to determine the maximum loads at the structure interfaces, Figure 2.17. Here we consider a highly idealized McPherson strut front suspension with a horizontal lower control arm, and we determine the maximum loads at the lower control arm attachment in the vertical, lateral, and fore-aft directions, A_V, A_L, A_{FA}, and similarly for the strut attachment, S_V, S_L, S_{FA}.

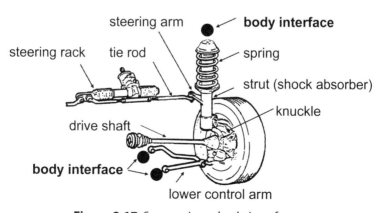

Figure 2.17 Suspension – body interfaces.

For the maximum lateral tire patch load during rollover mode, Figure 2.18:

$$A_L = \left(\frac{r-c}{d}+1\right)L_0$$

$$S_L = \left(\frac{r-c}{d}\right)L_0$$

(2.5)

where:

A_L = Lateral lower control arm load at the body attachment

S_L = Lateral load at the strut attachment

r, c, d = Dimensions shown in Figure 2.18

L_O = Lateral load at the tire patch

Lateral tire patch load, L_0, predominantly
through lower control arm attachment, A_L

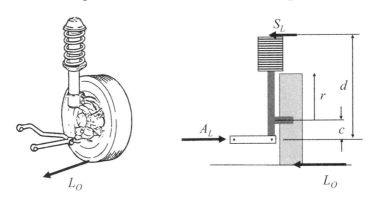

Figure 2.18 Suspension lateral loading.

For the maximum fore-aft tire patch load during braking mode, Figure 2.19a, the loads at the lower control arm ball joint are:

$$A_{FA} = \left(\frac{r-c}{d}+1\right)F_F$$

$$S_{FA} = \left(\frac{r-c}{d}\right)F_F$$

(2.6)

where:

A_{FA} = Fore-aft lower control arm load at ball joint

S_{FA} = Fore-aft load at the strut attachment

r, c, d = Dimensions shown in Figure 2.19a

F_F = Fore-aft load at the tire patch

Fore-aft tire patch load, F_F, predominantly
through lower control arm ball joint, A_{FA}

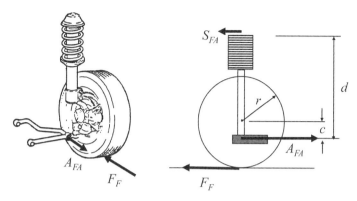

Figure 2.19a Suspension fore-aft loading at ball joint.

These ball joint loads are reacted at the lower control arm front and rear bushings, Figure 2.19b, with the specific breakdown of loads depending on the relative bushing stiffness. Often, one of the bushings is designed to be very stiff and take most of the fore-aft load applied at the ball joint as shown in Figure 2.19b.

Reaction of lower control arm ball joint load,
A_{FA}, at lower control arm attachment

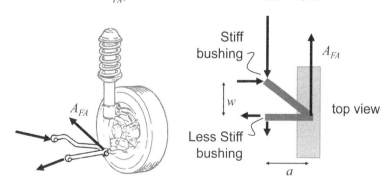

Figure 2.19b Suspension fore-aft loading at lower control arm attachment.

For the maximum vertical tire patch load during bump mode, Figure 2.20:

$$A_V = 0$$
$$S_V = R_F$$

(2.7)

where:

A_V = Vertical lower control arm load at body attachment

S_V = Vertical load at the strut attachment

R_F = Vertical load at the tire patch

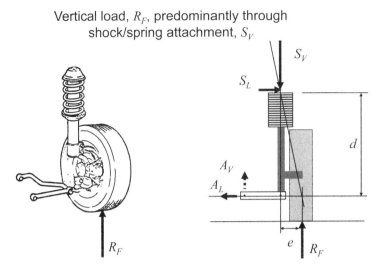

Figure 2.20 Suspension vertical bump load.

Numerical values for these loads become the strength requirements for the respective body structure interfaces. Note that the above maximum load requirements for body structure are for the specific suspension type shown in Figure 2.17. Other types will have different maximum loads applied to the body given the same tire patch loads. For example, an alternative suspension geometry is the Short-and-Long-Arm (SLA) suspension shown in Figure 2.21.

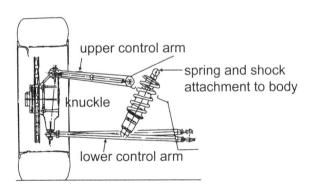

Figure 2.21 Short-and-long-arm suspension.

Looking, for example, at the maximum vertical load at the spring-shock attachment during a bump, we find, Figure 2.22:

$$A_V = \left(\frac{1}{\lambda} - 1\right)R_F$$

$$S_V = \frac{1}{\lambda}R_F$$

(2.8)

where:

A_V = Vertical lower control arm load at body attachment

S_V = Vertical load at the spring/shock attachment

λ = Lever ratio for the spring/shock attachment to lower control arm

R_F = Vertical load at the tire patch

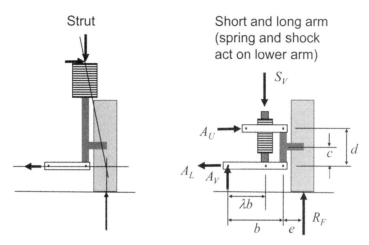

Figure 2.22 Vertical bump load for SLA suspension.

Thus, for a typical SLA lower control arm with lever ratio of $\lambda=0.7$, the spring-shock maximum loads are 1.4 times that of a McPherson strut, Figure 2.23.

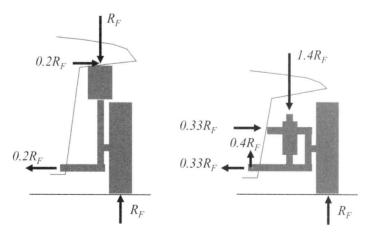

Figure 2.23 Comparison of loads: Strut and SLA suspensions.

The above analysis resulted in strength requirements at the suspension interfaces. For stiffness requirements we look to the *vibration isolation mode*. We would like seven to ten times the suspension bushing stiffness at each structure interface. With structure stiffness at this ratio, effective tuning of the bushing is enabled and also effective utilization of the bushing's damping qualities, as will be developed in the chapter on vibration.

Interfaces for other vehicle subsystems may be analyzed in a similar way to provide the requirements for the local structure at each subsystem interface. Figure 2.24 shows a sampling of interface requirements for a particular vehicle. Each point represents the dominant stiffness and strength requirement for an interface. Shown in this way, those interfaces which are stiffness dominant (upper left portion) and those that are strength dominant (lower right) can be seen. This distinction will be used later to select the most efficient structure to meet requirements.

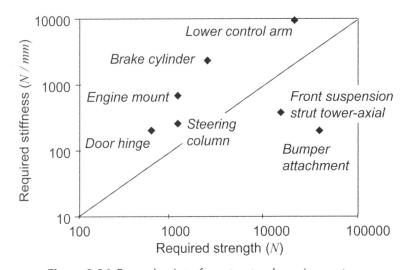

Figure 2.24 Examples: Interface structural requirements.

Although demanding some insight in identifying the critical modes-of-use, this process is straightforward. One challenge is deciding what modes are within reasonable customer usage, and which are clearly abusive and beyond the function of the product. Determining structural requirements also calls for considerable insight into which particular modes-of-use will result in the dominant requirement set for an interface. Here, by dominant, we mean the relatively small set of requirements which are the most stringent and will define the structure configuration and size. For preliminary design, the challenge is to identify the precious few defining requirements rather than creating an exhaustive list. Generally, the interfaces which define body load paths are those of the suspension and powertrain; influence of other interfaces is localized.

Although *locate and retain* addresses a very basic and important function of the body structure, there are other global structural requirements which must be met for a refined vehicle. To define these additional requirements we need to look at overall vehicle functions and see what role the body structure must play for those functions.

2.4 Flow Down of Requirements from Vehicle-Level Functions

A foundational principle of systems analysis can be stated as follows: to understand a system, one must look at the role it plays in the larger system which contains it, Figure 2.25 [7]. The system we would like to understand is the automobile body structure, so we must determine its role in making the larger containing system—the vehicle—successful, Figure 2.26. Many of the vehicle functions impose some needed structural performance from the body. To develop structure requirements then, we analyze the role the body structure plays in each of the functions which the vehicle provides, Figure 2.27.

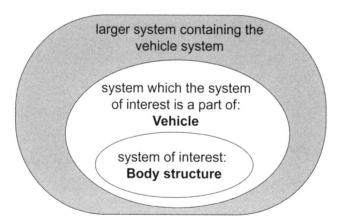

Figure 2.25 Expanding system boundaries.

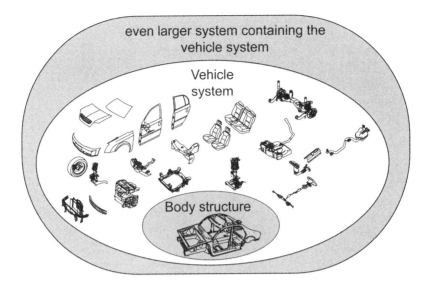

Figure 2.26 Automobile system boundaries.

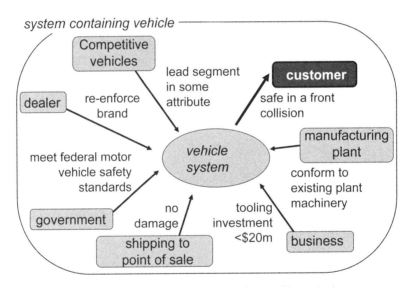

Figure 2.27 Some functions performed by vehicle.

Figure 2.28 lists many of the vehicle functions satisfying customer needs. The body structure plays an important role in achieving the functions of *noise and vibration, ride and handling, durability, safety, styling, energy use, package, human factors,* and *thermal environment*. Besides these functions directed at the ultimate customer, the vehicle interacts with many other entities and must also satisfy constraints from the manufacturing plant, government, dealer network, business financials, shipping, etc.

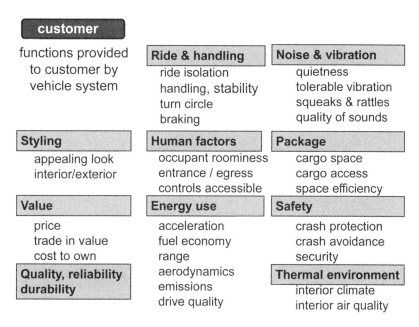

Figure 2.28 Functions supporting the customer.

By analyzing the role the body structure must play in satisfying each of these entities, we determine the functional requirements and constraints. The steps for this analysis are:

1. Identify the vehicle function to be accomplished.

2. Define how the vehicle subsystems will work together to provide that function— *function strategy*. This working together often includes a sharing of the applied loads between subsystems. Such load sharing can increase mass efficiency. An example of load sharing is given in the following example.

3. Analyze the role of the body structure in accomplishing the strategy (that is, define loads and allowable deflections in the categories of stiffness, strength, and energy absorption).

4. Flow down the overall body structure requirements to requirements on the structural subsystems and elements.

Example: Front impact structural requirements

Consider the vehicle-level function of safety in a front impact.

1. *Identify the vehicle function to be accomplished:* In this case, we will focus on minimizing injury in a *30 mph* front barrier.

2. *Define how the vehicle subsystems will work together to provide that function—function strategy*: To provide this function, a particular vehicle system strategy is defined, Figure 2.29. An example of such a strategy is: Decelerate the cabin by absorbing energy of impact with no cabin distortion, and with a cabin deceleration less than *20 g*. Air Cushion Restraint System (ACRS) deploys at *10 msec*, with optimal spatial relationship to seated occupant. Interior elements which the occupant may impact are crushable. Elements in motor compartment translate longitudinally with maximum crushable space.

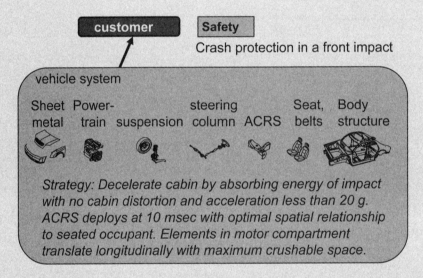

Figure 2.29 Strategy for crash protection in a front impact.

3. *Analyze the role of the body structure in accomplishing the strategy (that is, define loads and allowable deflections in the categories of stiffness, strength, and energy absorption):* To meet this strategy, several vehicle subsystems must work together in a coordinated way. These include the front-end sheet metal, front suspension and wheels, powertrain, steering column, ACRS, seat, and body structure, Figure 2.30. For example, applied barrier loads will be shared by the body structure and powertrain.

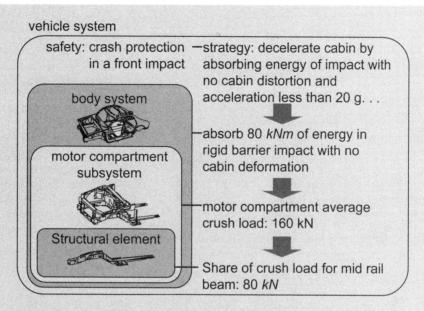

Figure 2.30 Vehicle requirements flow down to body structure requirements.

Now consider the role of the body structure in this strategy. It is one of the subsystems which will decelerate the cabin by absorbing energy of impact with no cabin distortion, and with a cabin deceleration less than *20 g*. From this, we can determine the amount of energy to be absorbed by the body structure using basic physics. This results in a body structure requirement in the energy-absorbing category, Figure 2.29.

4. *Flow down the overall body structure requirements to requirements on the structural subsystems and elements:* We can then flow this body requirement down to a structural subsystem requirement for the motor compartment, Figure 2.30 – the average load to be generated as the compartment crushes. We can further flow this *structural subsystem* requirement to a *structural element* requirement for the midrail beam – the average crush load for that beam.

We now have a procedure for identifying structural requirements based on vehicle functions. Applying this procedure to a particular vehicle will result in a large number of requirements and, as was the case for the locate-and-retain function, we again look for that small set of requirements which define the structure topology. We will suggest that small set as we discuss design for bending, torsion, crashworthiness, and vibration in subsequent chapters. In preparation for that, in the next chapter we will look at the behavior of the body structural elements.

References

1. Pahl, G. and Beitz, W., *Engineering Design*, Springer, Berlin, 1997.

2. Reinertsen, D., *Managing the Design Factory*, The Free Press, NY, 1997, p. 178.

3. Torenbeek, E., *Synthesis of Subsonic Airplane Design*, Delft University Press, The Netherlands, 1988.

4. Brown, J.C., Robertson, A.J., Serpento, S.T., *Motor Vehicle Structures*, SAE International, Warrendale, PA, 2001, p. 27.

5. Pawlowski, J., *Vehicle Body Engineering*, Business Books, London, 1969, Chapter 14.

6. Kamal M. and Wolf, J., *Modern Automotive Structural Analysis*, Van Nostrand Reinhold, NY, 1982, p. 50.

7. Kikuchi, N. & Malen, D., Course notes for ME513: Fundamentals of Body Engineering, University of Michigan, Ann Arbor, MI, 2007.

Chapter 3
Automotive Body Structural Elements

A unique characteristic of automobile body structure is the use of thin-walled structural elements. If a typical structural element from civil engineering—an I beam—is compared to a typical element from an automobile structure—a rocker beam—the difference in section proportions becomes apparent, Figure 3.1. If we compare the proportion of each section, the width to the thickness ratio, we see that it is relatively large for the automobile section. Further, we see that, unlike the I beam, the auto section is non-symmetrical, and that it is a fabrication of several formed pieces spot welded together. All of these differences lead to important differences in physical behavior, which we will cover in this chapter.

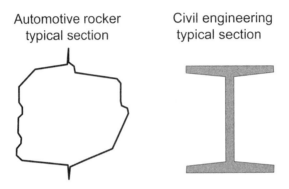

Figure 3.1 Beam sections.

3.1 Overview of Classical Beam Behavior

As a foundation for automotive beam behavior, we begin with classical beam behavior. This behavior is best embodied in a long straight beam with an I beam section; our assumptions are that the section is symmetric, the applied forces are down the axis of symmetry for the section, the section will not change shape upon loading, the deformation of the beam will be in the plane and in the direction of the applied load, the internal beam stresses vary in direct proportion with the strain, and failure is defined as yielding of the outermost fiber.

3.1.1 Static equilibrium at a beam section

The behavior of a beam under loading depends on the resulting bending moments at sections along the beam. To determine these moments, we can imagine the beam, Figure 3.2a, cut at some distance x from the end, Figure 3.2b. Looking at the left portion of the beam, an upward shear load, V, and a counter clockwise bending moment, M, act on the cut section. By setting the beam portion into static equilibrium we can find both $V(x)$ and $M(x)$. These are related by, Figure 3.2c, or the bending moment at a location x is equal to the area under the shear diagram between 0 and x.

$$M(x) = \int_0^x V dx \qquad (3.1)$$

where:

$M(x)$ = Bending moment on a section at x

V = Shear load on section at x

x = Coordinate along length of beam

In Equation 3.1 and subsequent equations where the bending moment, $M(x)$, appears, a positive bending moment is defined to be one which bends the beam such that the radius of curvature is above the beam.

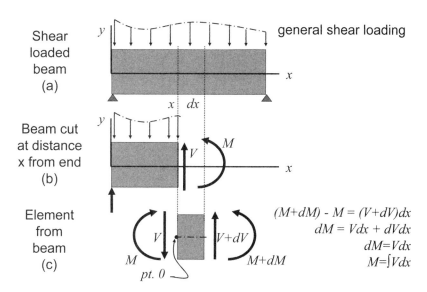

Figure 3.2 Beam equilibrium.

3.1.2 Stress over a beam section

At any cut of a beam, the direct stress varies linearly with the distance from the neutral axis, z, according to

$$\sigma = -\frac{Mz}{I} \tag{3.2}$$

where:

z = vertical distance from point of interest to the section neutral axis. (Defined to be positive in the upward direction.)

$M(x)$ = Bending moment on the section

σ = Direct stress at point of interest

The *moment of inertia, I,* is given by:

$$I = \int_{SECTION} z^2 dA \tag{3.3}$$

3.1.3 Beam deflection

The deflected shape of a beam may be described by a function, $y = f(x)$, where y is the lateral deflection from the initial shape at a distance x from the end. The curvature of the deflected beam at the distance x can be approximated by the second derivative, y''. This curvature is directly proportional to the bending moment acting at that section:

$$y'' = \frac{M(x)}{EI} \tag{3.4}$$

where:

   $M(x)$ = Bending moment on a section at x

   y'' = Curvature of the beam at x

   E = Young's modulus

   I = Moment of inertia

The deflection of a beam results by integrating Equation 3.3 twice and applying specific boundary conditions for the beam. Application of this equation provides useful results for stiffness for several typical beam conditions, Figure 3.3.

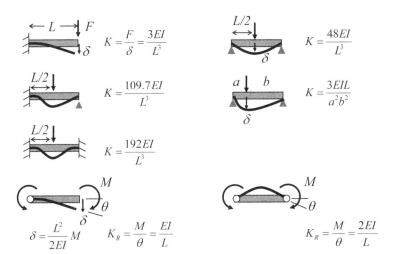

Figure 3.3 Beam stiffness equations.

Example: Cross member beam

The front motor compartment cross member holds the hood latch, Figure 3.4. Under use, aerodynamic loading places a vertical load of *1000 N* at the center of this beam. We wish to have no yielding (σ_y=210 N/mm²) in the cross member under this load, and a maximum linear deflection at the hood latch of *3 mm*. The size of the steel section is given as *30 mm* square with a thickness of *t=1.0 mm* and length *1000 mm*. We need to find if this section size will meet both the above requirements.

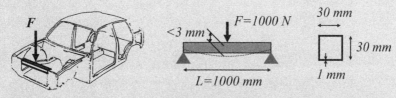

Requirements:
Stress less than $\sigma_{DESIGN}=210\ N/mm^2$
Deflection less than $3\ mm$

Figure 3.4 Motor compartment cross member.

The cross member may be modeled as a simply supported beam (in which the end restraints only prevent deflection at the ends and do not apply moments). We can take an arbitrary cut at x units from the end and between the left end and the center, Figure 3.5a. At this cut a shear force, V, acts and we place the left side of the beam into equilibrium to determine the value of V.

$$-500 + V = 0$$
$$V = 500N$$
for $0<x<L/2$

Similarly, for a cut between the center and right end:

$$-500 + 1000 + V = 0$$
$$V = -500N$$
for $L/2<x<0$

The value for V is plotted as the shear diagram, Figure 3.5b. Taking the area under this curve up to position x and using Equation 3.1, the moment diagram results, Figure 3.5b. The maximum bending moment is at $x=L/2$ and is

$$M(L/2) = \int_0^{L/2} Vdx = (500N)(1000mm/2) = 250000Nmm$$

Examining Equation 3.2 for stress, we see the maximum stress occurs along the length of the beam where bending moment is maximum, and at a point on the section which farthest from the neutral axis, $z = \pm15\ mm$.

Calculating the moment of inertia, I, as shown in Figure 3.5c:

For the two side walls, the moment of inertia is

$$I_1 = I_2 = \frac{tb^3}{12} = \frac{(1mm)(30mm)^3}{12} = 2250mm^4$$

For the top or bottom wall, the moment of inertia is the inertia about the center of area for that wall plus the transfer inertia—the area of the wall times the square of the distance from the section's neutral axis:

$$I_3 = I_4 = \frac{wt^3}{12} + wt\left(\frac{b}{2}\right)^2 \cong (30mm)(1mm)\left(\frac{30mm}{2}\right)^2 = 6750mm^4$$

Where we have neglected the first term, as it is very small for thin-walled sections where t is small relative to w or b. The total section moment of inertia is,

$$I = I_1 + I_2 + I_3 + I_4 = 2(2250 + 6750)mm^4 = 18,000mm^4$$

Substituting values into Equation 3.2

$$\sigma = -\frac{Mz}{I} = \frac{250,000 Nmm(\pm 15mm)}{18000mm^4} = \pm 208 N/mm^2$$

where the tensile stress (+) is at the lower cap *(z=−15 mm)* and the compressive stress (−) is at the upper cap *(z=15 mm)*.

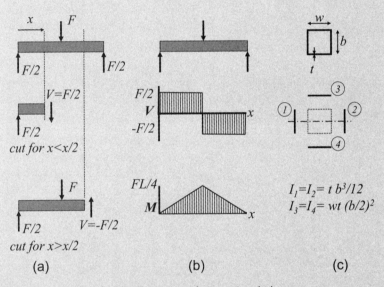

Figure 3.5 Cross member as simple beam.

Now looking at the deflection at the center of the span, we can use a result of Equation 3.4 contained in Figure 3.3:

$$k = \frac{applied\ load}{deflection} = \frac{48EI}{L^3}$$

$$deflection = \frac{L^3}{48EI}(applied\ load)$$

$$deflection = \frac{(1000mm)^3}{48(207000 N/mm^2)(18000mm^4)}(1000N) = 5.6mm$$

So for this example, the strength requirement is met *(208 N/mm² < 210 N/mm²)*, but the deflection requirement is not *(5.6 mm > 3 mm)*. The section size or thickness must be increased to satisfy the deflection requirement.

Note that the assumptions of the classical equations are met, as the beam is symmetrical and symmetrically loaded. Also, due to its thick-walled section proportions, the beam failure mode is by yield.

With this summary of classical beam results, we can now look at the unique behavior of automotive structural elements. In this discussion of body structural elements, we will divide them into two general types: 1) frameworks constructed of beams, and 2) panels, Figure 3.6. In the next section we will develop tools to understand the unique behavior of automotive beams.

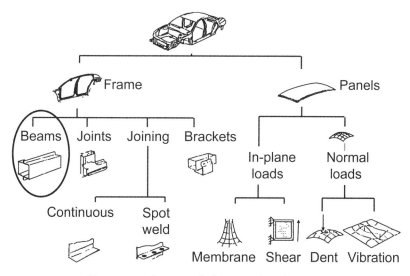

Figure 3.6 Structural element classification.

3.2 Design of Automotive Beam Sections

Several characteristics of automotive beams require analytical tools beyond classical beam theory. Below we will treat 1) the non-symmetrical nature of automotive beams, 2) local distortion of the section at the point of loading, 3) twisting of thin-walled members, 4) the effect of spot welds on structural performance.

3.2.1 Bending of non-symmetric beams

Beam sections in automotive applications are typically non-symmetrical, and we must develop the ability to predict deflection and stress for these sections. A symmetrically loaded section will deflect in the same direction as the applied load, Figure 3.7a. The deflection of a beam with a non-symmetrical section in general will not be colinear with the applied load, Figure 3.7b.

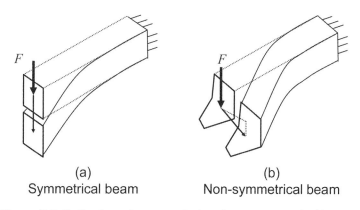

(a)
Symmetrical beam

(b)
Non-symmetrical beam

Figure 3.7 Deflection of symmetrical and non-symmetrical beams.

However, for any arbitrary non-symmetric section there is a specific axis through its centroid for which a load applied along that axis will produce a deflection colinear with that load [1]. This axis is defined as a *principle axis*. Perpendicular to this axis is a second principle axis with the same load-deflection property, Figure 3.8.

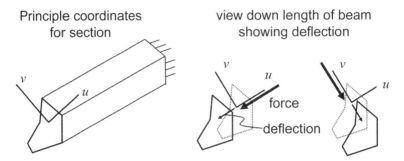

Figure 3.8 Non-symmetrical beam principle axes.

This suggests a way to predict deflections in a beam with a non-symmetric section and loaded in some arbitrary direction, Figure 3.9:

1. Resolve the load into components along each principle axis.

2. Solve for the resulting deflection for each of these components using the equations of Figure 3.3. Note that the moment of inertia is taken about the axis perpendicular to the load. Each of these deflections will be along the respective principle axis.

3. Take the vector sum of the two deflections to determine the magnitude and direction of the total deflection.

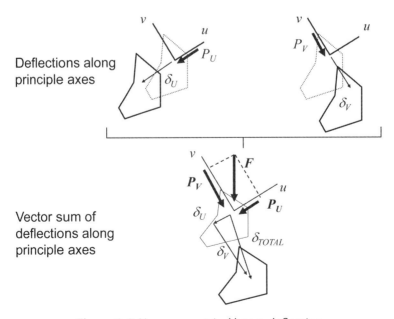

Figure 3.9 Non-symmetrical beam deflection.

For stress, the steps are similar after determining the bending moment vector acting on the section, Figure 3.10:

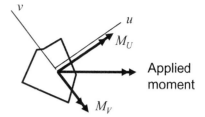

1. Resolve the moment into components along each principle axis.

2. For each component of the moment, solve for the resulting stress using Equation 3.2. Note that the dimension z is the distance to the point of interest from the axis which is colinear with the moment vector, and the moment of inertia is about the same axis.

Figure 3.10 Moments acting on non-symmetrical beam.

3. Take the algebraic sum of the two stresses for the resultant stress.

Example: Non-symmetric beam

Consider the steering column mounting beam, Figure 3.11. We can view this beam with non-symmetrical section as a cantilever with a downward vertical tip load, Figure 3.12a, and we are interested in determining the tip deflection and also the stress at a specific point A where the beam joins the restraining structure.

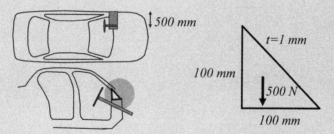

Figure 3.11 Steering column support beam.

The section properties can be determined using section analysis software [2], with the resulting orientation of the two principle axes shown along with the moments of inertia about these axes, Figure 3.12b.

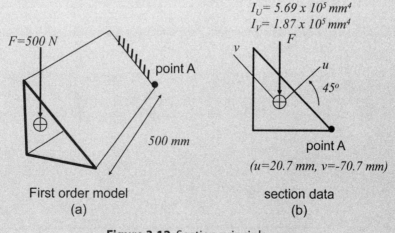

First order model
(a)

section data
(b)

Figure 3.12 Section principle axes.

45

To determine deflection, we resolve the applied load into a component in the u direction and a component in the v direction.

$$F_V = F\cos(\theta) = 500\cos(45^o) = 353.5N$$
$$F_U = F\sin(\theta) = 500\sin(45^o) = 353.5N$$

Then we can use the cantilever equation from Figure 3.3 to determine the deflections along each of these axes, Figure 3.13.

$$\delta_U = \frac{F_U l^3}{3EI_V}, \quad \delta_V = \frac{F_V l^3}{3EI_U}$$

$$\delta_U = \frac{500N(0.707)(500mm)^3}{3(207\ x10^3\ N/mm^2)(1.87\ x10^5\ mm^4)} = 0.38mm$$

$$\delta_V = \frac{500N(0.707)(500mm)^3}{3(207\ x10^3\ N/mm^2)(5.69\ x10^5\ mm^4)} = 0.125mm$$

Adding the deflection vectors yields the total deflection, Figure 3.13b.

$$\delta_{TOTAL} = \sqrt{\delta_U^2 + \delta_V^2} = 0.4mm$$

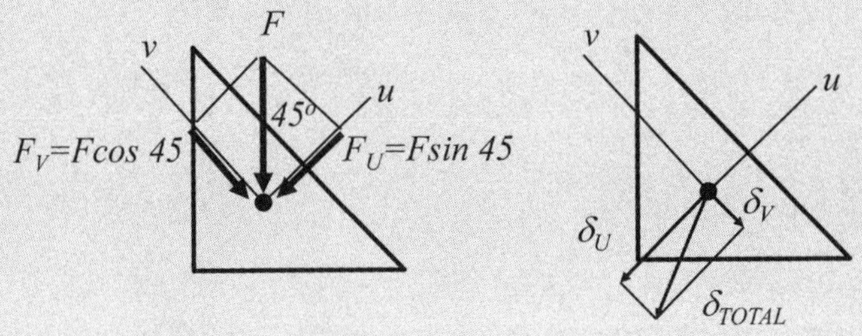

Force components
in principle coordinate system
(a)

Vector sum of
deflections
(b)

Figure 3.13 Beam deflection calculation.

To determine stress at point A, we first identify the bending moment at the section of interest. The bending moment for the section at the wall is a counterclockwise moment, and using the right-hand rule is a horizontal vector pointing to the right, Figure 3.14. Next, take the components of the applied bending moment along the u and v axes. We can see that the moment vector down the u axis would place the bending radius of curvature below the beam, and similarly for the moment along the v axis; this gives both moment components a negative sign, Figure 3.15. Solving for the stress from each of these moments using Equation 3.2 (The distance from the neutral axis to the point, z, is shown here as v and u.):

$$\sigma_U = -\frac{M_U v}{I_U}$$

$$\sigma_U = -\frac{(-500N \cdot 500mm \cdot \cos 45^o)(-70.7mm)}{5.69\ x10^5\ mm^4}$$

$$\sigma_U = -22N/mm^2$$

$$\sigma_V = -\frac{M_V u}{I_V}$$

$$\sigma_V = -\frac{(-500N \cdot 500mm \cdot \sin 45^o)(20.7mm)}{1.87 \times 10^5 mm^4}$$

$$\sigma_V = +19.57 N / mm^2$$

$$\sigma_{POINT\ A} = \sigma_U + \sigma_V$$

$$\sigma_{POINT\ A} = (-22 + 19.57)N / mm^2$$

$$\sigma_{POINT\ A} = (-2.43)N / mm^2$$

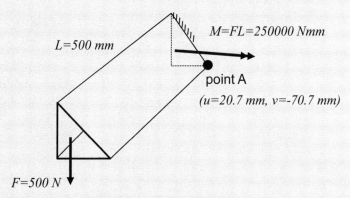

Figure 3.14 Moment vector.

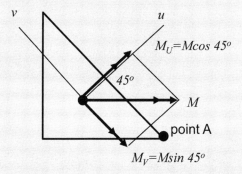

Figure 3.15 Moment resolved into principle components.

3.2.2 Point loading of thin-walled sections

In the development of beam theory for thin-walled sections, we have assumed that the applied point loads only influence the beam through the resulting bending moments. In practice, with thin-walled sections the point load also distorts the beam section in the vicinity of the load, Figure 3.16. This undesirable distortion leads to a reduced apparent beam stiffness and increased local stress. In this section, we will develop tools to predict the degree of local distortion and strategies to minimize local distortion.

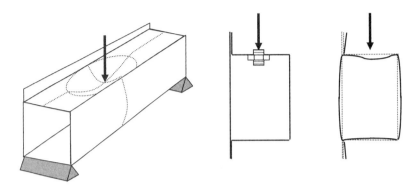

Figure 3.16 Local deformation under point load.

To investigate the effects of point loading of a thin-walled section, consider the latch structure of a van, Figure 3.17. We will idealize this structure as a simply supported beam with rectangular section. The latch applies a point load at the center of the section, and we are interested in the resulting stiffness measured at this point. Under such a load, the deformation we would observe is shown in Figure 3.18. In addition to the idealized beam deformation, we would see a local distortion of the section near the point load. In effect, the thin-walled point-loaded beam would be two springs in series: the idealized beam stiffness, and the stiffness of the local distortion of the section, Figure 3.18. The idealized beam stiffness is summarized in Figure 3.3. Now we will look at a way to predict the local distortion.

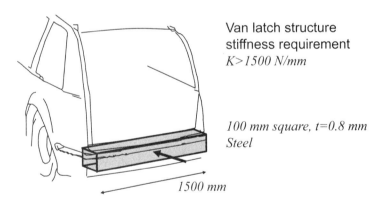

Van latch structure
stiffness requirement
K>1500 N/mm

100 mm square, t=0.8 mm
Steel

1500 mm

Figure 3.17 Van cross member.

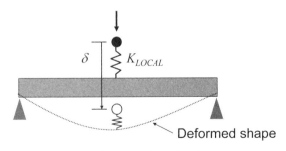

Figure 3.18 Deflection of van latch structure.

Looking again at the distorted beam, we can separate out the idealized beam deflection by imagining the beam supported along its neutral axis, Figure 3.19a. The point load is reacted by a deformed patch on the top surface of the beam which we denote as the *affected zone* (we take the length and width of the affected zone to be the section width, *b*). The affected zone can now be imagined to be cut into slices of unit length with each slice acting as a spring supporting the load. Figure 3.19b. As such, the distorted sections act as an *elastic foundation* for the point load.

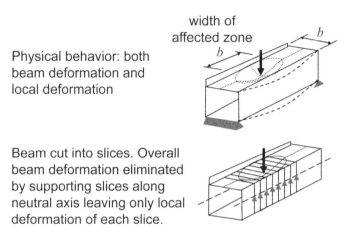

Physical behavior: both beam deformation and local deformation

Beam cut into slices. Overall beam deformation eliminated by supporting slices along neutral axis leaving only local deformation of each slice.

Figure 3.19a Idealized beam.

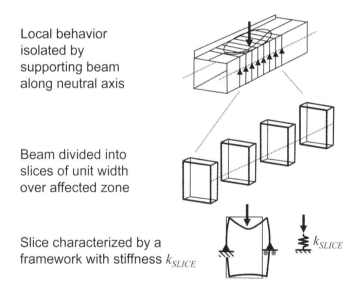

Local behavior isolated by supporting beam along neutral axis

Beam divided into slices of unit width over affected zone

Slice characterized by a framework with stiffness k_{SLICE}

Figure 3.19b Idealized beam.

Consider the foundation stiffness to be k_{SLICE} N/mm/mm and assume the deflected shape for the top plate centerline is, Figure 3.20:

$$y = -\Delta \sin \frac{\pi x}{b}, \quad 0 < x < \pi$$

where:

y = Normal deflection of top plate at centerline

b = Plate width

Δ = Plate deflection at point of load application

x = Coordinate along length of plate

The strain energy, de, in one slice may be integrated over the length of the affected zone to arrive at the energy of distortion for the affected zone:

$$de = \frac{1}{2}(k_{SLICE}\, dx)(y)^2$$

$$e = \frac{1}{2}\int_0^b k_{SLICE}(y)^2\, dx$$

$$e = \frac{1}{2}\int_0^b k_{SLICE}(-\Delta \sin \frac{\pi x}{b})^2\, dx = \frac{1}{2}k_{SLICE}\Delta^2\frac{b}{2}$$

We can equate this distortion energy, e, to the work done by the external force, $1/2F\Delta$. Equating work to strain energy yields:

$$\frac{1}{2}F\Delta = \frac{1}{2}k_{SLICE}\Delta^2\frac{b}{2}$$

$$\frac{F}{\Delta} = \frac{1}{2}k_{SLICE}b$$

$$K_{LOCAL} = \frac{1}{2}k_{SLICE}b \tag{3.5}$$

So the stiffness as seen by the point load, k_{LOCAL}, is directly related to the stiffness of a slice of the section, k_{SLICE}. For a rectangular section h high, b wide and t thick, Figure 3.21, the stiffness of a unit slice is:

$$k_{SLICE} = \frac{16Et^3(h+b)}{b^3(4h+b)} \tag{3.6}$$

so substituting Equation 3.6 into 3.5 for a point-loaded thin-walled rectangular beam:

$$K_{LOCAL} = \frac{8Et^3(h+b)}{b^2(4h+b)}$$

where:

K_{LOCAL} = Local stiffness of section under a central point load

t = Section thickness

h = Section height

b = Section width

E = Young's Modulus

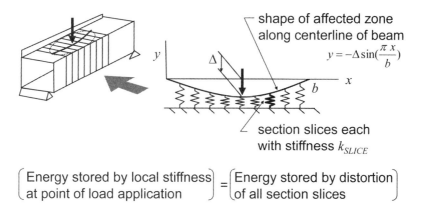

shape of affected zone
along centerline of beam

$$y = -\Delta \sin(\frac{\pi x}{b})$$

section slices each
with stiffness k_{SLICE}

$$\begin{pmatrix} \text{Energy stored by local stiffness} \\ \text{at point of load application} \end{pmatrix} = \begin{pmatrix} \text{Energy stored by distortion} \\ \text{of all section slices} \end{pmatrix}$$

Figure 3.20 Idealized beam analysis.

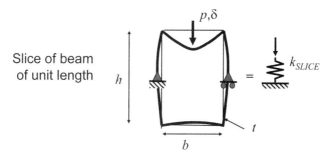

Slice of beam
of unit length

$$= k_{SLICE}$$

Figure 3.21 Rectangular section under point load.

Example: Local deformation

As a numerical example, consider the van cross member of Figure 3.17. Looking first at the idealized beam deflection and using the stiffness equations of Figure 3.3:

$$K_{IDEAL\ BEAM} = \frac{48EI}{l^3}$$

$$K_{IDEAL\ BEAM} = \frac{48(207000\frac{N}{mm^2})(5.33 x10^5\ mm^4)}{(1500mm)^3}$$

$$K_{IDEAL\ BEAM} = 1569\frac{N}{mm}$$

For the dimensions given, the local stiffness is:

$$K_{LOCAL} = \frac{8Et^3(h+b)}{b^2(4h+b)} = \frac{8(207000\frac{N}{mm^2})(0.8mm)^3(100mm+100mm)}{(100mm)^2(4\cdot100mm+100mm)}$$

$$K_{LOCAL} = 33.9\frac{N}{mm}$$

Now consider $K_{IDEAL\,BEAM}$ and K_{LOCAL} to be two springs in series:

$$K_{SYSTEM} = \frac{K_{IDEAL\,BEAM}K_{LOCAL}}{K_{IDEAL\,BEAM} + K_{LOCAL}}$$

$$K_{SYSTEM} = \frac{(33.9)(1569)}{33.9 + 1569} \frac{N}{mm}$$

$$K_{SYSTEM} = 33.2 \frac{N}{mm}$$

So, because of local distortion, the stiffness at the point of load application is only 2% of the idealized beam stiffness. The large beam section is not being effectively used.

The above example demonstrates why an untreated thin-walled section performs so poorly under a point load. In practice there are several strategies to reduce this local distortion so that the idealized beam stiffness is more fully utilized. Fundamentally the point load must load the shear web of the section directly, Figure 3.22. This can be achieved by moving the load point to align with the web, Figure 3.23a. When the load point cannot be physically aligned with the web, Figure 3.23b, a stiff structural element, which reacts the load to the webs, can be added to the section. Finally, a through-section attachment with bulkhead can be used to achieve this transfer of load to the web, Figure 3.23c. Two examples of this strategy are shown in Figure 3.24.

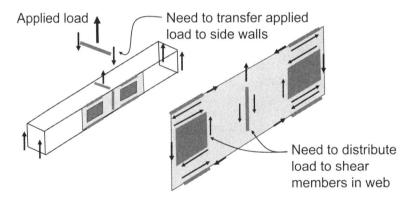

Figure 3.22 Shear web.

3.3 Torsion of Thin-Wall Members

Elementary treatment of twisting of bars is limited to solid circular sections. For solid circular bar, the basic equations [3] for angular deflection and for shear stress are given by:

$$\theta = \frac{TL}{GJ} \tag{3.7a}$$

$$\tau = \frac{Tr}{J} \tag{3.7b}$$

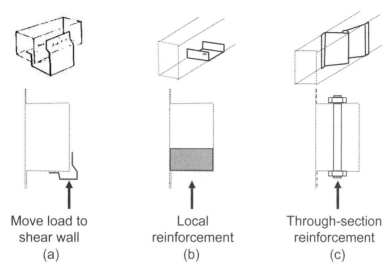

Move load to shear wall
(a)

Local reinforcement
(b)

Through-section reinforcement
(c)

Figure 3.23 Loading the shear web.

Through-section engine mount attachment

Through-section door hinge attachment

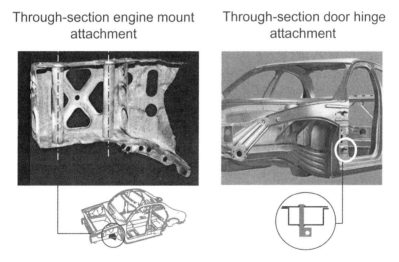

Figure 3.24 Example sections reacting a point load. (Courtesy of A2Mac1.com Automotive Benchmarking, and the American Iron and Steel Institute, UltraLight Steel Auto Body)

where:

T = Applied torque

θ = Resulting angle of twist

J = Polar moment of inertia

G = Shear modulus

L = Length of the bar

r = Radius of the bar

τ = Maximum shear stress (at the surface of the bar)

3.3.1 Torsion of members with closed section

For the thin-walled hollow sections we see in automotive body structures, Equations 3.7 are not valid, and we will now develop the equivalents for predicting angular deflection and shear stress for *thin-walled closed* sections:

$$\theta = \frac{TL}{GJ_{EFF}} \tag{3.8a}$$

$$\tau = \frac{T}{2At} \tag{3.8b}$$

where:

θ = angle of rotation

T = Applied torque

L = Beam length

τ = Shear stress

A = Area enclosed by the section

t = Thickness

J_{EFF} = Thin-wall torsion constant

For a thin-walled closed section with *constant thickness*:

$$J_{EFF} = \frac{4A^2t}{S} \tag{3.9}$$

where:

J_{EFF} = Thin-wall closed-section torsion constant with constant thickness

S = Section perimeter.

A = Area enclosed by the section

t = constant thickness

For sections with *varying thickness*:

$$J_{EFF} = \frac{4A^2}{\sum_i \frac{s_i}{t_i}} \tag{3.10}$$

where the section is divided into i segments each of uniform thickness, t_i, and length s_i.

We will now develop these equations [4]. Consider a thin-walled closed tube of arbitrary cross-section shape and length, L, Figure 3.25a. We apply equal and opposite torques at the ends of the tube and imagine the tube cut along its length, Figure 3.25b. All along one of the cut edges will be a shearing stress, τ_1, and we can look at the static equilibrium of a rectangular element adjacent to the cut, Figure 3.26a. Setting longitudinal forces into equilibrium gives $\tau_1 t_1 = \tau_2 t_2$. The quantity (τt) is the shearing force per unit length and we define this as *shear flow, q*. Further, setting moments on this element into equilibrium, Figure 3.26b, yields $q' = q$, or the shear flow on all sides of this element are equal. This shows that the shear flow, (τt), is uniform throughout the tube. Now we seek a relationship between shear flow and applied torque.

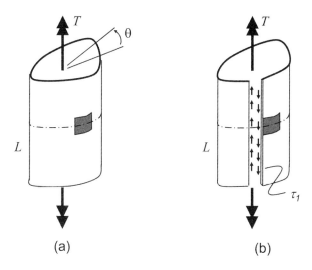

Figure 3.25 Thin-walled closed tube.

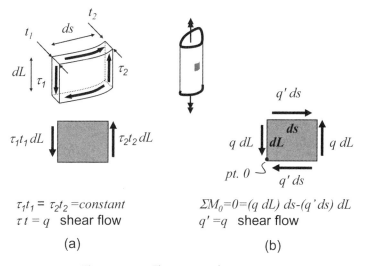

Figure 3.26 Element under torsion.

Looking at the end view of the tube, we can consider a small element of perimeter, dS. As shear flow is uniform throughout the tube, the force, dF, acting on this element is $dF=qdS$. This force is directed tangent to the section, and results in a small torque, dT, about some arbitrary point O contained within the interior of the section. The torque is equal to the force, dF, times the distance, r, from the line of action of the force to the point O, or $dT=rdF=rqdS$. To determine the total torque we must integrate around the perimeter:

$$T = \oint_{PERIMETER} rqdS$$

Since q is constant, it can be brought out of the integral:

$$T = q \oint_{PERIMETER} r\,dS$$

To solve this integral, note that the shaded area in Figure 3.27 is given by:

$$dA = \frac{1}{2}r\,dS$$

Substituting this relationship into the previous yields the desired relationship between stress and torque. (Note that A is the area enclosed by the tube perimeter and not the area of the material.):

$$T = 2q \oint_{PERIMETER} dA$$

$$T = 2qA$$

$$q = \frac{T}{2A}$$

$$\tau = \frac{T}{2tA} \tag{3.11}$$

where:

    T = Applied torque

    A = Area enclosed by the section

    t = Thickness

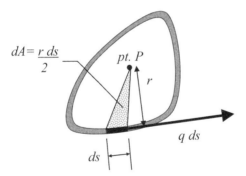

Figure 3.27 Tube section.

Now consider the angle of twist, θ, resulting from the application of the torque, T, Figure 3.25. We will equate the work done by the external torque to the strain energy stored within the twisted tube.

$$\text{external work from applied torque } = \frac{1}{2}T\theta$$

$$\text{strain energy} = \int \frac{1}{2}\gamma\tau\,dV = \int \frac{1}{2}\frac{\tau}{G}\tau\,dV = \int \frac{1}{2}\frac{\tau^2}{G}\,dV$$

$$\text{strain energy} = \iint \frac{1}{2}\frac{\tau^2}{G}t\,dL\,dS$$

using Equation 3.8 to substitute for shear stress, τ:

$$\text{strain energy} = \iint \frac{1}{2G}\left(\frac{T}{2At}\right)^2 t \, dL \, dS$$

$$\text{strain energy} = \frac{1}{2G}\frac{T^2}{(2A)^2}\iint \frac{1}{t}\, dL \, dS$$

$$\text{strain energy} = \frac{1}{2G}\frac{LT^2}{(2A)^2}\oint \frac{dS}{t}$$

Equating work to strain energy:

$$\frac{1}{2}T\theta = \frac{1}{2G}\frac{LT^2}{(2A)^2}\oint \frac{dS}{t}$$

$$\theta = \frac{TL}{4GA^2}\oint \frac{dS}{t}$$

Rearranging yields:

$$\theta = \frac{TL}{\left[G\left[\dfrac{4A^2}{\oint \dfrac{dS}{t}}\right]\right]} = \frac{TL}{GJ_{EFF}},$$

(3.12)

$$\text{where} \quad J_{EFF} = \frac{4A^2}{\oint \dfrac{dS}{t}}$$

which agrees with Equation 3.8a where the variables are defined. Note for the case where the thickness is constant, the effective torsional constant is

$$J_{EFF} = \frac{4A^2 t}{S}$$

(3.13)

Example: Torsion of beam with closed section

Consider again the steering column mounting beam, Figure 3.28 but now under torsional loading. We are interested in the angle of rotation and the shear stress under a pure torque of $T=25x10^4$ Nmm.

Equation 3.13 gives the effective torsion constant:

$$J_{EFF} = \frac{4A^2}{\oint \dfrac{dS}{t}} = \frac{4tA^2}{S} = \frac{4(1mm)\left(\dfrac{1}{2}(100mm)(100mm)\right)^2}{(100mm+100mm+141.4mm)} = 2.929 \, x10^5 \, mm^4$$

Equation 3.12 results in the angle of twist:

$$\theta = \frac{TL}{GJ_{EFF}} = \frac{(25x10^4 \, Nmm)(500mm)}{(78x10^3 \, N/mm^2)(2.929 \, x10^5 \, mm^4)} = 5.47 \, x10^{-3} \, rad \, (0.312^o)$$

Finally, Equation 3.11 results in the shear stress:

$$\tau = \frac{T}{2tA} = \frac{(25x10^4 \, Nmm)}{2(1mm)\left(\frac{1}{2}(100mm)(100mm)\right)} = 25N / mm^2$$

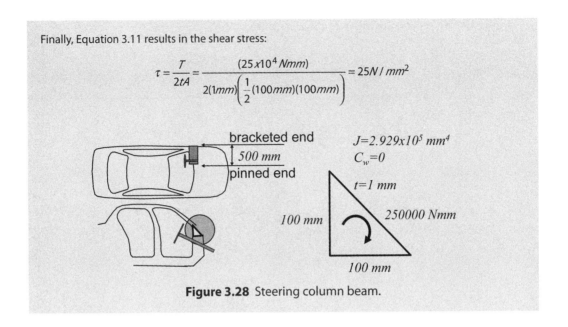

Figure 3.28 Steering column beam.

3.3.2 Torsion of members with open section

The above discussion concerned tubes with closed section. For open sections, the development of the equations requires application of the *theory of elasticity* and is beyond the scope of this book. However, we state the results [5] below, Figure 3.29, for a *thin-walled open* section of constant thickness:

$$J_{EFF} = \frac{1}{3}t^3S \tag{3.14a}$$

$$\tau = \frac{Tt}{J_{EFF}} \tag{3.14b}$$

where:

J_{EFF} = Torsion constant for an open section

S = Developed section length

t = Thickness

T = Applied torque

τ = Shear stress

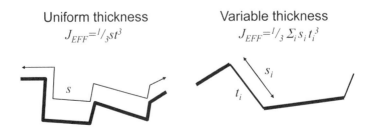

Figure 3.29 Torsion of open sections.

For open sections with non-uniform thickness:

$$J_{EFF} = \frac{1}{3} \sum_i t_i^3 s_i \qquad (3.15)$$

where the section is divided into i segments each of uniform thickness, t_i, and length s_i.

Note the somewhat unexpected result that the shape of the open section does not enter the expressions, and in general open sections are much more flexible than a closed section of the same perimeter.

Example: Torsion of beam with open section

Consider again the steering column mounting beam as in the previous example, but now, rather than a closed section, there is a very thin slot as shown in Figure 3.30. Applying Equation 3.14a gives:

$$J_{EFF} = \frac{1}{3} t^3 s = \frac{1}{3}(1mm)^3(100mm + 100mm + 141.4mm) = 113.8mm^4$$

From Equation 3.8a:

$$\theta = \frac{TL}{GJ_{EFF}} = \frac{(25 \times 10^4 \, Nmm)(500mm)}{(78 \times 10^3 \, N/mm^2)(113.8mm^4)} = 14.1 rad \ (803^o)$$

From Equation 3.14b:

$$\tau = \frac{Tt}{J_{EFF}} = \frac{(25 \times 10^4 \, Nmm)(1mm)}{113.8mm^4} = 2197 N/mm^2$$

Clearly both of these values are beyond the linearity assumptions; however, we learn that not only is the section *very* flexible, but the shear stress is quite high compared to the closed section of the previous example.

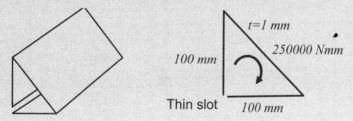

Figure 3.30 Slotted beam section.

3.3.3 Warping of open sections under torsion

The above examples for torsion of open and closed sections showed the open section to be relatively highly flexible. The reason open sections perform so poorly in torsion is that the section has considerable warping in the longitudinal direction, Figure 3.31. In Equations 3.14a and 3.14b, we assumed that this warping is unrestrained in any way. However, if we rigidly hold an end of the open tube and prevent warping, the stiffness of the tube can be markedly increased. In Figures 3.32 and 3.33, we provide formulae for prediction of angle of rotation when one or both ends of a tube are restrained from warping [6]. Note the use of the warping constant, Cw, for the section. Warping depends only on the geometry of the section, with large Cw values indicating greater out-of-plane deformation, and $Cw=0$ indicating the section remains planar under torque loading. For general sections, the constant Cw may be obtained using section analysis software, or for simple sections it may be calculated using the formulae shown in Figure 3.34 [6].

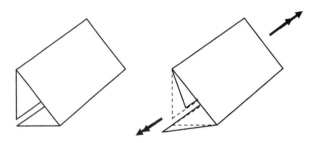

Figure 3.31 Open-section warping.

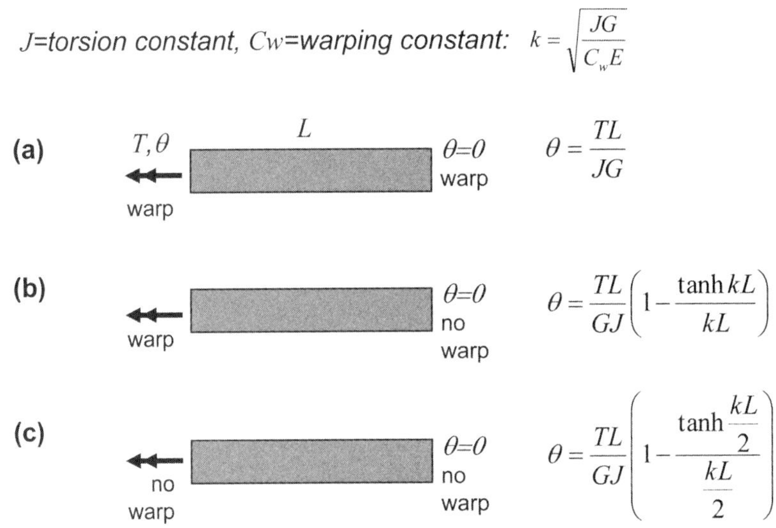

J=torsion constant, Cw=warping constant: $\quad k = \sqrt{\dfrac{JG}{C_w E}}$

(a) T, θ $\quad$ L $\qquad$ $\theta = 0$ warp $\qquad$ $\theta = \dfrac{TL}{JG}$

warp

(b) warp $\qquad$ $\theta = 0$ no warp $\qquad$ $\theta = \dfrac{TL}{GJ}\left(1 - \dfrac{\tanh kL}{kL}\right)$

(c) no warp $\qquad$ $\theta = 0$ no warp $\qquad$ $\theta = \dfrac{TL}{GJ}\left(1 - \dfrac{\tanh \dfrac{kL}{2}}{\dfrac{kL}{2}}\right)$

Figure 3.32 Formulae for twist of end-loaded warping tubes.

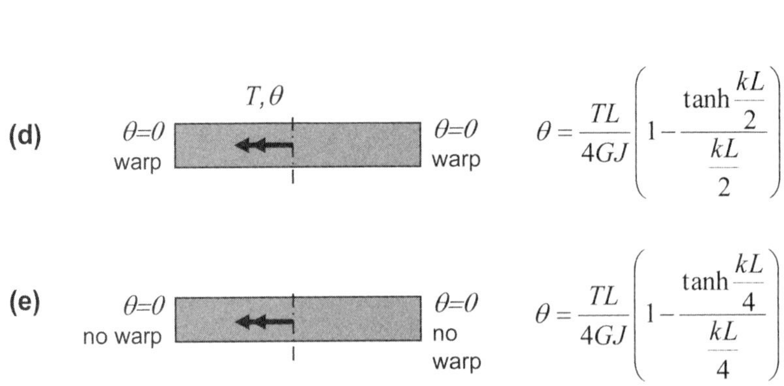

(d) $\theta = 0$ warp $\qquad T, \theta \qquad \theta = 0$ warp $\qquad$ $\theta = \dfrac{TL}{4GJ}\left(1 - \dfrac{\tanh \dfrac{kL}{2}}{\dfrac{kL}{2}}\right)$

(e) $\theta = 0$ no warp $\qquad \theta = 0$ no warp $\qquad$ $\theta = \dfrac{TL}{4GJ}\left(1 - \dfrac{\tanh \dfrac{kL}{4}}{\dfrac{kL}{4}}\right)$

Figure 3.33 Formulae for twist of center-loaded warping tubes.

$$Cw = \frac{ta^4b^3}{6}\left(\frac{4a+3b}{2a^3-(a-b)^3}\right)$$

$$Cw = \frac{2tr^5}{3}\left(\alpha^3 - 6\frac{(\sin\alpha-\alpha\cos\alpha)^2}{\alpha-\sin\alpha\cos\alpha}\right)$$

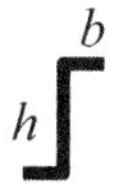

$$Cw = \frac{th^2b^3}{12}\left(\frac{2h+b}{h+2b}\right)$$

$$Cw = 0$$

$$Cw = \frac{th^2b^3}{12}\left(\frac{2h+3b}{h+6b}\right)$$

Figure 3.34 Formulae for warping coefficients.

Example: Warping beam

Look again at the steering column mounting beam with a very thin slot from the preceding example, Figure 3.30. Previously we allowed unrestrained warping in the calculation of angle of rotation, now we will constrain both ends from out-of-plane deflection by adding a very stiff mounting bracket to each end, Figure 3.35. Referring to Figure 3.32, the boundary conditions for this example match case *(c)*, and we use the relationship:

$$\theta = \frac{TL}{GJ}\left[1 - \frac{\tanh(kL/2)}{kL/2}\right]$$

where:

θ = Angle of twist with warping

T = Applied torque

G = Shear modulus

L = Beam length

J = Torsion Constant

$$k = \sqrt{\frac{JG}{C_wE}}$$

E = Young's Modulus

Cw = Warping constant

For this section, $J=113.8\,mm^4$ from the previous example, and we are given $CW=1.4x10^9\,mm^6$. Substituting into the above:

$$k = \sqrt{\frac{JG}{C_wE}} = \sqrt{\frac{(113.8mm^4)(78\,x10^3\,N/mm^2)}{(1.4x10^9\,mm^6)(207\,x10^3\,N/mm^2)}} =$$

$$k = 1.75013x10^{-4}\,/mm$$

and

$$\theta = \frac{TL}{GJ}\left[1 - \frac{\tanh(kL/2)}{kL/2}\right] =$$

$$\theta = \frac{25 \times 10^4\, Nmm(500mm)}{(78 \times 10^3\, N/mm^2)(113.8mm^4)}\left[1 - \frac{\tanh\left((1.75013 \times 10^{-4}/mm)(500mm)/2\right)}{(1.75013 \times 10^{-4}/mm)(500mm)/2}\right]$$

$$\theta = 8.98 \times 10^{-3}\, rad$$

Compare this result with $\theta = 14.1\, rad$ for the unconstrained warping case, and $\theta = 5.47 \times 10^{-3}\, rad$ for the closed-section case. We can see that restraining warping is highly effective in stiffening an open tube. However, this comes at the penalty of the increased structure (and mass) needed to prevent warping, Figure 3.35.

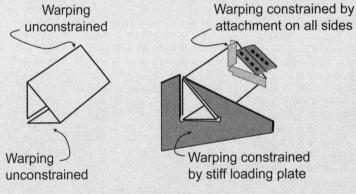

Figure 3.35 Constraining warping.

The physical means of restraining warping is often challenging to achieve in practice. In Figure 3.36, other boundary conditions are shown for the steering column beam for the cases (a), (b), (c) of Figure 3.32.

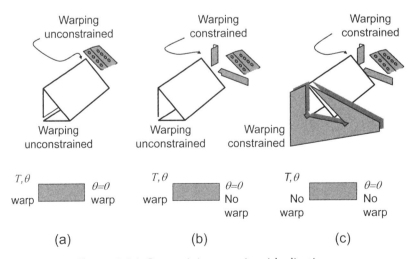

Figure 3.36 Constraining warping: Idealization.

Note that the effectiveness of restraining warping depends on the tube length; the stiffening influence of restraining warping diminishes as the tube becomes longer. Intuitively this is because the warping restraint primarily influences the region around tube end, and as the tube becomes longer this region becomes proportionately smaller.

Example: Effect of length of warping beam

As an example of the effect of beam length on angle of rotation, we continue with the example of the steering column mount tube and use the equations in Figure 3.32 to determine the angle of rotation for an open section tube with (a) unrestrained ends, (b) with one end restrained, (c) with two ends restrained, and a closed section (no warping). Figure 3.37 shows the resulting angle of rotation per unit length as we vary length from *400* to *1000 mm* for these cases. The trend of increasing angle of twist as length increases can be seen for cases (b) and (c) where we restrain warping. For the shorter tubes, *L=400 mm*, the both-ends-restrained open section, (c), is nearly as stiff (93%) as the closed section; for the longer tubes, *L=1000 mm*, the both-ends-restrained open section, (c), is only *15%* as stiff as the closed section.

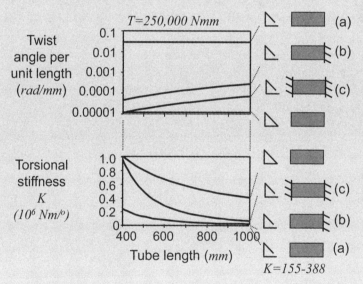

Figure 3.37 Twist angle: Varying length and end conditions.

3.3.4 Effect of spot welds on structural performance

Automotive body sections are fabrications of several formed elements, usually spot welded together. Spot welding is a robust and cost-efficient process for mass production, Figure 3.38, but does present some unique structural challenges. One of these is the addition of shear flexibility in the section which is most evident during torsion of fabricated sections. In this section we will develop tools to predict the degree of shear flexibility, the implications on section stiffness, and strategies to minimize the flexibility.

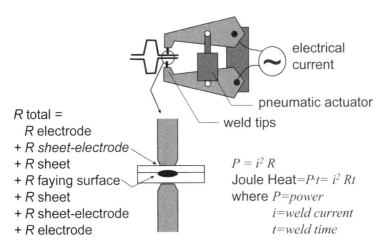

R total =
  R electrode
  + R sheet-electrode
  + R sheet
  + R faying surface
  + R sheet
  + R sheet-electrode
  + R electrode

$P = i^2 R$
Joule Heat $= P \cdot t = i^2 Rt$
where $P = power$
  $i = weld\ current$
  $t = weld\ time$

Figure 3.38 Spot welder.

3.3.4.1 Peel versus shear loading condition

Let us look at a single spot weld loaded by equal and opposite tensile loads, Figure 3.39. This condition is referred to as *shear loading*. The loads act on the centerline of the flange thickness, and the offset between centerlines creates a moment at the weld. The moment bends the flange and creates high stress in the area of the weld, as shown in Figure 3.40. Even the small offset of loads under shear loading creates stress concentrations, which reduce fatigue performance. Figure 3.41 shows this effect for a mild steel sample loaded in shear with the fatigue limit being reduced from the base material—in this case, by a factor of seven. When an adhesive is used in addition to the spot weld, the stresses are more evenly distributed and the fatigue performance is enhanced.

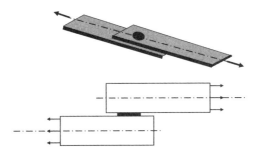

Figure 3.39 Shear-loaded spot-welded joint.

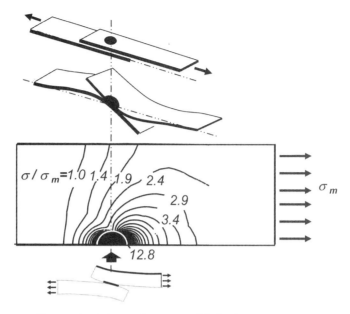

Figure 3.40 Stress in spot-welded joint.

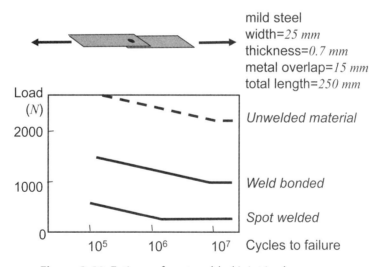

Figure 3.41 Fatigue of spot-welded joint in shear.

Another loading configuration—*peel loading*—increases this detrimental offset even more, Figure 3.42. Comparing weld strength for an individual weld, Figure 3.43, shows the effect of increasing the loading offset, *x*, beyond the sheet thickness as in shear loading. Because of this effect, good design practice is to use part geometry to put welds into *shear loading* rather than *peel loading*. Figure 3.44 and 3.45 show auto body examples for welds in peel and the preferred design with the weld in shear. Note that in each case, we assume a tensile load within the plane of the thin-wall material, and attempt to minimize the offset of this tensile load from the weld.

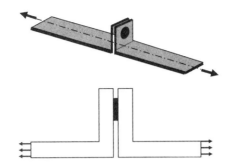

Figure 3.42 Peel loading of spot-welded joint.

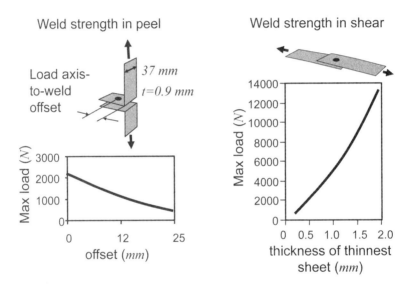

Figure 3.43 Offset effect on spot-welded joint strength.

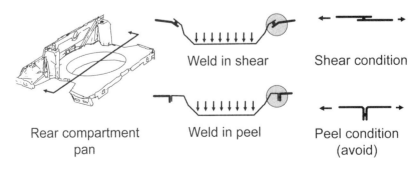

Figure 3.44 Examples of rear compartment joints in shear and peel.

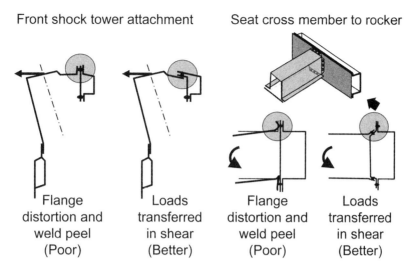

Front shock tower attachment   Seat cross member to rocker

Flange distortion and weld peel (Poor) | Loads transferred in shear (Better) | Flange distortion and weld peel (Poor) | Loads transferred in shear (Better)

Figure 3.45 Examples of joints in shear and peel.

3.3.4.2 Longitudinal stiffness of a shear-loaded weld flange

Even with a well-designed spot-welded joint under shear loading, local deformations, which will reduce the apparent stiffness of a section, will occur. We will now look at a welded flange and determine how it elastically deforms under the action of shear loading. Isolating one spot weld, the distortion under a shear load is a rotation with the center at the interface of the weld, Figure 3.46. This rotation causes a relative longitudinal deflection between the two flanges. Now looking at a series of welds under shear loading, we can expect the deformation shown in Figure 3.47.

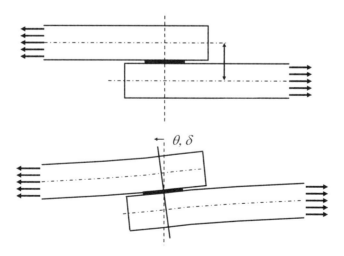

θ, δ

Figure 3.46 Distortion at spot weld.

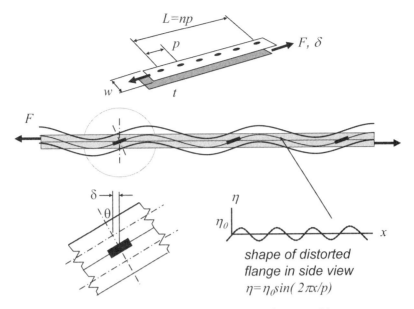

Figure 3.47 Distortion for a series of spot welds.

At each weld the same small rotation occurs and the result is a relative longitudinal deflection at the end of the flanges. Let the deflected shape of the flange be given as:

$$\eta = \eta_0 \sin \frac{2\pi x}{p}$$

At each weld, the slope is θ at $x=0, p, 2p \ldots$:

$$\frac{d\eta}{dx} = \eta_0 \frac{2\pi}{p} \cos \frac{2\pi x}{p}$$

$$\frac{d\eta}{dx}(at\ x = 0,\ p,\ 2p, \ldots) = \eta_0 \frac{2\pi}{p} = \theta$$

From the geometry:

$$\delta = \frac{t}{2}\theta \text{ or } \theta = \frac{2}{t}\delta$$

Substituting into the above gives:

$$\eta_0 \frac{2\pi}{p} = \frac{2}{t}\delta$$

$$\eta_0 = \frac{p}{t\pi}\delta$$

where:

η_0 = Amplitude of the deformed flange sine wave (Figure 3.47)

t = Thickness for upper and lower flange

p = Spot weld pitch (distance between spot welds)

δ = Lateral deflection of the flange from the original position

We can now use energy principles and equate work done by the external shearing force to the strain energy stored in bending the flange into the sinusoidal shape.

The work done by an external elastic shearing force acting through distance δ is:

$$work = \frac{1}{2}F\delta = \frac{1}{2}(qL)\delta = \frac{1}{2}(qnp)\delta$$

where:

q = Shear flow at the welded flange

n = Number of spot welds in the length

p = Weld pitch (distance between spot welds)

δ = Deflection down the length of the welded flange

The bending strain energy in the distorted flange with moment of inertia I is:

$$e = \int_0^L \frac{EI(\eta")^2}{2}dx \text{ and } \eta" = -\eta_0\left(\frac{2\pi}{p}\right)^2 \sin\frac{2\pi x}{p}$$

$$e = \frac{EI}{2}\int_0^{L=np}\left[-\eta_0\left(\frac{2\pi}{p}\right)^2\sin\frac{2\pi x}{p}\right]^2 dx$$

$$e = \frac{EI}{2}\eta_0^2\left(\frac{2\pi}{p}\right)^4\int_0^{L=np}\sin^2\left(\frac{2\pi x}{p}\right)dx$$

Substituting the identity and integrating:

$$\sin^2\left(\frac{2\pi x}{p}\right) = \frac{1-\cos 2\left(\frac{2\pi x}{p}\right)}{2}$$

$$e = \frac{EI}{2}\eta_0^2\left(\frac{2\pi}{p}\right)^4\frac{np}{2}$$

$$\text{from above } \eta_0 = \frac{p}{t\pi}\delta \text{ so}$$

$$e = \frac{4EI\pi^2}{t^2}\frac{n\delta^2}{p}$$

Now equating work with strain energy gives:

$$\frac{1}{2}(qnp)\delta = \frac{4EI\pi^2}{t^2}\frac{n\delta^2}{p}$$

$$q = \frac{8EI\pi^2}{(pt)^2}\delta$$

For a flange width, w, and thickness, t, $I=wt^3/12$ and:

$$\delta = \frac{3p^2}{2E\pi^2wt}q \tag{3.16}$$

where:

δ = The deflection down the length of the welded flange

p = Weld pitch (distance between spot welds)

q = Shear flow across welded flanges (force per unit length)

E = Young's Modulus

w = Flange width

t = Flange thickness

Equation 3.16 provides some insight into the added flexibility when a spot-welded flange is present in a section. When a shear load is introduced at the weld interface, a longitudinal deflection will occur. This deflection is proportional to the square of the weld pitch. The deflection occurs whenever a shear across the weld is present, during both section bending or in torsion. To illustrate the influence on section stiffness, consider a tube, closed by a single spot-weld flange, Figure 3.48.

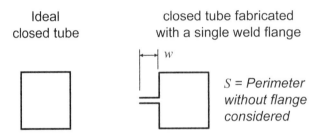

Figure 3.48 Closed tube with single weld flange.

The tube of Figure 3.48 is twisted by torque, T, and we are interested in the angle of deflection [7]. In our previous development of the closed-tube twisting behavior, we used a balance of work performed by the external torque to the internal shear strain energy within the tube. We can consider the same balance, but now the internal energy is also contained within the distorted weld flange.

External work = (shear strain energy in tube wall) + (strain energy in distorted flange)

$$\frac{1}{2}T\theta = \frac{SL}{2Gt}q^2 + \frac{1}{2}(qL)\delta$$

Using the result of Equation 3.16 for δ:

$$\frac{1}{2}T\theta = \frac{SL}{2Gt}q^2 + \frac{1}{2}(qL)\left(\frac{3p^2}{2E\pi^2 wt}q\right)$$

Which can be simplified using the relationship, $\dfrac{G}{E} = \dfrac{1}{2(1+\mu)}$ to:

$$\frac{T}{\theta} = \frac{(stiffness\ of\ closed\ tube\ without\ weld\ flange)}{\left[1 + \dfrac{3}{4\pi^2(1+\mu)}\dfrac{p}{wS}\right]}$$

(3.17)

where:

(T/θ) = Torsional stiffness of the spot-welded tube

p = Weld pitch

μ = Poisson's ratio

w = Flange width

S = Perimeter of the section without flange considered

This expression gives an estimate of the reduced stiffness in a twisted section when a single spot-welded flange is present.

Example: Spot-weld flange

Consider once again the slotted steering column mounting section, Figure 3.30. Let us close the section with a single spot-welded flange, Figure 3.49. From Equation 3.17, the factor

$$\psi = \frac{1}{1 + \dfrac{3}{4\pi^2(1+\mu)}\dfrac{p^2}{wS}}$$

is the fraction of the closed tube stiffness which will remain when the section is closed by a spot welded flange. Choosing typical values for the flange; $w=8\,mm$, $t=1\,mm$ and varying the weld pitch over the range $40\,mm < p < 60\,mm$, we see the tube stiffness with the weld is reduced to between 85% to 96% of the closed wall stiffness, Figure 3.50.

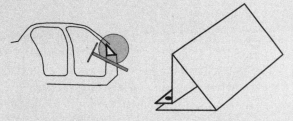

Figure 3.49 Steering column section with spot weld.

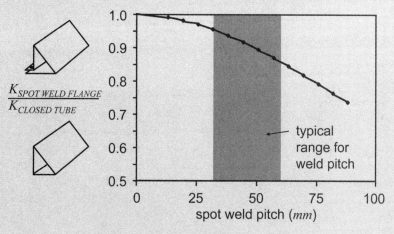

Figure 3.50 Spot weld spacing effect.

3.4 Thin-Wall Beam Section Design in Automobiles

In the previous sections, we have been discussing the unique behavior of the thin-wall sections used in the automotive body. It is helpful to ask a fundamental question: Why are automotive sections so often thin walled? We will use a simple thought experiment to answer this question. Consider a steel cantilever beam with a tip load. In this experiment the cross-section area is fixed (and therefore the beam mass is fixed), and we are free to choose the cross-section shape to maximize strength and stiffness. From basic beam theory we know for a cantilever beam that:

$$K = \frac{3EI}{L^3} \quad \text{and} \quad F_{MAX} = \frac{I\sigma_{DESIGN}}{Lc} \tag{3.18}$$

where:

K = Stiffness in bending

F_{MAX} = Maximum load (strength)

E = Young's Modulus

L = Beam length

c = Distance from the neutral axis to the outer fiber

I = Moment of inertia, $I = \int z^2 dA$

σ_{DESIGN} = Allowable design stress

In this experiment we will take the material yield stress as the design stress, $\sigma_{Design} = \sigma_Y$.

First let us imagine a square cross section of unit area as our base for stiffness and strength performance, Figure 3.51a. Now observe from Equations 3.18 that both stiffness and strength increase with moment of inertia, and as section material is moved away from the neutral axis the moment inertia is increased. This observation leads to an I beam shape with increased moment of inertia for constant cross-section area, yielding improved strength *and* stiffness. For example, Figure 3.51b shows an I beam of the same mass as the base square, resulting in 17 times the base stiffness. With no apparent difficulties with this design approach, we can move the material further from the neutral axis to result in the I beam of Figure 3.51c and a performance of 115 times the base stiffness. Given our assumptions, this approach could be continued until a *very* tall I beam of paper thickness resulted in presumably a *very* high stiffness and strength.

If we now move from a strictly thought experiment to testing two of the designed sections, the thick-walled I beam of Figure 3.51b and the thin-walled I beam of Figure 3.51c, we would find that while very stiff, the thin-walled section fails at an unexpectedly low load. The difference between expected performance and actual performance for the thinner-walled section is the existence of a new failure mode—*elastic plate buckling* of the compressive elements of the section.

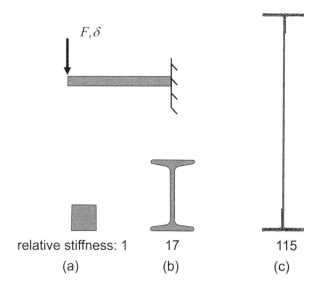

Figure 3.51 Relative bending stiffness for equal cross-section area.

The general behavior of a compressively loaded plate is to react loads by direct stress. However, if the plate is sufficiently thin, it bifurcates into the buckled shape, Figure 3.52. The compressive stress at which this occurs depends on the plate width-to-thickness ratio, b/t. The design stress for a thin plate under compressive loading is then the lower of yield stress or plate buckling stress, Figure 3.53.

Thus, in section design we deal with a trade-off: Thick-walled sections with higher strength but lower stiffness performance, or thin-walled sections with higher stiffness but lower strength performance due to plate buckling. Selection of the best section proportion then depends on the relationship of strength requirement to stiffness requirement for the section. This trade-off can best be illustrated with a specific automobile body design example.

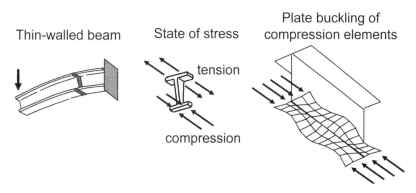

Figure 3.52 Deflection behavior of thin-walled members.

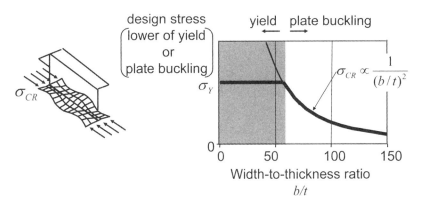

Figure 3.53 Plate buckling stress.

Example: Rocker sizing in convertible

Consider a highly idealized convertible, Figure 3.54 [8]. The bending performance of this vehicle is provided by the two rocker sections, which act as center-loaded simply supported beams. The specified bending stiffness requirement for this vehicle is *3335 N/mm* and strength requirement is *3335 N*. (The origin of global structure requirements will be discussed in subsequent chapters.) These requirements may be visualized graphically, Figure 3.55, with required performance shown as the shaded acceptable region. We wish to determine the mild steel rocker section width, *b*, and thickness, *t*, which meets both strength and stiffness requirements at a minimum mass. For this example we arbitrarily hold the section height at *1½* times the width.

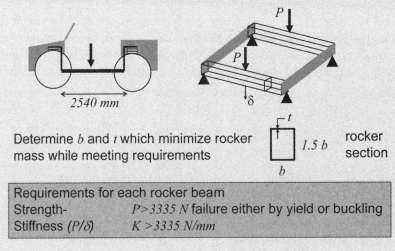

Figure 3.54 Convertible rocker sizing for bending.

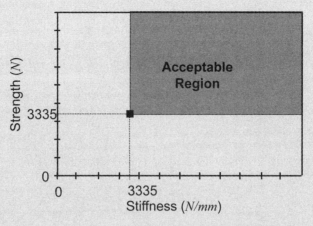

Figure 3.55 Convertible bending requirements.

Again using basic beam equations but now recognizing plate buckling as a possible failure mode, we have:

$$K = \frac{48EI}{L^3} \text{ and } F_{MAX} = \frac{4I\sigma_{DESIGN}}{Lc},$$ (3.19 a, b)

$$\text{where } \sigma_{DESIGN} = Min\begin{pmatrix} \sigma_Y = 207N/mm \\ \sigma_{CR} = \frac{748355}{(b/t)^2}N/mm \end{pmatrix}$$ (3.19c)

(We will develop the above relationship for plate buckling stress later.) First, let us find when the failure mode will be yield and when it will be plate buckling of the compressive cap of the section. By equating yield stress to buckling stress in Equation 3.19c, the *(b/t)* ratio is found below which failure is by yield, and above which failure is by buckling:

$$\sigma_Y = \sigma_{CR}$$
$$207N/mm = \frac{748355}{(b/t)^2}N/mm$$
$$(b/t) \cong 60$$

We can now hold the cross-section area constant (constant mass rocker), and then vary *(b/t)* until a section meeting the requirements is found. Using section geometry, we can relate thickness and width to cross-section area, *A*, and width-to-thickness ratio, *(b/t)*:

$$t = \frac{b}{(b/t)} \quad \text{and} \quad b = \sqrt{\frac{A}{5}\left(\frac{b}{t}\right)}$$ (3.20)

Now we can choose a specific value for cross-section area and then vary width-to-thickness ratio. Using Equations 3.20 to calculate *b* and *t*, we can then calculate I and, using Equations 3.19 a & b, calculate the resulting body stiffness and strength. Let us first follow this procedure for sections which will fail by yield by using a width-to-thickness ratio range of *5<(b/t)<60*. Doing this for a cross-section area of *250 mm²* gives the performance shown in Figure 3.56. Each data point represents a section of cross-section area *250 mm²* (and mass *10.4 kg*) with the width-to-thickness ratios shown. As no section design for this cross-section area meets the requirements, we must choose a larger cross-section area. After some iteration, Figure 3.57 shows the result for a section with minimum area, *A=348 mm²* (and mass *14.5 kg*) and with *(b/t) = 60*, which satisfies the strength requirement, but not the stiffness requirement. We must continue to increase the area to *A=1232 mm²* (and mass *51 kg*) to meet both requirements with *(b/t)=60*. Notice that while just meeting the stiffness requirement, this section greatly exceeds the strength requirement. This is due to the constraints imposed by the Equations 3.19 with a design stress of yield.

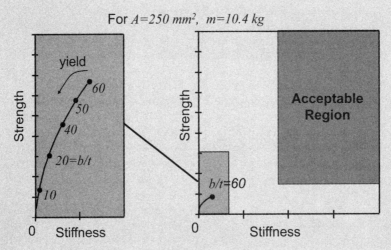

Figure 3.56 Thick-walled section performance.

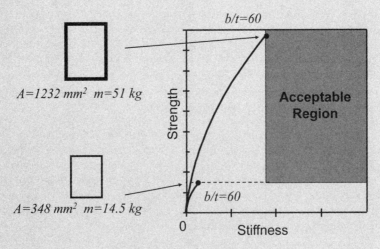

Figure 3.57 Thick-walled section performance meeting requirements.

Now let us allow the section to fail by plate buckling, and use a width-to-thickness ratio range of *60<(b/t)<160*. After some iteration, we find a section area of *A=832 mm²* (and mass *34.5 kg*) and *(b/t)=135*, which meets both requirements simultaneously, Figure 3.58. Comparing the mass for these beams, Figure 3.59, we see that using the thin-wall section results in a considerable mass savings and a more efficient structure. Note that it is the specific relationship of strength requirement to stiffness requirement that leads to the optimal section proportion.

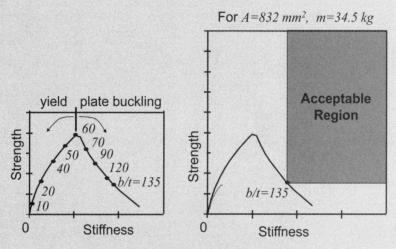

Figure 3.58 Thin-walled section performance.

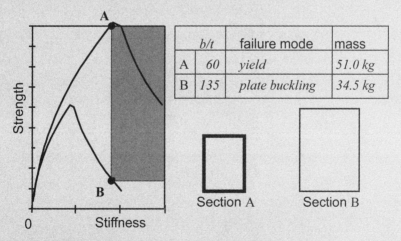

Figure 3.59 Mass savings of thin-walled section.

In the above example, we examined the motivation for using thin-walled sections for the case of global bending. Looking back at structural requirements for body subsystems, Figure 3.60, we can see that many of the subsystems, but not all, are stiffness dominated, and thin-wall sections are most efficient for those elements.

Understanding the relationship between section proportion and dominant requirement, we can look at an actual section and infer the dominant requirement for the section proportion. Figure 3.61 shows a typical automotive body structure with the measured

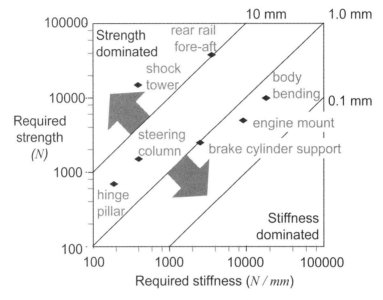

Figure 3.60 Dominant structural requirement.

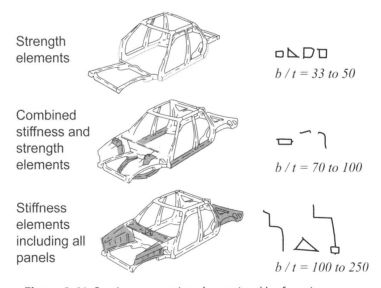

Figure 3.61 Section proportion determined by function

(b/t) ratios for various sections. For those sections with lower *(b/t)*, strength is the dominant requirement, and these sections are associated with reacting loads in a crash—roof crush, side impact, maintaining cabin integrity. For those sections with moderate *(b/t)*, a balance of strength and stiffness is required, and these sections are associated with major subsystem attachment points—suspension attachments and powertrain attachment. For those sections with high *(b/t)*, the stiffness requirement is clearly dominant, and these sections are associated with the overall stiffness of the body.

We have shown that many structural elements in automobile body design are dominated by stiffness requirements. A mass-effective means to design for stiffness performance is to use thin-walled sections which have a failure mode of buckling of the compressive section elements. In the next section, we will look at the governing equations for important modes of plate buckling.

3.5 Buckling of Thin-Walled Members

Perhaps the most significant difference between automotive sections and others is the failure modes caused by plate buckling of section elements. In this section we will develop tools to estimate plate buckling stress in section elements, and to estimate the strength of a buckled section.

3.5.1 Plate buckling

In the convertible rocker sizing example above, we used the expression:

$$\sigma_{CR} = \frac{748355}{(b/t)^2} N/mm$$

to describe the buckling stress for the compressive cap of the rocker section. We will now look at the physics of plate buckling and develop this and other buckling relationships.

Consider a flat plate of length a and width b where $a \geq b$, Figure 3.62. The plate lies in the x–y plane and is loaded by compressive stress, f_X, along the edges as shown. In the buckled state, the plate will have deflections, $w(x,y)$, normal to the plate in the z direction. For this analysis, we will consider the plate to have simply supported boundary conditions (no deflection and zero bending moment) along all edges.

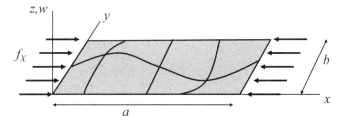

Figure 3.62 Flat plate.

Looking at a small element of this plate, we identify the loads acting, Figure 3.63. On the shaded y face of the element acts a bending moment per unit of length M_Y, which results in direct bending stresses σ_Y and a twisting moment, M_{XY}, which results in a

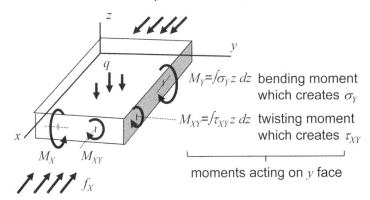

Compressive stress in x direction: f_X
Moments per unit length: M_X, M_Y, M_{XY}
Distributed normal load per unit area: q

$M_Y = \int \sigma_Y z\, dz$ bending moment which creates σ_Y

$M_{XY} = \int \tau_{XY} z\, dz$ twisting moment which creates τ_{XY}

moments acting on y face

Figure 3.63 Plate element loading.

shear stress, τ_{XY}. Similar moments, M_X and M_{XY}, are applied to the x face. Also acting on the element is a normal load per unit of area, q, and a compressive stress, f_X.

By applying 1) static equilibrium of the element under these loads, 2) compatibility of deformations within the plate, and 3) the material stress-strain relationship, the plate equations result. For the development of these equations, the reader is referred to texts on the theory of elasticity [9, 10, 11].

$$\frac{\partial^4 w}{\partial x^4} + 2\frac{\partial^4 w}{\partial x^2 \partial y^2} + \frac{\partial^4 w}{\partial y^4} + \frac{f_X t}{D}\frac{\partial^2 w}{\partial x^2} + \frac{q}{D} = 0 \qquad (3.21)$$

$$M_X = -D\left(\frac{\partial^2 w}{\partial x^2} + \mu\frac{\partial^2 w}{\partial y^2}\right)$$

$$M_Y = -D\left(\mu\frac{\partial^2 w}{\partial x^2} + \frac{\partial^2 w}{\partial y^2}\right)$$

$$M_{XY} = -D(1-\mu)\left(\frac{\partial^2 w}{\partial x \partial y}\right)$$

where $D = \dfrac{Et^3}{12(1-\mu^2)}$ is the plate flexural rigidity and $w(x,y)$ is the out-of-plane deflection of the plate which satisfies all these relationships.

To solve the plate equation, a function for the plate deflection is guessed and then substituted into Equations 3.21 to test its validity. Not only must the deflection satisfy the first of Equations 3.21, but the moments at the edges must satisfy the boundary conditions. For our case of a simply supported plate those boundary conditions are:

$$M_X(x=0, y) = 0 \quad \text{and} \quad M_X(x=a, y) = 0$$
$$M_Y(x, y=0) = 0 \quad \text{and} \quad M_Y(x, y=b) = 0$$

By observation of buckled plates, a reasonable guess at the deflected shape is:

$$w = A_{mn} \sin\left(\frac{m\pi x}{a}\right)\sin\left(\frac{n\pi y}{b}\right), \quad \text{where } m, n = 1, 2, 3, \ldots.$$

which satisfies the boundary conditions. (An example of this shape is shown in Figure 3.62 for $m=2$, $n=1$.) Substituting this assumed deformation into the first of Equations 3.21 with $q=0$ gives:

$$\left[\pi^4\left(\frac{m^2}{a^2}+\frac{n^2}{b^2}\right)^2 - \frac{f_X t}{D}\frac{m^2\pi^2}{a^2}\right]A_{mn}\sin\left(\frac{m\pi x}{a}\right)\sin\left(\frac{n\pi y}{b}\right) = 0 \qquad (3.22)$$

where $n, m = 1, 2, 3, \ldots$

Since the term $A_{mn}\sin\left(\frac{m\pi x}{a}\right)\sin\left(\frac{n\pi y}{b}\right)$ is not in general zero, the term in brackets must be equal to zero, and solving for the compressive stress, f_X, yields:

$$f_X = \frac{D\pi^2}{tb^2}\left[m\left(\frac{b}{a}\right)+\frac{n^2}{m}\left(\frac{a}{b}\right)\right]^2 \qquad (3.23)$$

where:

$f_X = \sigma_{CR}$, the critical plate buckling stress with all edges simply supported

D = Plate flexural rigidity $D = \dfrac{Et^3}{12(1-\mu^2)}$

t = Plate thickness

a = Plate length

b = Plate width (for the compressively loaded edge)

m = Number of half sine waves in buckled plate along x axis, $m=1, 2, 3 \ldots$

n = Number of half sine waves in buckled plate along y axis, $n=1, 2, 3 \ldots$

We would like to find the minimum compressive stress which would induce this deflected shape. This occurs at the minimum for the term in brackets. First observe that this term is, for all values of m, lowest when $n=1$. Physically, this occurs when the deflected shape across the plate is one half sine wave. To investigate the bracketed term with $n=1$, we plot it against plate length-to-width ratio, a/b, Figure 3.64. This plot shows for long plates where $a/b>1$ a minimum value of four results. Thus the critical compressive plate buckling stress for a simply supported plate is then:

$$\sigma_{CR} = \frac{D\pi^2}{tb^2}4 = 4\frac{E\pi^2}{12(1-\mu^2)(b/t)^2} \qquad (3.24)$$

Where variables are those of Equation 3.23. For plates where $a/b<1$ use $m=1$, $n=1$ in Equation 3.23 and:

$$\sigma_{CR} = \frac{D\pi^2}{tb^2}\left[\left(\frac{b}{a}\right)+\left(\frac{a}{b}\right)\right]^2 \qquad (3.25)$$

Where variables are those of Equation 3.23. However, for most practical applications in automobile sections, $a/b>1$ and Equation 3.24 is used when the boundary conditions are simply supported.

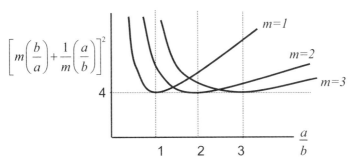

Figure 3.64 Buckling constant as function of (a/b).

Example: Plate buckling-1

To relate (*b/t*) ratio to critical stress, substituting values for mild steel *(E=207,000 N/mm², μ=0.3)* into Equation 3.24 yields:

$$\sigma_{CR} = 4\frac{(207000 N/mm^2)\pi^2}{12(1-.3^2)(b/t)^2} = \frac{748355}{(b/t)^2} N/mm^2$$

which is the relationship used earlier in the convertible example.

The critical stress of Equation 3.24 was developed for simply supported boundary conditions. With different boundary conditions, the form of Equation 3.24 may still be used, but with a plate buckling coefficient, *k*, replacing the value of 4.

$$\sigma_{CR} = k\frac{E\pi^2}{12(1-\mu^2)(b/t)^2} \tag{3.26}$$

Figure 3.65 provides values for the plate buckling coefficient for several boundary conditions found in practice [11, 12, 13]. Note the inclusion of shear and bending

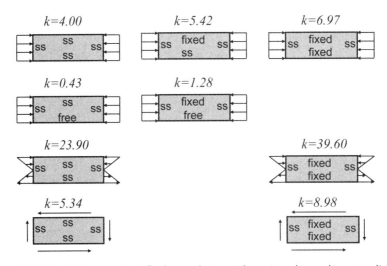

Figure 3.65 Buckling constant for long plates with various boundary conditions.

loading in this table. Whenever a plate is under a compressive state, it is subject to buckling. In the bending stress case, the plate is subject to buckling only in the compressive half of the web. In the shear case, Mohr's circle can be used to look at the stresses in an element rotated $45°$, Figure 3.66. It can be seen that, while the external loading is shear, on this rotated element one of the principle stresses is compression. Figure 3.67 shows the diagonal waves typical of a shear buckled plate.

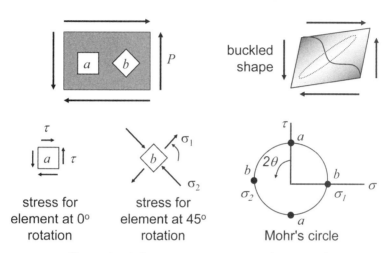

stress for element at 0° rotation   stress for element at 45° rotation   Mohr's circle

Figure 3.66 Compressive stress in a shear panel.

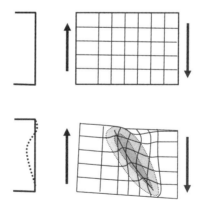

Figure 3.67 Shear buckled panel.

3.5.2 Identifying plate boundary conditions in practice

Our discussion of plate buckling has been in the context of automotive beam sections in which the plates are elements of the beam section. Given the complexity of automotive beam sections, it is at times difficult to choose the most appropriate boundary conditions and plate size. Our choices of simply supported, fixed, or free are highly idealized over the true physical constraint. Some general rules to help in this selection are shown in Figure 3.68a & b:

1. Each bent corner forms a simply supported boundary condition when the bent angle is greater than 40°. The plate width is then the distance from corner to corner.

2. When a plate is supported by a flange of thickness equal to or greater than the plate thickness, the boundary condition is fixed. The plate width is measured from the center line of the weld.

3. When an edge is unconnected, the boundary condition is free.

Figure 3.69 provides an example application of these rules for a typical automotive section.

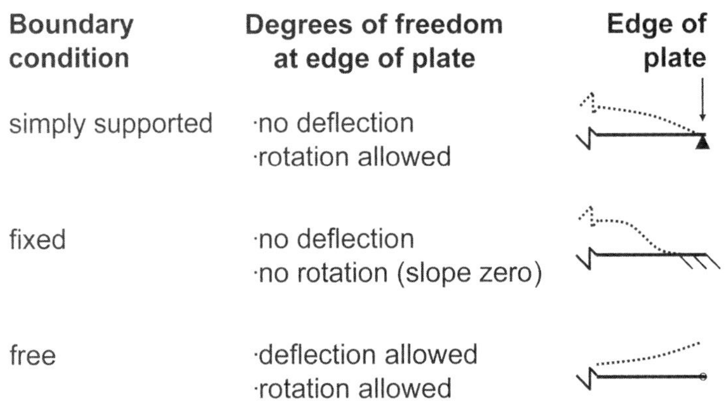

Boundary condition	Degrees of freedom at edge of plate	Edge of plate
simply supported	·no deflection ·rotation allowed	
fixed	·no deflection ·no rotation (slope zero)	
free	·deflection allowed ·rotation allowed	

Figure 3.68a Identifying plate edge conditions.

simply supported	fixed	free
ss ss for $\theta<40°$ for $\theta>40°$	$t_{FLANGE}=t_{PLATE}$ fixed ss $t_{FLANGE}<t_{PLATE}$ fixed $t_{FLANGE}>t_{PLATE}$	free
Each bent corner with angle >40° is ss. For angle less <40°, plate extends to next corner	When a plate is supported by a flat flange where $t_{FLANGE}\geq t_{PLATE}$ the boundary condition is fixed. Otherwise boundary is ss.	Un-connected edge is free
plate size: corner to corner	plate size: from center line of weld	plate size: from edge

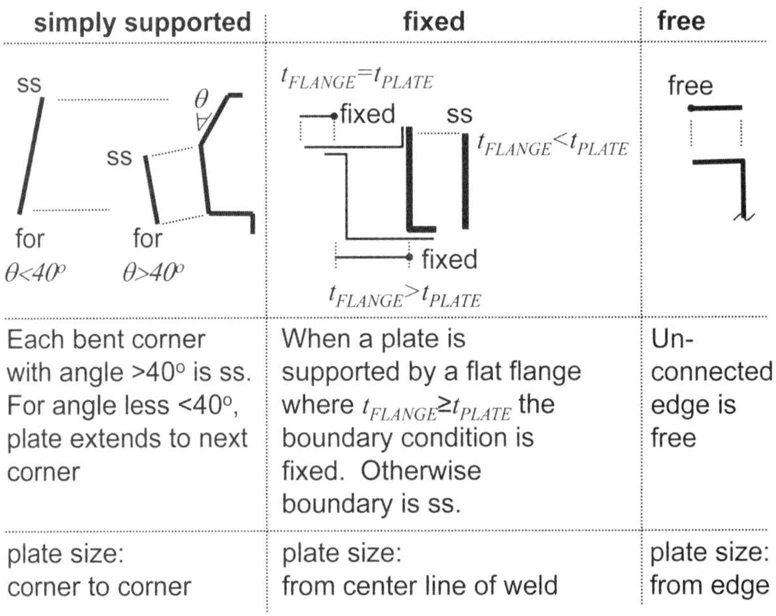

Figure 3.68b Identifying plate edge conditions.

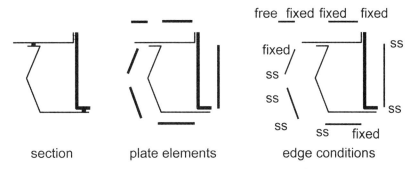

free fixed fixed fixed

fixed

ss

ss

ss

ss

free fixed fixed fixed

fixed

ss

ss

ss

ss fixed

section   plate elements   edge conditions

Figure 3.69 Example of plate edge conditions.

Example: Plate buckling-2

A *100-mm*-square thin-walled steel beam of thickness *0.86 mm* is loaded in compression. Determine the compressive load which will cause plate buckling in the section walls.

For this section, each of the four sides has a corner with a plate of the same thickness. The most appropriate boundary conditions are simply supported, with the plate uniformly loaded in compression. Using Figure 3.65, the buckling coefficient is *k=4*. The resulting critical stress using Equation 3.26 is

$$\sigma_{CR} = 4 \frac{(207000N \, / \, mm^2)\pi^2}{12(1-.3^2)(100mm \, / \, 0.86mm)^2}$$

$$\sigma_{CR} = 55.35N \, / \, mm^2$$

The load to cause this buckling on all four sides is

$$P_{CR} = \sigma_{CR}A = (55.35N \, / \, mm^2) \cdot 4(100mm \cdot 0.86mm)$$

$$P_{CR} = 19040N$$

Example: Plate buckling-3

A spot weld flange *(t=0.9 mm and b=15 mm)* is under uniform compressive stress down its length. See Figure 3.68b right side. Determine the stress at which the flange will buckle.

The flange has simply supported boundary conditions on three sides and is free on the fourth side. For this condition, the buckling coefficient is *k=0.43* from Figure 3.65. Using Equation 3.26

$$\sigma_{CR} = 0.43 \frac{(207000N \, / \, mm^2)\pi^2}{12(1-.3^2)(15mm \, / \, .9mm)^2}$$

$$\sigma_{CR} = 290N \, / \, mm^2$$

3.5.3 Post buckling behavior of plates

In the previous section we developed the ability to calculate the critical plate buckling stress for an automotive section, as well as the load at which this buckling will occur. We now will look at the behavior of plates in which the plate buckling stress is exceeded.

In the earlier example of the convertible rocker section, we made an implicit assumption that the onset of plate buckling of the rocker cap represents *excessive deflection* and therefore failure of the beam. For certain structural elements which the customer sees (quarter panel, roof, door panel), this is a valid assumption; a structural element with the out-of-plane buckling deformation would appear to have failed even though this deformation is elastic and will reverse when the load is removed. However, there are also non-visual structural elements for which the elastically buckled plate does not represent excessive deflection. For these elements the plate may be loaded beyond critical plate buckling before ultimate failure occurs. We now investigate the physical behavior of plates loaded beyond their critical stress, σ_{CR}.

To understand the post buckling behavior of a panel, first consider a slender beam with compressive loads at each end, Figure 3.70a. If the load is gradually increased, we will reach a critical load, $P_{CR}=\pi^2EI/L^2$, and the beam will undergo Euler buckling (not plate buckling). If we now attempt to increase the end load beyond the critical load, the beam will collapse. It has no ability to react increased load above the critical beam buckling load. Compare that behavior to that of a buckled plate, Figure 3.70b. Gradually increasing the compressive edge load, we will reach the plate buckling stress and the plate will bifurcate into the buckled pattern of deformation. However, unlike the slender beam, the load can continue to increase without collapse of the plate. Eventually an ultimate compressive load, Figure 3.70b, will be reached.

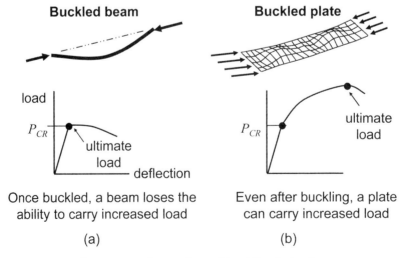

Buckled beam

load

P_{CR}

ultimate load

deflection

Once buckled, a beam loses the ability to carry increased load

(a)

Buckled plate

P_{CR}

ultimate load

Even after buckling, a plate can carry increased load

(b)

Figure 3.70 Comparison of buckling behavior.

For those cases where the mere visual detection of a buckled pattern does not represent excessive deformation, we would like to use this additional load-carrying capacity of the buckled plate. We begin by looking at a simply supported plate with an increasing compressive edge load, Figure 3.71a, and consider the stress across the plate at a section some distance from the applied load. Below the plate buckling stress, the stress distribution across the plate will be uniform and is given by $\sigma = P/(bt)$. As the load is increased to plate buckling, the stress remains uniform with $\sigma_{CR} = P/(bt)$, Figure 3.71a. Continuing to increase the load above critical plate buckling, the stress distribution across the plate becomes nonlinear. The stresses at the edges of the plate increase more rapidly than the stress at the center of the plate, Figure 3.71b. Eventually, by further increasing the load, the stress at the edges of the plate reaches the material yield stress and the plate exhibits an ultimate failure.

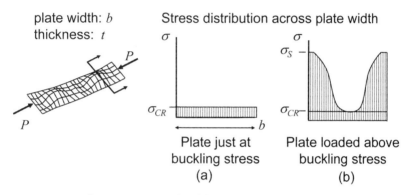

Figure 3.71 Post buckling stress distribution.

The difference between the load where a plate buckles and the load where the buckled plate yields can be considerable, and for efficient structures we must take advantage of this post buckled load carrying ability. In the next section, we will develop tools to do this.

3.5.4 Effective width

In many plate applications, we need to know the relationship between the applied compressive load, P, and maximum stress in the plate, σ_S, after elastic plate buckling has occurred. A convenient means to calculate the P-σ_S relationship is to replace the real plate with an imaginary plate having an *effective width, w*, which is less than the real plate width, Figure 3.72 [14]. The effective plate is assumed to have a uniform stress of σ_S across it, Figure 3.73b.

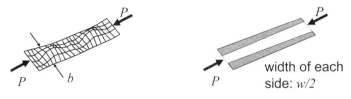

Physical buckled plate with
maximum stress: σ_S
thickness t
and width b

Effective unbuckled plate with
uniform stress: σ_S
thickness t
and width w

width of each
side: $w/2$

Figure 3.72 Effective width concept.

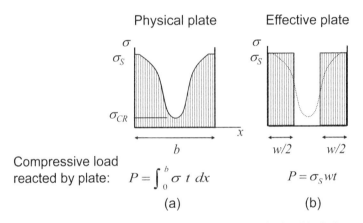

Physical plate   Effective plate

Compressive load
reacted by plate: $\quad P = \int_0^b \sigma\, t\, dx \qquad P = \sigma_S wt$

(a)   (b)

Figure 3.73 Determining effective width of a buckled plate.

For this effective plate, with currently unknown width w, the relationship between load and maximum stress is:

$$P_{\text{EFFECTIVE PLATE}} = \sigma_S(wt) \tag{3.27}$$

where:

$P_{EFFECTIVE\ PLATE}$ = Load on plate

σ_S = Maximum stress on plate

w = Effective width of plate

t = Thickness of plate

We can use Equation 3.27 as a convenient way to calculate the applied load, P, given σ_S, providing we know the effective width, w. In the following, we develop a means to calculate the effective width given a maximum plate stress, σ_S.

Looking at the physical plate with non-uniform stress distribution and maximum stress, σ_S, Figure 3.73a, the applied compressive load is given by:

$$P_{\text{REAL PLATE}} = \int_0^b \sigma(tdx) \tag{3.28}$$

Looking at the general shape of the stress distribution for a buckled plate, Figure 3.73a, we take for the stress distribution across the plate:

$$\sigma = \left(\frac{\sigma_S + \sigma_{CR}}{2}\right) + \left(\frac{\sigma_S + \sigma_{CR}}{2}\right)\cos\frac{2\pi x}{b} \tag{3.29}$$

where $\sigma_S > \sigma_{CR}$

This function gives us a stress of σ_{CR} at the center of the plate and σ_S at the edges.

Now substituting Equation 3.29 into 3.28 and integrating yields:

$$P_{\text{REAL PLATE}} = tb\left(\frac{\sigma_S + \sigma_{CR}}{2}\right) \tag{3.30}$$

as the force applied to the buckled plate which results in a maximum stress σ_S.

Equating the load for each plate shown in Figure 3.73 and solving for the effective width, w, gives the desired result:

$$P_{\text{REAL PLATE}} = P_{\text{EFFECTIVE PLATE}}$$

$$tb\left(\frac{\sigma_S + \sigma_{CR}}{2}\right) = \sigma_S(wt)$$

$$w = \frac{1}{2}\left(1 + \frac{\sigma_{CR}}{\sigma_S}\right)b \tag{3.31}$$

where:

   w = Effective width of plate under the stress σ_S

   σ_{CR} = Buckling stress for plate

   σ_S = Maximum stress on plate

   b = Actual width of plate

The above relationship gives us the width of an imaginary effective plate which has a uniform stress of σ_S across it. When we are interested in the load to induce a maximum stress of $\sigma_S > \sigma_{CR}$ in a buckled plate, we can use Equation 3.31 to determine the effective plate and then Equation 3.27 to determine the load, P, to induce that maximum stress.

Example: Effective width-1

Consider that the earlier example of the *100-mm--square thin-walled steel beam of thickness 0.86 mm* is loaded in compression. We determined the critical buckling stress for each side plate to be *55.35 N/mm²*, and the resulting compressive load to cause plate buckling to be *19040 N*. Now let us ask what load will cause a maximum stress of *111 N/mm²* in each plate.

In this example σ_S=*111 N/mm²* and σ_{CR}=*55.35 N/mm²*. Since $\sigma_S > \sigma_{CR}$ we are in the post buckling behavior for the plates, and we use Equation 3.31 to find the effective width of each of the four plates:

$$w = \frac{1}{2}\left(1 + \frac{55.35 N/mm^2}{111 N/mm^2}\right)100mm = 75mm$$

Now calculating the load on each plate using Equation 3.27

$$P_{\text{EFFECTIVE PLATE}} = 111 N/mm^2 (75mm \cdot 0.86mm)$$
$$P_{\text{EFFECTIVE PLATE}} = 7160 N \text{ per side}$$
and for all four sides
$$P = 28640 \text{ N}$$

Example: Effective width-2

A *100-mm--square thin-walled steel beam of thickness 0.86 mm* is loaded by a bending moment, M_x, in the x direction using the right-hand rule. Under this moment, the maximum compressive stress in the top plate is *111 N/mm²*. What is the effective moment of inertia, $I_{EFFECTIVE}$, for the section under this moment loading?

In the previous example, we determined that *w=75 mm* at σ_S=*111 N/mm²*. The effective section is then shown in Figure 3.74. We replace the top plate with the effective plate; since the other three plates have not buckled, they are fully effective. We can now use standard tools to determine the moment of inertia for this effective section. We find that the centroid is *46.65 mm* up from bottom of section and $I_{EFFECTIVE}$=*5.16x10⁵ mm⁴* (90% of the nominal unbuckled section moment of inertia

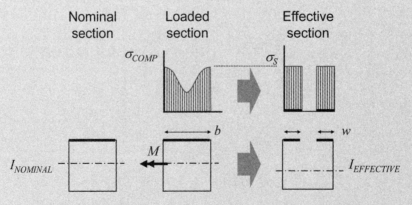

Figure 3.74 Reduced moment of inertia for beam with buckled element.

Our selection for the true stress distribution across the buckled plate, Equation 3.29, is based on reasonable fit to the observed plate stress distribution and also computational convenience. Alternative empirical relationships for effective width shown below give more accurate results in practice. These relationships are compared in Figure 3.75:

$$w = 0.894b\sqrt{\frac{\sigma_{CR}}{\sigma_S}}$$

$$w = b\sqrt{\frac{\sigma_{CR}}{\sigma_S}}\left(1 - 0.22\sqrt{\frac{\sigma_{CR}}{\sigma_S}}\right)$$

(3.32a, b)

where:

w = Effective width of plate under the stress σ_S

σ_{CR} = Buckling stress for plate

σ_S = Maximum stress on plate

b = Actual width of plate

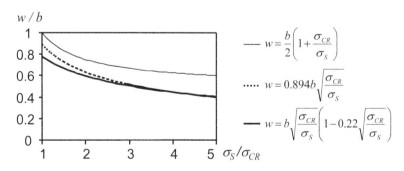

Figure 3.75 Comparison of effective width relationships.

3.5.5 Thin-walled section failure criteria

With the ability to calculate the compressive load acting on a buckled plate when given the maximum stress, we can now consider the ultimate failure load for the thin-walled plate. Extensive testing has been done on the ultimate load-carrying ability for a thin-wall section [15]. In these tests, a load was applied to a thin-walled square steel section as shown in Figure 3.76, and the peak load, P_{ULT}, was identified. Sections with a range of (b/t) ratio were tested from moderately thick walled, $(b/t)=50$, to highly thin walled, $(b/t)=200$. The measured ultimate load, P_{ULT}, was compared with five calculated values, Figure 3.77. These values include 1) the onset of yield at the outer fiber, P_Y; 2) fully plastic yield, P_p; 3) the onset of critical plate buckling, P_{CR}; 4) the onset of yield for the effective buckled section, P_{YE}; and 5) the onset of yield for a U section in which the compressive cap has been discarded, P_{YU}.

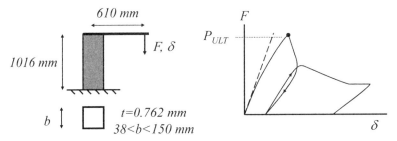

Figure 3.76 Behavior of square thin-walled sections. (Courtesy of SAE International)

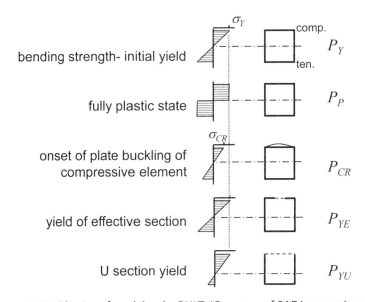

Figure 3.77 Physics of peak load – PULT. (Courtesy of SAE International)

The important result of this study is shown in Figure 3.78. For a wide range of (b/t) seen in automotive construction, the best predictor of ultimate load is P_{YE}, the onset of yield for the effective buckled section.

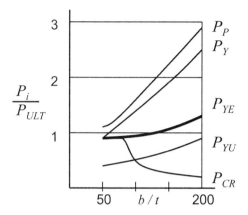

Figure 3.78 Ultimate load predicted by yield of effective section. (Courtesy of SAE International)

Example: Failure of a thin-walled beam

A Z section beam is part of a bumper reaction structure, Figure 3.79a. Calculate the ultimate compressive load for this section. The beam is stubby (i.e. will not Euler buckle as a beam), and the material is steel with σ_y=207 N/mm².

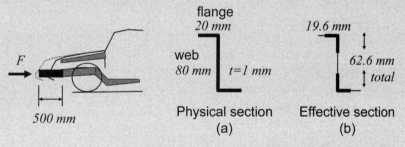

Figure 3.79 Bumper extension.

The ultimate load is given by P_{YE}, the onset of yield for the effective buckled section. First we find the plate buckling stress of each of the three plates which make up the Z section. Each of the two identical flanges have simply supported boundary conditions on three sides and are free on the fourth, so from Figure 3.65, k=0.425. Using Equation 3.26

$$\sigma_{CR} = 0.425 \frac{(207000 N/mm^2)\pi^2}{12(1-0.3^2)(20mm/1mm)^2} = 199 N/mm^2$$

for the flanges, and for the simply supported web k=4:

$$\sigma_{CR} = 4 \frac{(207000 N/mm^2)\pi^2}{12(1-0.3^2)(80mm/1mm)^2} = 117 N/mm^2$$

At a maximum stress equal to the material yield stress, $\sigma_s = \sigma_y$, and from Equation 3.31, the effective width for each plate is:

$$w = \frac{1}{2}\left(1 + \frac{199 N/mm^2}{207 N/mm^2}\right) 20mm = 19.6mm, \text{ for each flange}$$

$$w = \frac{1}{2}\left(1 + \frac{117 N/mm^2}{207 N/mm^2}\right) 80mm = 62.6mm, \text{ for the web}$$

The effective section, Figure 3.79c, is visualized with a uniform stress at yield. Using Equation 3.27, the ultimate compressive load is:

$$P_{YE} = (207 N/mm^2)(19.6mm \cdot 1mm +$$
$$19.6mm \cdot 1mm + 62.6mm \cdot 1mm)$$
$$P_{YE} = 21073 N$$

3.5.6 Techniques to inhibit buckling

From the above example, it can be seen that the existence of plate buckling reduces the ultimate load-bearing capacity of a section under compression. We may increase the strength of the section if somehow we can increase the critical plate buckling stress, σ_{CR}. In this section we look at ways to do this.

The plate buckling relationship, Equation 3.26, can be grouped as shown:

$$\sigma_{CR} = (k)\left(\frac{E\pi^2}{12(1-\mu^2)}\right)\left(\frac{1}{(b/t)^2}\right) \tag{3.33}$$

This grouping suggests means to increase σ_{CR} by increasing any of the three bracketed terms. The first enclosed term relates to boundary conditions, the second to normal stiffness of the plate, and the third to the width-to-thickness ratio. Several practical means to increase these terms will be discussed:

- Boundary conditions: flange curls and flanged holes
- Normal stiffness of the plate: material, curved elements, foam filling
- Width-to-thickness ratio: reducing width with beads and added edges.

The plate width may be reduced by adding beads to a section. By adding a longitudinal bead to a section, Figure 3.80, the plate width is effectively halved and the critical stress increased by a factor of four. Adding corners, Figure 3.81, can reduce plate width as shown, but in this case with some reduction in nominal moment of inertia as we are moving material closer to the neutral axis. Therefore, this technique is often used in sections under axial load as will be shown in the chapter on crashworthiness. Figure 3.82 is an example of adding corners to reduce plate width while maintaining moment of inertia.

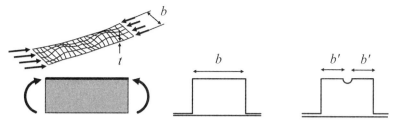

Figure 3.80 Reducing plate width by adding a bead.

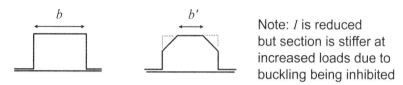

Note: I is reduced but section is stiffer at increased loads due to buckling being inhibited

Figure 3.81 Reducing plate width by chamfering corners.

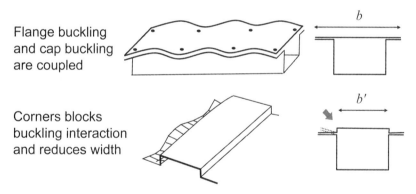

Flange buckling and cap buckling are coupled

Corners blocks buckling interaction and reduces width

Figure 3.82 Reducing plate width by adding corners.

While we have focused on plates as elements of beam sections, the idea of increasing critical stress using added corners or beads also applies to panels where the loading is in shear, Figure 3.83. Note that for crossing bead patterns it is important that the beads do not fully intersect. Such intersections act as hinge points which reduce plate normal stiffness considerably.

An example of altered boundary condition is a flange with the addition of a flange curl, Figure 3.84. This changes the boundary conditions from free to simply supported with the buckling coefficient increasing from *0.425 to 4.0*. Similar increases in buckling stress are seen in panels with flanged holes, Figure 3.85.

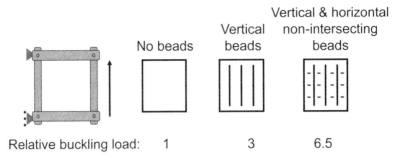

Figure 3.83 Reducing plate width using beads-shear case.

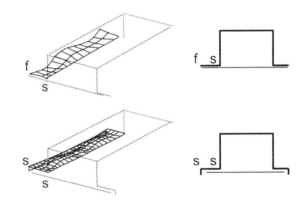

Figure 3.84 Changing boundary condition with a flange curl.

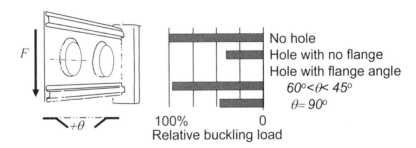

Figure 3.85 Flanged hole effect on shear buckling.

Plate normal stiffness can also be increased by curving the plate into a cylindrical element, Figure 3.86. This simple geometry change can increase critical stress dramatically [16, 17]. Yet another method to increase normal stiffness is by filling the beam section with foam. An example of the effects of foam filling is included in the section below on initial imperfections in plates.

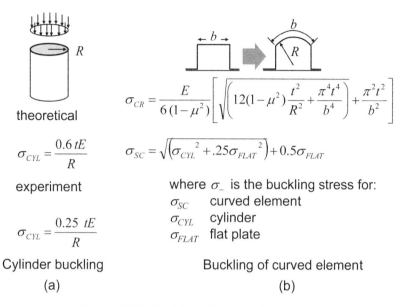

$$\sigma_{CR} = \frac{E}{6(1-\mu^2)}\left[\sqrt{\left(12(1-\mu^2)\frac{t^2}{R^2} + \frac{\pi^4 t^4}{b^4}\right)} + \frac{\pi^2 t^2}{b^2}\right]$$

theoretical

$$\sigma_{CYL} = \frac{0.6\,tE}{R}$$

$$\sigma_{SC} = \sqrt{\left(\sigma_{CYL}^2 + .25\sigma_{FLAT}^2\right)} + 0.5\sigma_{FLAT}$$

experiment

where σ_- is the buckling stress for:
σ_{SC} curved element
σ_{CYL} cylinder
σ_{FLAT} flat plate

$$\sigma_{CYL} = \frac{0.25\ tE}{R}$$

Cylinder buckling

Buckling of curved element

(a)

(b)

Figure 3.86 Buckling of curved elements.

Example: Inhibiting buckling

Consider again the Z section under compressive loading, but now with two buckling inhibiting techniques; a flange curl, and a central bead on the web, Figure 3.87. Again we are interested in the ultimate load.

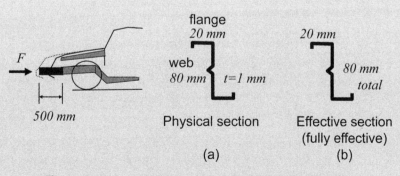

Figure 3.87 Bumper extension with section modifications.

With a flange curl, the boundary conditions for the flange are simply supported on all sides, with a resulting $k=4$. The plate buckling stress is for the stiffened flange:

$$\sigma_{CR} = 4\frac{(207000N/mm^2)\pi^2}{12(1-.3^2)(20mm/1mm)^2}$$

$$\sigma_{CR} = 1873N/mm^2$$

For the web with a central bead, we now have two plates of width $b=40\,mm$. With simply supported boundary conditions, $k=4$, and the plate buckling stress for the stiffened web:

$$\sigma_{CR} = 4\frac{(207000N/mm^2)\pi^2}{12(1-.3^2)(40mm/1mm)^2}$$

$$\sigma_{CR} = 468N/mm^2$$

Since both these critical stresses are above the material yield stress, the section is now fully effective and the ultimate load is:

$$P_{YE} = (207N/mm^2)(20mm \cdot 1mm + 20mm \cdot 1mm +$$
$$+ 40mm \cdot 1mm + 40mm \cdot 1mm)$$

$$P_{YE} = 24840N$$

Compare to *21,073 N* for the unstiffened section.

3.5.7 Note on the use of high-strength steel

High-strength steels now make up a large portion of the steel used in automobile body structures. The motivation for this use is to gain higher strength and energy absorption at a reduced mass. Often this change in steel grade is done by direct substitution; only material and thickness is changed while keeping existing section geometry unchanged. It is important to recognize that because of plate buckling, the change in the strength is not directly proportional to yield stress. Take the previous Z section example with dual-phase steel, $\sigma_Y = 650\ N/mm^2$ *(94,000 psi)* substituted for mild steel, $\sigma_Y = 207\ N/mm^2$ *(30,000 psi)*. This substitution does not achieve a *(650/207)=3.14* times increase in strength because the buckling stress of the web, $\sigma_{CR}=468\ N/mm^2$ *(68,000 psi)* is now below yield and the section is no longer fully effective as it was with mild steel. When substituting with higher-strength materials, it is important to reexamine the effective properties of a section and add buckling inhibitors which may not have been necessary with lower-strength steel.

3.5.8 Note on bifurcation and initial imperfection

The mathematical development of plate buckling considers an ideally flat plate which snaps instantly—bifurcates—from a flat plate to the buckled deformation. In practice, the plates making up the elements of an automotive section are imperfect, containing as-fabricated out-of-plane geometry.

To see the effect of this imperfection on buckling behavior, consider an edge view of a plate, Figure 3.88. Here we will greatly simplify our model to understand the most basic physics, and will consider an analogy to the plate—a beam supported by an elastic foundation of stiffness k_f *(N/mm/mm)*, Figure 3.89. The foundation stiffness represents, for example, the additional rigidity of foam filling the section.

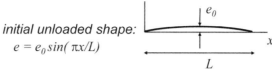

side view of top plate element at center line

initial unloaded shape:
$e = e_0 sin(\pi x/L)$

Figure 3.88 Initial imperfection in plate.

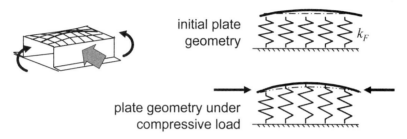

initial plate geometry k_F

plate geometry under compressive load

Figure 3.89 Elastic foundation model of initial imperfection.

The beam is given an initial unloaded 'imperfect' shape characterized by a half sine wave with amplitude e_0, Figure 3.90 top.

$$e = e_0 sin(\pi x/L) \tag{3.34a}$$

The foundation springs in this initial condition are unstressed. The beam may deflect from this shape a distance $\eta(x)$ under the action of compressive end loads, P. We take as the deflected shape a half sine wave with amplitude η_0, Figure 3.90 center,

$$\eta = \eta_0 sin(\pi x/L) \tag{3.34b}$$

Cutting the beam at a distance x from the end, Figure 3.90 bottom, the bending moment at the cut is

$$M(x) = -P(\eta + e) - \int_0^x k_F \eta(\xi)(x - \xi)d\xi \tag{3.34c}$$

substituting the expressions for e, Equation 3.34a, and η, Equation 3.34b, and integrating gives:

$$M(x) = \left[-P(\eta_0 + e_0) + k_F \eta_0 \left(\frac{L}{\pi} \right)^2 \right] sin \frac{\pi x}{L}$$

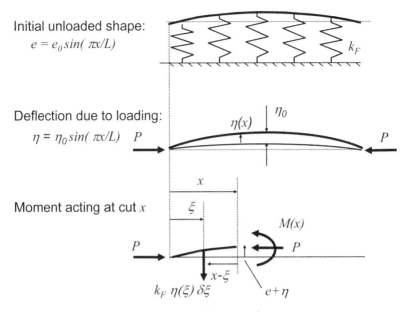

Initial unloaded shape:
$e = e_0 \sin(\pi x/L)$

k_F

Deflection due to loading:
$\eta = \eta_0 \sin(\pi x/L)$ P

$\eta(x)$ η_0

P

x

Moment acting at cut x

ξ

$M(x)$

P P

$x-\xi$

$k_F \, \eta(\xi) \, \delta\xi$ $e+\eta$

Figure 3.90 Equilibrium at beam cut.

Substituting Equations 3.34b and c into the beam equation:

$$\frac{\partial^2 \eta}{\partial x^2} = \frac{M}{EI}$$

gives the following result:

$$\frac{\eta_0}{e_0} = \frac{1}{\dfrac{P_{CR}}{P}\left(1 + \dfrac{k_F}{EI}\right) - 1} \tag{3.34d}$$

where:

P = Applied compressive load

P_{CR} = Euler buckling load, $P_{CR}=EI(\pi/L)^2$

e_0 = Initial out-of-plane imperfection

η_0 = Out-of-plane deflection

EI = Bending stiffness of plate

k_F = Stiffness per unit length which resists the out-of-plane deflection

Equation 3.34d gives us the deflection at the center of the beam, η_0, normalized to the initial imperfection magnitude, e_0. The deflection depends on the applied load, P, normalized to the critical Euler buckling load, P_{CR}.

First consider the behavior without any foundation stiffness, $k_F=0$. Equation 3.34d with $k_F=0$ yields

$$\frac{\eta_0}{e_0} = \frac{1}{\dfrac{P_{CR}}{P} - 1}$$

$$\frac{P}{P_{CR}} = \frac{\dfrac{\eta_0}{e_0}}{\dfrac{\eta_0}{e_0} + 1} \qquad (3.34e)$$

Figure 3.91 plots the applied normalized compressive load, P/P_{CR}, against the lateral normalized deformation, η_0/e_0, at the center of the beam, Equation 3.34e. It can be seen that the existence of an initial imperfection tends to 'ease' the beam into the buckled shape rather than a sharp snap-over into the deflected shape at P_{CR}. While this model describes an imperfect *beam*, the physical behavior is analogous to a real plate with an initial imperfection.

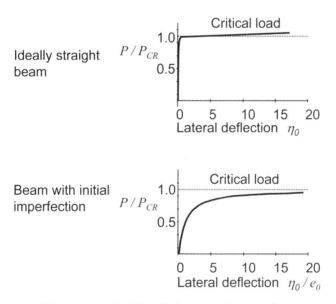

Figure 3.91 Buckling behavior with imperfection.

In the earlier section on inhibiting buckling, the technique of increasing the normal stiffness of the plate was mentioned. One embodiment of this technique is to foam fill a section. To buckle, a plate element of the section must also deflect the foam, and this action inhibits the onset of buckling by effectively increasing the normal stiffness of the plate. This can be modeled, Figure 3.89, as a beam (the plate) on an elastic foundation (the foam) by using Equation 3.34d. In this case, the factor $k_F/(EI)$ is the ratio of foam compressive stiffness to beam stiffness. Figure 3.92 shows the results of applying Equation 3.34d, and illustrates how the foam stiffness tends to increase the buckling load above the critical load for the unstiffened beam.

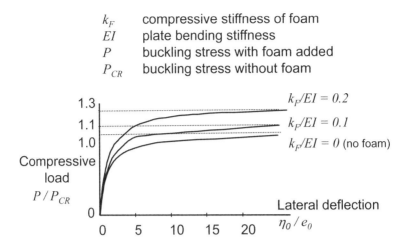

k_F compressive stiffness of foam
EI plate bending stiffness
P buckling stress with foam added
P_{CR} buckling stress without foam

Figure 3.92 Plate buckling of a foam-filled section.

3.6 Automobile Body Panels: Plates and Membranes

A panel is a flat or curved surface with thin thickness and, along with the beam, is a primary structural element in the automobile body, Figure 3.93. The bending stiffness of thin-walled automotive panels is quite low, while the in-plane stiffness is quite high. In this sense, a valid analogy for an automotive panel is a sheet of paper: little resistance to bending, but if laid flat and shearing loads applied in the plane of the paper, it provides considerable resistance. Only when a panel is highly curved does it begin to provide much stiffness to out-of-plane loads.

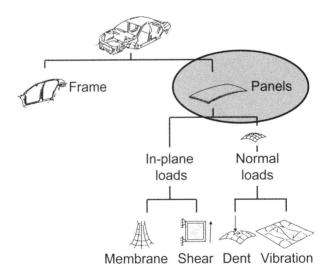

Figure 3.93 Loading classification for panels.

We will therefore divide the treatment of panels by the type of loads acting on the panel: normal loading of curved panels, and in-plane loading of flat or curved panels, Figure 3.93. In the next section we will look at loading of curved panels by a *point* load normal to the surface. Panels may also be loaded by a *distributed* normal load, for example by inertia during panel vibration; this topic will be treated later in the chapter on vibration.

3.6.1 Curved panel with normal loading

The exterior panels of an automobile are largely influenced by the overall styling of the vehicle. Because of this, structural performance is not the shape-defining function. However, a small set of structural requirements are used to screen valid shapes for curved exterior panels. These involve the reaction to normal point loading, which we will now discuss.

A curved panel loaded normal to the surface at a point exhibits complex load-deflection behavior, Figure 3.94. Initially, the load-deformation relationship is linear and elastic and can be characterized by a stiffness, K. At some higher load, the panel may begin to invert its curvature, Figure 3.94. This inversion may be either *soft*, where the surface stays in contact with the load applicator, or *hard*, where the surface snaps over and looses contact with the load applicator. The load where a hard snap-over occurs is referred to as the critical oil-canning load, P_{CR} (from the behavior of the bottom of an oiling can). A customer may judge the solidness of a panel by pushing with a thumb, and both K and P_{CR} relate to the customer perception of panel quality, with higher values being preferred.

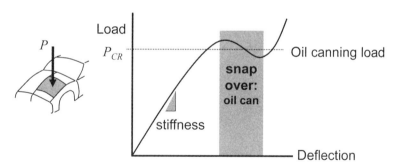

Figure 3.94 Curved panel behavior under normal load.

Dent resistance, another case of normal loading of panels, is measured by the kinetic energy of a dart, directed normal to a surface, which leaves a permanent dent in the panel. Again referring to Figure 3.94, this energy is proportional to the area under the load-deflection curve up to the point where permanent deformation occurs.

The properties of panel stiffness, oil-can load, and dent resistance for a curved panel are difficult to predict, and the equations below are a combination of analytical and empirical considerations. All equations assume simply supported boundary conditions for the panel [18, 19].

3.6.2 Normal stiffness of panels

In modeling the normal stiffness of automotive panels, we assume the panel is large and the distortion under a point normal load is localized and does not extend to the boundaries. First consider the theoretical normal stiffness of a partial spherical shell under a concentrated load:

$$K = \frac{B}{R} \frac{Et^2}{\sqrt{1-\mu^2}} \tag{3.35}$$

where:

R = Spherical radius $(R >> t)$

B = Constant: $B=2.309$ for a shallow shell, [6]

E = Young's Modulus

μ = Poisson's ratio

t = Panel thickness

Automotive panels are generally doubly curved—curvature not the same in two orthogonal directions—and Equation 3.35 can be generalized by replacing the spherical curvature $(1/R)$ with:

$$\frac{1}{R_{EQ}} = \frac{\left(\frac{L_1^2}{R_1}\right) + \left(\frac{L_2^2}{R_2}\right)}{2L_1L_2} = \frac{1}{2}\left[\frac{1}{R_1}\left(\frac{L_1}{L_2}\right) + \frac{1}{R_2}\left(\frac{L_2}{L_1}\right)\right] \tag{3.36}$$

where:

R_{EQ} = Equivalent radius for a doubly curved panel

L_1, L_2 = Rectangular panel dimensions

R_1, R_2 = Radii of curvature in orthogonal directions, shown in Figure 3.96

A convenient means to express panel curvature is using the crown height H_C, where:

$$H_C = \left(\frac{L_1^2}{8R_1}\right) + \left(\frac{L_2^2}{8R_2}\right) \tag{3.37a}$$

Substituting Equation 3.37a into Equation 3.36, gives:

$$\frac{1}{R_{EQ}} = \frac{4H_C}{L_1L_2} \tag{3.37b}$$

To calculate normal stiffness for a doubly curved panel, use Equation 3.36 or 3.37 to determine the equivalent spherical radius, R_{EQ}, then substitute into Equation 3.35. The constant, B, in Equation 3.35 ranges from the shallow spherical value, $B=2.309$ to $B=2.96$ for a shallow doubly curved panel $(H_C/t=4)$ to $B=3.618$ for a deeper doubly curved panel $(20<H_C/t<60)$ [18].

3.6.3 Oil-canning resistance

The critical buckling load, P_{CR}, at which the panel will bifurcate and reverse its curvature is given by

$$P_{CR} = \frac{CR_{CR}\pi^2 Et^4}{L_1 L_2 (1-\mu^2)}$$

(3.38)

where:

$C = 0.645 - 7.75x\ 10^{-7} L_1 L_2$ for $L_{1,2}$ in millimeters

$R_{CR} = 45.929 - 34.1832 + 6.397\lambda^2$

$$\lambda = .5\sqrt{\frac{L_1 L_2}{t}}\sqrt{\frac{12(1-\mu^2)}{R_1 R_2}}$$

valid over the range
$$\begin{cases} \dfrac{R_1}{L_1}\ and\ \dfrac{R_2}{L_2} > 2 \\[2mm] \dfrac{1}{3} < \dfrac{L_1}{L_2} < 3 \\[2mm] L_1 L_2 < 0.774 m^2 \end{cases}$$

3.6.4 Dent resistance

The denting resistance, *W*, is the minimum energy to dent the surface. Here the definition of a dent is a *0.025 mm (0.001 in)* permanent deformation in the panel. Dent resistance depends on σ_{YD}, the yield at a *dynamic* strain rate (rather than the more typical *static* value). The dynamic strain rate corresponding to denting (*10 to 100 /sec*) is many orders of magnitude higher than the typical static tensile test strain rate (*0.001 /sec*). The ratio of dynamic to static yield strength of steels at strain rates of up to *100 /sec* is shown in Figure 3.95. The denting energy, based on empirical curve fit, is [18]:

$$W = 56.8\frac{(\sigma_{YD}t^2)^2}{K}$$

(3.39)

where:

K = Panel normal stiffness (Equation 3.35)

t = Panel thickness

σ_{YD} = Yield strength at a dynamic strain rate

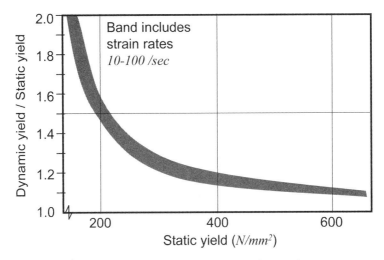

Figure 3.95 Dynamic yield stress for steel.

Example: Curved panel behavior

A steel automobile hood outer panel is shown in Figure 3.96. A portion of the panel, with the given dimensions, is supported at its periphery by the hood inner structure, which provides simply supported boundary conditions. The dynamic yield stress is $\sigma_{YD}=298\ N/mm^2$. What is the panel stiffness, oil-can load, and denting energy?

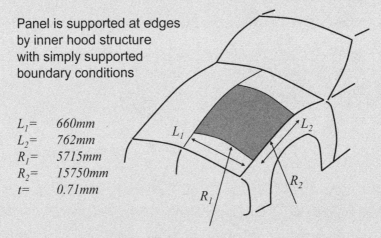

Panel is supported at edges
by inner hood structure
with simply supported
boundary conditions

$L_1=$ $660mm$
$L_2=$ $762mm$
$R_1=$ $5715mm$
$R_2=$ $15750mm$
$t=$ $0.71mm$

Figure 3.96 Automobile hood panel.

To determine stiffness, we will use Equation 3.35, but first need the panel crown height, H_C, given below Equation 3.37a,

$$H_C = \left(\frac{(660mm)^2}{8(5715mm)} \right) + \left(\frac{(762mm)^2}{8(15750mm)} \right) = 14.14mm$$

Now using Equation 3.37b to arrive at the equivalent spherical curvature for the panel:

$$\frac{1}{R_{EQ}} = \frac{4(14.14mm)}{(660mm)(762mm)} = 0.0001124\,/\,mm$$

Finally using Equation 3.35 with $B=2.309$:

$$K = 2.309(0.00011246\,/\,mm)\frac{(207000N\,/\,mm^2)(0.71mm)^2}{\sqrt{1-0.3^2}} = 28.39N\,/\,mm$$

Oil can load is given by Equation 3.38, where:

$$C = 0.645 - 7.75 \times 10^{-7}(660mm)(762mm) = 0.255$$

$$\lambda = .5\sqrt{\frac{660mm \cdot 762mm}{0.71mm}}\sqrt{\frac{12(1-0.3^2)}{5715mm \cdot 15750mm}} = 7.854$$

$$R_{CR} = 45.929 - 34.183\lambda + 6.397\lambda^2 = 172$$

Then substituting into Equation 3.38:

$$P_{CR} = \frac{(.255)(172)\pi^2(207000N\,/\,mm^2)(0.71mm)^4}{660mm \cdot 762mm(1-0.3^2)} = 50N$$

The denting energy is given by Equation 3.39 where we have calculated the panel stiffness, K, above:

$$W = 56.8\frac{(298N\,/\,mm^2(0.71mm)^2)^2}{28.39N\,/\,mm}$$

$$W = 4.515 \times 10^4\,Nmm$$

3.6.5 In-plane loading of panels

In this section we will look at panels where the applied loads are within the plane of the panel, Figure 3.93. This includes membrane panels which are curved, but over a small element of the surface, the loads are tangent to the surface. A second case of in-plane loading—that of shear loading of panels—will be treated later in the chapter on body torsion.

3.6.5.1 Membrane shaped panels

Many of the structural elements of the underbody are panels, including floor pan, motor compartment sides, dash, and wheelhouse inner panel. Unlike the exterior panels, these underbody panels are largely shaped by structural requirements. Therefore, our orientation in this section is that of design rather than analysis. The question we would like to answer is: Given a set of loads applied to a panel, what is the best panel shape to react to those loads? (*Best* here is defined as stiffest, strongest, and lightest.) We begin with a simple observation: consider a small element taken from a loaded panel and viewed from its edge, Figure 3.97. If the loading on the panel is pure bending, the stress distribution is linear with zero stress at the neutral axis, Figure 3.97a. Intuitively, much of the material of the section is stressed at a relatively low level yet we are 'paying' for the mass of that material.

The panel stiffness in bending, $D = \dfrac{Et^3}{12(1-\mu^2)}$, is very low since the thickness in automotive panels is very small. Now consider a panel loaded by pure membrane loads, Figure 3.97b. All the material of the section is uniformly stressed across the thickness and contributes to reacting the applied load. Given this fully stressed quality of a membrane, we would like to design the panel *shape* such that only membrane loading is present.

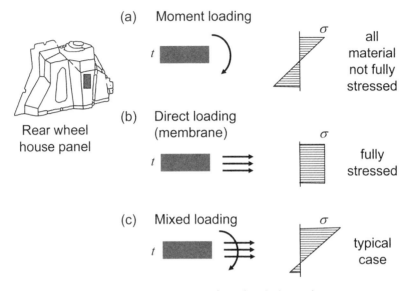

Figure 3.97 Element from loaded panel.

For simplicity in understanding basic structural behavior, we will consider axially symmetrical membranes, Figure 3.98. The membrane surface is defined by rotating a *generating line* about the axis of symmetry. Consider a small element on the surface. Two radii of curvature define the geometry of the element. A longitudinal radius, R_L,

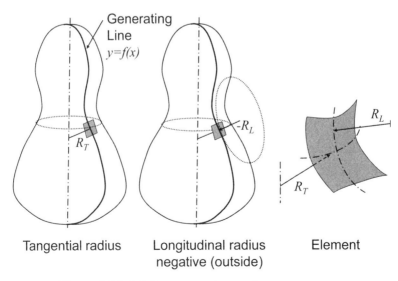

Figure 3.98 Axially symmetric membrane geometry.

which defines the radius of a longitudinal section of the element, and a tangential radius, R_T, which defines the radius of a tangential section of the element. The longitudinal radius is defined by the curvature of the generating line at the point where the element sits. The tangential radius is a line which is normal to the element and extends to the axis of the membrane.

Acting on the element are tangential stresses, σ_{TAN}, and longitudinal stresses, σ_{LONG}, as shown in Figure 3.99, as well as an outward pressure, p. These stresses are uniform across the thickness. The element must be in static equilibrium, and balancing forces in the direction normal to the element, Figure, 3.99:

$$p(ds)(ds') - 2\sigma_{TAN}(tds')\frac{d\theta}{2} + 2\sigma_{LONG}(tds)\frac{d\phi}{2} = 0$$

from geometry, $ds'=R_T\,d\theta$ and $ds=R_L\,d\phi$, and substituting into the above yields the first equation a membrane surface must satisfy:

$$\frac{p}{t} = \frac{\sigma_{TAN}}{R_T} + \frac{\sigma_{LONG}}{R_L} \tag{3.40}$$

where:

p = Pressure normal to membrane

t = Thickness of membrane

σ_{TAN} = Tangential stress

σ_{LONG} = Longitudinal stress

R_T = Radius of curvature in tangential direction

R_L = Radius of curvature in longitudinal direction

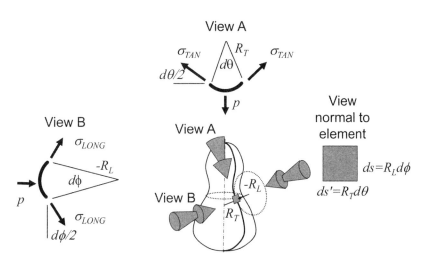

Figure 3.99 Loads on membrane element.

Now cutting the membrane perpendicular to the axis, Figure 3.100, the cut portion may be placed into equilibrium in the axial direction. The net upward force on the cut section is F, and along the cut the longitudinal stress, σ_{LONG}, acts. At the cut, the normal to the surface makes an angle α with the axis as shown. Static equilibrium yields the second equation a membrane surface must satisfy:

F=(component of stress along axis)/(Area on which stress acts)

$$F = (\sigma_{LONG} \sin\alpha)(2\pi t\, x) \text{ or } F = (\sigma_{LONG} \sin\alpha)(2\pi t\, R_T \sin\alpha)$$

$$F = 2\pi t\, x(\sigma_{LONG} \sin\alpha) \text{ or } F = 2\pi t\, R_T \sigma_{LONG} \sin^2\alpha \tag{3.41}$$

where:

F = Net force acting on a cut portion of membrane

α = Angle made between normal to the surface with the axis

t = Thickness of membrane

σ_{LONG} = Longitudinal stress

R_T = Radius of curvature in tangential direction

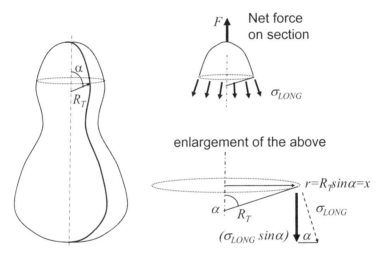

Figure 3.100 Cut perpendicular to membrane axis.

Now look at the *generating line, y=f (x)*, for the membrane, Figure 3.101. The longitudinal and tangential radii may be defined at any point, *(x,y)*, on the line using analytical geometry [20]:

$$R_T = \frac{x}{\sin\alpha}, \quad R_L = \frac{\left[1+\left(\dfrac{dy}{dx}\right)^2\right]^{\frac{3}{2}}}{\dfrac{d^2y}{dx^2}}$$

$$\sin(\alpha) = \frac{\dfrac{dy}{dx}}{\left[1+\left(\dfrac{dy}{dx}\right)^2\right]^{\frac{1}{2}}}, \quad \cos(\alpha) = \frac{1}{\left[1+\left(\dfrac{dy}{dx}\right)^2\right]^{\frac{1}{2}}} \tag{3.42}$$

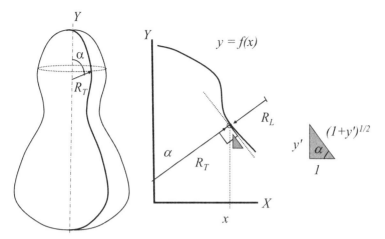

Figure 3.101 Membrane generating line.

With the set of Equations 3.40, 3.41, and 3.42 we can now determine *the* membrane shape, defined by the generating line, $y=f(x)$, for a specific loading. The sequence we will follow is:

1. Define the loading

2. Substitute the loading into the force balance, Equation 3.41

3. Substitute into the resulting equation the geometric relationships of Equations 3.42

4. A differential equation for the generating line will result, which we solve for $y=f(x)$ to define the shape of the membrane panel.

Example: Uniform load on circular membrane

Consider a panel with a uniform pressure load of p and a circular boundary. Such a load may represent an automobile rear compartment load floor. What is the shape of a panel which will react this loading with only membrane stress?

The total load $F=p(Area)=p\pi x^2$ substituting into Equation 3.41

$$p\pi x^2 = 2\pi t x(\sigma_{LONG} \sin\alpha)$$

and substituting for $sin\alpha$ from Equations 3.42:

$$x = 2t \frac{\sigma_{LONG}}{p} \frac{\frac{dy}{dx}}{\left[1 + \left(\frac{dy}{dx}\right)^2\right]^{\frac{1}{2}}}$$

$$\beta = 2t \frac{\sigma_{LONG}}{p}$$

$$\frac{dy}{dx} = \frac{x}{\left[\beta^2 - x^2\right]^{\frac{1}{2}}}$$

$$y = \int \frac{x}{\left[\beta^2 - x^2\right]^{\frac{1}{2}}} dx$$

which may be integrated to determine the generating line resulting in:

$$x^2 + y^2 = \beta^2, \quad \beta = 2t \frac{\sigma_{LONG}}{p} \tag{3.43}$$

which gives a circle of radius β as a generating line, and when swept about an axis, gives a spherical surface as the panel shape with $R_T = \beta$ and $R_L = \beta$. Substituting the value for β into Equation 3.40:

$$\frac{p}{t} = \frac{\sigma_{TAN}}{\left(2t \frac{\sigma_{LONG}}{p}\right)} + \frac{\sigma_{LONG}}{\left(2t \frac{\sigma_{LONG}}{p}\right)}$$

$$\sigma_{TAN} = \sigma_{LONG} = \frac{p\beta}{2t} \tag{3.44}$$

So as designers of the panel loaded as above, if we shape the panel as a spherical surface, the loaded panel will have no bending stress and will have a uniform design stress across the thickness given by $\sigma = p\beta/2t$. As the material is fully stressed, the panel will be stiffer than alternative shapes.

Example: Circular membrane loaded by rigid ring

Consider a panel with a load, F, applied to a rigid ring, Figure 3.102a. Again the boundary of the panel is circular. Such a panel is an idealization of a suspension attachment point. What is the shape of a panel which will react this loading with only membrane stress?

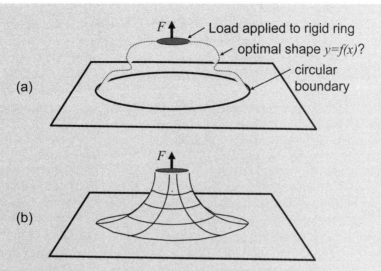

Figure 3.102 Circular membrane loaded by rigid ring.

Equation 3.41 gives:

$$F = (2\pi x)t\,\sigma_{LONG}\sin\alpha$$

$$F = 2\,\pi t\,\sigma_{LONG}\,\frac{x\dfrac{dy}{dx}}{\left[1+\left(\dfrac{dy}{dx}\right)^{2}\right]^{\frac{1}{2}}}$$

letting $\rho = \dfrac{F}{2\,\pi t\sigma_{LONG}}$

$$\frac{y}{\rho} = \int \frac{1}{\left[\left(\dfrac{x}{\rho}\right)^{2}-1\right]^{\frac{1}{2}}}\,d\!\left(\frac{x}{\rho}\right)$$

$$\frac{x}{\rho} = \cosh\!\left(\frac{y}{\rho}\right) \tag{3.45}$$

which gives a horn-shaped surface as the panel shape. From Equation 3.40 we find

$$R_{T} = R_{L} \tag{3.46}$$

and each element on the surface is a saddle shape as in the patch shown in Figure 3.98. Again both stresses are equal:

$$\sigma_{TAN} = \sigma_{LONG} = \frac{F}{2\,\pi t\rho} \tag{3.47}$$

As the designer of this panel, we shape the panel into the horn shape described by Equation 3.45 and the loaded panel will have no bending stress, and we expect the panel to be stiffer than alternative shapes.

3.6.5.2 Membrane analogy

In the above examples, we let the applied loads define the ideal shape for the panel with the notion that a uniformly stressed membrane panel will result in a light, stiff panel. We have used the special case of an axially symmetric membrane to demonstrate basic principles and to ease computation. However, rarely in automotive panel design practice will we have symmetric boundary conditions. For general loading and boundary conditions, closed-form analytic solutions are typically not available. However, much insight may be gained by visualizing panel shape using the *membrane analogy technique* [21].

In this technique the boundary constraints and loading for the panel are accepted, and we imagine a thin rubber membrane or a soap film [22] stretched under the action of the loads and constraints. Both the rubber membrane or soap film are structures with zero bending stiffness, so they must react to loads in a pure membrane state. An example of this analogy is shown in Figure 3.103 [23]. Here the objective is to define the shape for a building roof when loaded by uniform dead weight. A model of the roof is visualized using a membrane analogy—a light fabric cloth saturated with plaster. The desired four restraints are applied at the corners, and the weight of the plaster provides the uniform downward load on the inverted structure. The shape taken by the fabric is that of a membrane having uniform stress (no bending) across the structure's thickness.

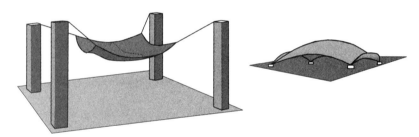

Figure 3.103 Membrane shaped roof structure.

Example: Rear load floor panel

Vibration of the rear load floor panel is a source of noise in the automobile. Generally a high resonance frequency for the panel is desirable, and this implies a high normal stiffness. The inertial loads during vibration approximate a uniformly distributed normal load across the panel. To maximize panel stiffness in this condition, we would like to define the membrane shape for a uniform downward load satisfying the irregular panel boundary, Figure 3.104a.

A heated plastic film was constrained at the boundary and loaded by its weight, simulating the uniformly distributed load. In the heated state, the film had only the ability to generate direct stress and behaved as a membrane. Upon cooling, the film held its shape, shown in Figure 3.104b. The shape is similar to the spherical membrane in *Example 2* above, but accommodates the irregular boundary conditions.

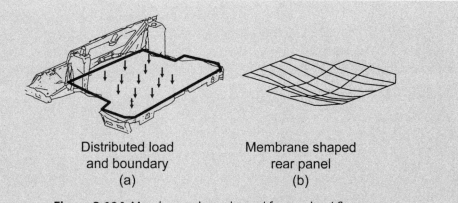

Distributed load
and boundary
(a)

Membrane shaped
rear panel
(b)

Figure 3.104 Membrane shaped panel for rear load floor.

Example: Suspension strut attachment panel

The suspension strut loads are often reacted by the wheelhouse panel. These strut loads are point loads directed along the axis of the strut. Figure 3.105a shows a typical strut attachment for the front suspension and Figure 3.105b for the rear suspension. The boundary restraints for the wheelhouse panels are generally irregular and non-planar, Figure 3.106a & b. High panel stiffness at the point of load attachment is desirable, and we would like to determine a membrane shape under these loads which satisfy the irregular boundary conditions.

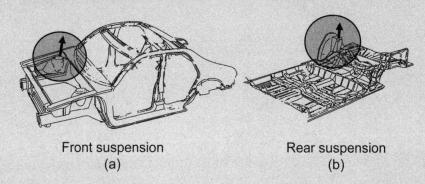

Front suspension
(a)

Rear suspension
(b)

Figure 3.105 Suspension strut attachment panels.

Using the same technique as used in example 1 above, a heated plastic film was constrained at the boundary and loaded in the direction of the strut axis through a small rigid disk representing the strut mounting bracket. For the motor compartment panel, Figure 3.107 shows the rigid boundary, the unloaded membrane, and finally the resulting loaded membrane shape. Figure 3.108 shows the rear panel membrane shape. Note the similarity to the horn membrane shape in the earlier axis-symmetrical example, Figure 102, but now accommodating the irregular boundary conditions. Note also that any small element on this panel also has a saddle shape, which is required by Equation 3.40 whenever the distributed load p is zero.

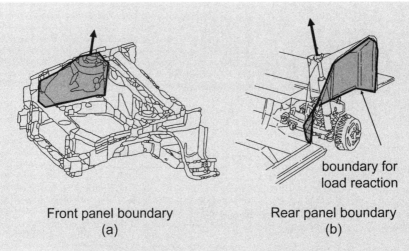

Front panel boundary
(a)

boundary for
load reaction

Rear panel boundary
(b)

Figure 3.106 Boundaries for strut attachment panels.

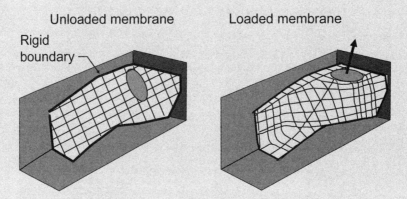

Unloaded membrane

Loaded membrane

Rigid
boundary

Figure 3.107 Forming motor compartment panel as a membrane.

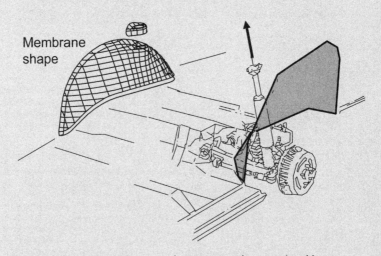

Membrane
shape

Figure 3.108 Rear strut attachment-membrane wheel house.

Once a panel has this more efficient membrane shape, the predominant internal loads act tangent to the surface at any point. In restraining such a panel, the weld flanges must also be tangent to the surface and the panel must blend smoothly into the supporting boundary structure to realize the stiffness offered by the membrane shape, Figure 3.109.

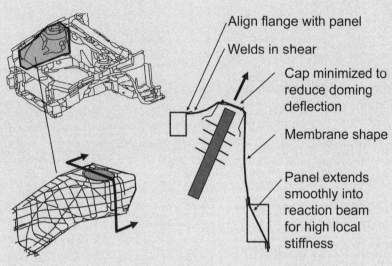

Figure 3.109 Reacting tensile loads in a membrane panel.

Modeling sheet metal structures as ideal membranes is certainly an approximation. Comparing some attributes between actual sheet metal and ideal membrane behavior:

Idealized Membrane Behavior: Can accept either tensile or compressive stress

Actual Sheet Metal Behavior: Buckling can occur under compressive stress

This is a serious limitation to the practical application of membrane panels. Proper design must seek to generate tensile loads only, or assure the buckling stress will not be exceeded. In the case of load floors or strut towers, tensile stresses are assured due to the unidirectional nature of the load.

Idealized Membrane Behavior: All stress lies in plane of surface (direct stress only)

Actual Sheet Metal Behavior: Bending stresses can be reacted

This is a highly beneficial property of real panels. Because membrane geometry can balance only the set of loads used to define it, the panel should collapse when the load deviates even slightly as it does in practice. That this does not occur is due to the ability to generate bending moments in the real panel.

Idealized Membrane Behavior: Boundary conditions are ideal (boundary loads are tangent to surface)

Actual Sheet Metal Behavior: Boundary conditions do not agree completely with theoretical requirements

This provides a general design constraint on the geometry of the structure at the boundary of a membrane panel. The required tangent reactions can practically be achieved only imperfectly and require bending moments to be generated in the reaction structure.

Despite these cautions, membrane structures in the automotive body are highly efficient means to react both distributed and point loads.

3.7 Summary: Automotive Structural Elements

In this chapter we have looked at how automotive structural elements respond to loading, how they deflect, and how they fail. We developed equations to predict stiffness and strength given the section geometry, the material and the bending moment, torque, or applied force. This has given us a set of section design tools.

However, to apply these tools the relationship between loads applied to the body system and the resulting loading on a particular section must be known. We need to *flow down* structural requirements from the global body level to the individual section. An example of this flow down for the B Pillar section is shown in Figure 3.110. In the subsequent chapters we will look at this flow down of requirements for several global body system cases including body bending, body torsion, crashworthiness, and vibration.

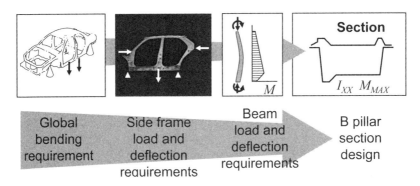

Figure 3.110 Flow of body strength and stiffness requirements.

References

1. Byars, E. and Snyder, R., *Engineering Mechanics of Deformable Bodies*, International Textbook Co, Scranton, PA, 1964.

2. "Geometrical Analysis of Sections," CARS 2008 Software, American Iron and Steel Institute, Southfield, MI, 2008.

3. Den Hartog, J. P., *Advanced Strength of Materials*, McGraw-Hill Co, NY, 1952.

4. Crandall, S., Dahl, N. and Lardner, T., *An Introduction to the Mechanics of Solids*, McGraw-Hill, NY, 1978.

5. Timoshenko, S. and Goodier, J., *Theory of Elasticity*, McGraw-Hill Co., NY, 1951.

6. Roark, R. and Young, W., *Formulas for Stress and Strain*, McGraw-Hill Co., NY, 1982.

7. Nisawa, J., Tomioka, N. and Yi, W., "Analytical Method of Rigidities of Thin Walled Beams with Spot Welding Joints and Its Application to Torsion Problems," JSAE Review, March 1985, pp. 76–83.

8. Kikuchi, N. & Malen, D., Course notes for ME513 Fundamentals of Body Engineering, University of Michigan, Ann Arbor, MI, 2007.

9. Timoshenko S., *Theory of Plates and Shells*, McGraw Hill, NY, 1959. pp. 444–447.

10. Timoshenko, S. and Gere, J., *Theory of Elastic Stability*, McGraw Hill, NY, 1961.

11. Yu, Wei-Wen, *Cold-Formed Steel Design*, John Wiley & Sons, NY, 1985.

12. Bloom, F. and Coffin, D., *Handbook of Thin Plate Buckling and Postbuckling*, Chapman & Hall/CRC, NY, 2001.

13. *Aluminum Construction Manual, Section 1-Specification for Aluminum Structures*, Aluminum Association, Washington, DC, 1970.

14. Cook, R. and Young, W., *Advanced Mechanics of Materials*, Macmillan Co., NY, 1985.

15. Lin, Kuang-Huei, "Stiffness and Strength of Square Thin-Walled Beams," SAE Paper No. 840734, SAE International, Warrendale, PA, 1984.

16. Parks, M. B. and Yu, W. W., "Structural Behavior of Members Consisting of Flat and Curved Elements," SAE Paper No. 870464, SAE International, Warrendale, PA, 1987.

17. Bruhn, E. F., *Analysis & Design of Flight Vehicle Structures*, Tri-State Offset Co., USA, 1973.

18. *Cold-Formed Steel Design Manual*, American Iron and Steel Institute, Washington, DC, 1986, Section 3.3.4.

19. Swenson, W. E. and Traficante, R.J., "The Influence of Aluminum Properties on the Design, Manufacturability and Economics of an Automotive Body Panel," SAE Paper No. 820385, SAE International, Warrendale, PA, 1985.

20. Goodman, A. W., *Analytic Geometry and the Calculus*, Mac Millan Co, NY 1963, pp. 269, 292.

21. Malen, D. E., *Empirical Method for Determining the Shape of a Vehicle Body Membrane Panel*, US Patent 4,581, 192, April 8, 1986.

22. Kato, T., Hoshi, K. and Umemura, E., "Application of Soap Film Geometry for Low Noise Floor Panels," *Proceedings of the 1999 Noise and Vibration Conference*, P-342, SAE 1999-01-1799, SAE International, Warrendale, PA, 1999.

23. Billington, D., *Heinz Isler as Structural Artist*, Catalog for exposition at Princeton University, April 1980.

Chapter 4
Design for Body Bending

In this chapter we consider the overall body structure supported and loaded similar to a single beam. The supporting points and loads are applied symmetrically to the vehicle center line; that is, loads on the right side are the same as loads on the left side. We will consider two types of body bending requirements: strength and stiffness.

4.1 Body Bending Strength Requirement

A most basic structure requirement is to locate and retain the vehicle subsystems in the correct positions. Powertrain, occupants, suspension, etc. must be supported by the body structure. Consider a vehicle at rest with the weight of the vehicle subsystems being supported by the body structure, which we will idealize as a beam in the side view, Figure 4.1. A requirement for this structure is that it does not fail under this loading condition.

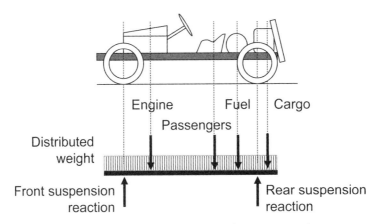

Figure 4.1 Body loaded by subsystem weight.

Let us look at the bending moments being applied to the structure under this condition. Using standard bending moment analysis techniques, we can identify the shear loads and moments being applied at any position along the length of the structure, Figure 4.2 [1]. A strength requirement for this body structure is to react these moments without failure.

More severe bending conditions than static weight loading can be imagined. The first is dynamic loading where the inertia loads of the subsystem exert larger forces during use than in the static condition. This condition can be addressed by multiplying the forces and moments of the static case by a dynamic acceleration factor. A typically used factor is 2-g loading—application of twice the static loads. A second condition is jacking or towing, where one support point is moved to an end of the vehicle, Figure 4.3. For this condition, we have taken an extreme case in which passenger loads are present. Although not typical, this represents a possible case for which the customer would not expect structural failure. Both front or rear jacking result in a larger maximum bending moment than that under static loading, as shown in Figure 4.4.

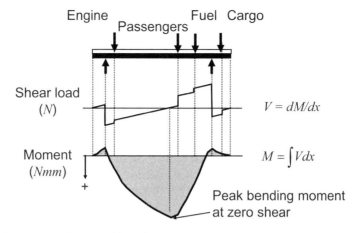

Figure 4.2 Shear and bending moment due to subsystem weight.

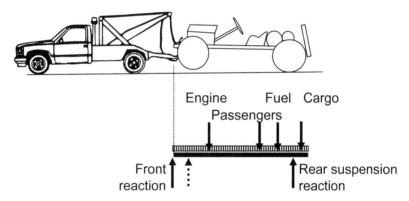

Figure 4.3 Front towing or jacking condition.

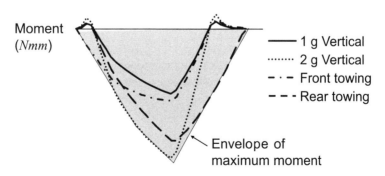

Figure 4.4 Envelope of moments.

Considering all the above conditions, we define an envelop of maximum bending moments which the body structure must react without excessive permanent deformation, Figure 4.4. To apply this requirement for practical design, we seek a simple test configuration which will produce an approximation to this bending moment envelope.

To define this test configuration, consider the body supported at the suspension points and loaded by just one or two loads (H point load) at the seating position. A beam loaded in this way will have the bending moment diagram shown in Figure 4.5a. We can superimpose the diagram for this case over the bending moments for the vehicle, Figure 4.5b. Now, by varying the magnitude of the H point loads, we can approximate the envelope of maximum moments. The resulting values for the loads become the bending strength requirement for that vehicle. This simpler condition is referred to as the H Point Bending Test because, in practice, the loads are applied at the seating location (H point).

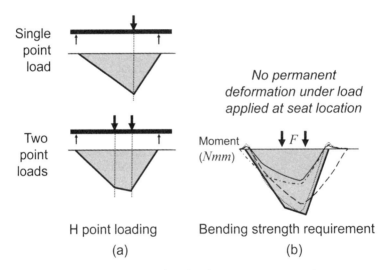

Figure 4.5 Equivalent load at passenger position.

The H Point Bending Test set-up is shown in Figure 4.6. The body is supported at the suspension attachments (usually the strut or spring attachment) with ball and socket conditions. The loads are applied in increments through a fixture loading the seat attachment points or adjacent rocker section. Vertical deflection measurements are taken along the longitudinal structural elements. For each load increment, the maximum deflection along the length is plotted on a load-deflection curve. The maximum load is increased in load-unload cycles to determine at what load there remains a permanent deformation of the structure which would affect vehicle performance. This load, F_s, in Figure 4.6, is the measured bending strength for the body, and can be compared to the bending strength requirement as calculated by the procedure shown in Figure 4.5b.

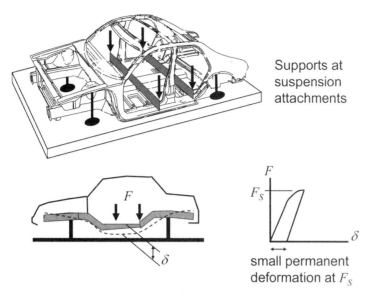

Figure 4.6 H point bending test convention.

Such a standardized bending test allows the comparison of competitive vehicles. Figure 4.7 shows the maximum bending moment for a sampling of 20 vehicles. Note that the body bending strength requirement depends on the bending moment analysis for the particular vehicle under consideration. Those bending moments depend on the placement of the subsystem mass, and the longitudinal dimensions of the vehicle, particularly the wheelbase.

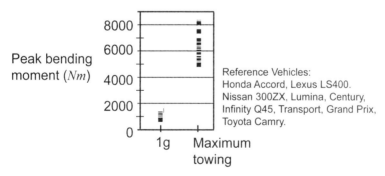

Figure 4.7 Bending test strength benchmarking. (Data courtesy of the American Iron and Steel Institute, UltraSteel Auto Body)

4.2 Body Bending Stiffness Requirement

From the H Point Bending test, we can measure bending stiffness—the slope of the load-deflection curve in the linear region. We now will develop the rationale for setting a bending stiffness requirement.

Consider the feeling of solidness as the vehicle drives over road irregularities. Solidness is a subjective feeling that the vehicle is "well put together," "vault-like," and not "loose" or "shaky." This subjective feel has been correlated to engineering parameters; one of the more significant is body vibration resonance.

The body structure acts like a vibrating beam with free end conditions, Figure 4.8a. As with a simple beam, the body has resonant frequencies for which a small dynamic force at the resonant frequency can cause large deformations. Although the number of resonant frequencies is infinite, we will concentrate on the lowest frequency of primary bending, Figure 4.8b (we will defer looking at torsional frequencies to a later chapter).

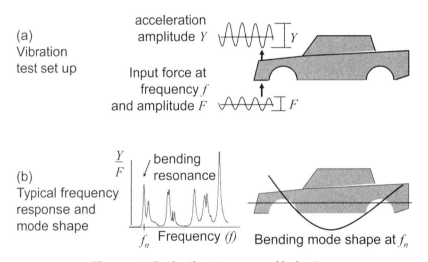

Figure 4.8 Body vibration test and behavior.

Figure 4.9 shows benchmark data for body bending resonant frequency for 47 vehicles. Three conditions are shown; Body shell (no components, doors, or glass), body shell with windshield, and full vehicle. The body shell frequencies with or without glass are similar, and are in general higher than the full vehicle condition. This is because the additional subsystem mass in the vehicle condition reduces the resonant frequency, as we will discuss below.

Customer testing in ride mules [2] where the bending resonant frequency could be varied, Figure 4.10, have shown that, to achieve the feeling of solidness, a desirable range for vehicle bending frequency is from *22–25 Hz*. This frequency range is relatively free from major exciting forces and responders, and is also in a range in which humans are less sensitive to vibration.

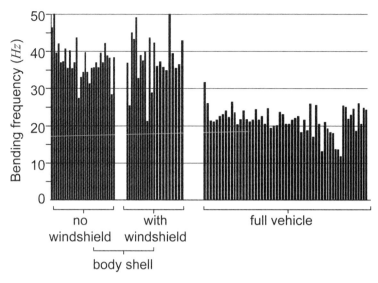

Figure 4.9 Bending resonant frequency benchmarking.

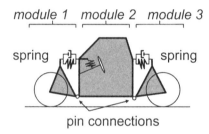

Figure 4.10 Customer evaluation mule for structural solidness.

We can now use this target of *22–25 Hz* for the vehicle bending frequency to establish a body bending stiffness target. Let us view the body structure as a uniform beam, Figure 4.11a. The primary bending frequency of a uniform beam [3] is related to its length, linear mass density, and section bending stiffness (*EI*):

$$\omega_n = \frac{22.4}{L^2}\sqrt{\frac{EIg}{w}}$$

$$\omega_n = 22.4L^{-\left(\frac{3}{2}\right)}\sqrt{\frac{EI}{M}}$$

(4.1)

where:

    w = Weight per unit length

    M = Total mass = *wL/g*

    ω_n = Bending resonant frequency

    L = Beam length

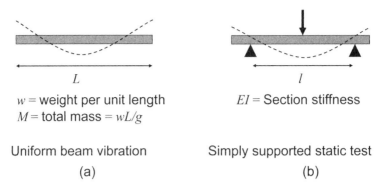

w = weight per unit length
M = total mass = wL/g

EI = Section stiffness

Uniform beam vibration

Simply supported static test

(a)

(b)

Figure 4.11 First-order model for bending frequency.

Now consider the same beam but loaded with a single static load at its center span, Figure 4.11b. The beam is supported at points representing the suspension points—the length between supports is the wheelbase. The stiffness of such a beam is related to the support length and section bending stiffness (EI):

$$K = \frac{48EI}{l^3}$$

$$EI = \frac{Kl^3}{48}$$

(4.2)

where:

l = Wheel base

M = Rigidly mounted mass

K = Required bending stiffness

We can now eliminate (EI) between Equations 4.1 and 4.2 by substituting Equation 4.2 for (EI) into Equation 4.1, and we are left with a relationship between bending resonant frequency and static bending stiffness, K.

$$\omega_n = 22.4L^{-\left(\frac{3}{2}\right)}\sqrt{\frac{Kl^3}{48M}}$$

(4.3)

$$\omega_n = \frac{22.4}{\sqrt{48}}\left(\frac{l}{L}\right)^{\frac{3}{2}}\sqrt{\frac{K}{M}}$$

where:

l = Wheel base

L = Overall length

M = Rigidly mounted mass

K = Required bending stiffness of the body

ω_n = Desired bending resonant frequency for the vehicle (rad/sec)

Thus given a target for vehicle bending frequency, ω_n, the lengths l and L, and the rigidly mounted mass, M, we can identify the required H point bending stiffness, K. Rigidly attached masses are those which participate fully in the vibration of the body structure and do not include those masses which are isolated with bushings. For preliminary design, the rigidly attached mass is taken as *0.4* to *0.6* times the vehicle curb mass, with the larger value relating to luxury vehicles.

Example: Developing a bending stiffness requirement

Consider a midsize vehicle in the early design stage having the following numerical values for key parameters:

$$l = 2700\,mm$$

$$L = 4550\,mm$$

M = rigidly attached mass ~0.4 to 0.6 times curb mass. For this case, the lower value for $M = 0.4 \times 1446 = 578\,kg$ and the upper value for $M = 0.6 \times 1446 = 868\,kg$

We desire to establish the bending stiffness requirement that would allow us to achieve a bending resonance for the vehicle in the *22–25 Hz* range. Substituting numerical values into Equation 4.3 we can investigate the relationship between resonant frequency, rigidly mounted mass, and bending stiffness. This analysis indicates that a bending stiffness requirement of approximately *7000 N/mm* will achieve the desired *22–25 Hz* vehicle frequency, Figure 4.12.

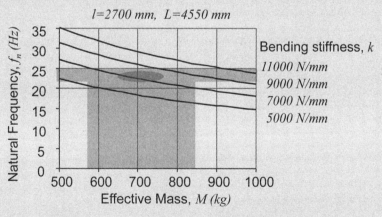

Figure 4.12 First order estimation of bending stiffness.

Note that the bending stiffness requirement for a specific vehicle depends on the parameters in Equation 4.3. Thus vehicles which have higher mass loading (highly optioned luxury cars for example) or cars with long overall length (four-door sedans vs. two-seat sport coupes, for example) will require higher static bending stiffness to achieve the same frequency target. Benchmark data for bending stiffness is shown in Figure 4.13.

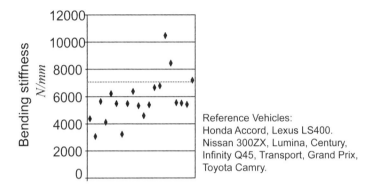

Figure 4.13 Bending stiffness test and benchmarking. (Courtesy of the American Iron and Steel Institute, UltraLight Steel Auto Body)

In addition to adequate bending stiffness being important for structural feel, a high bending stiffness is also significant in reducing the relative deformations which cause squeaks and rattles during normal use.

To summarize, we have shown how both strength and stiffness requirements may be established for body bending. Typical values for these requirements are shown in Figure 4.14 for a midsize vehicle. We will now look at the design of body structure to meet these requirements. The next section treats design for the strength requirement, followed by a section on design for the stiffness requirement.

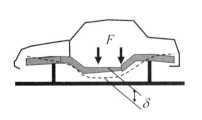

Bending strength

nominal value
$F = 6680\ N$
no permanent deformation

Bending stiffness

nominal value
$K = 7000\ N/mm$

restraints at suspension attachments

Figure 4.14 Typical bending requirements: Midsize vehicle.

4.3 Internal Loads During Global Bending: Load Path Analysis

First we will consider the design of structure to meet the body strength requirement. Our objective is to understand how the global body requirement flows down to loads on structure elements such as the beams in the side frame. Once we have loads on individual beams, we can then use the techniques discussed in Chapter 3 to design the appropriate beam sections.

We will idealize the body as a set of structural surface and bar elements. A structural surface [4] is a flat element which is loaded in shear along its edges. Loads normal to the surface, or bending moments, cannot be reacted, Figure 4.15a. A bar element is a linear element which can only react loads along its axis, either end loads or shearing loads along the length, Figure 4.15b. Using only structural surfaces and bars, we can construct the model for the auto body shown in Figure 4.16.

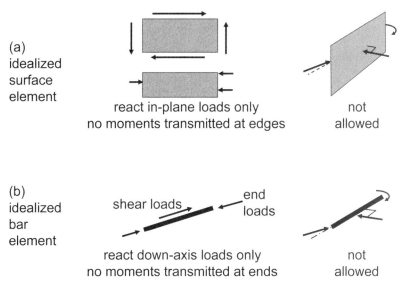

(a) idealized surface element

react in-plane loads only
no moments transmitted at edges

not allowed

(b) idealized bar element

shear loads   end loads

react down-axis loads only
no moments transmitted at ends

not allowed

Figure 4.15 Idealized structural elements.

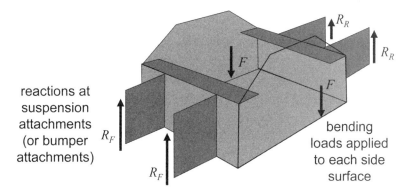

reactions at suspension attachments (or bumper attachments)

R_F

R_F

R_R

R_R

F

F

bending loads applied to each side surface

Figure 4.16 Structural surface and bar body model.

We can now load this model as in an H point bending test by applying a downward load F in the plane of the side structural surface, Figure 4.16, and supporting the structure with edge loads R_F and R_R. Note that this loading is symmetrical side-to-side.

The magnitude of the applied load, F, represents the bending strength requirement which we would like to react with just a small amount of permanent deformation. We would like to know the loads that are applied to each of the structural elements, a through f, in Figure 4.17. Once we know these maximum internal loads, we can then design each element for strength. This process is known as flow down of requirements, from a global requirement to a requirement for a specific structural element.

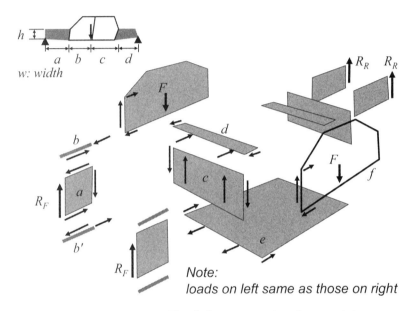

Figure 4.17 Internal loads for structural surface model.

By exploding the structure [5], we can use free-body analysis to identify these element loads, Figure 4.17. We begin by considering equilibrium of the full body, (see Figure 4.17 upper left corner) and solve for the unknown reaction forces R_R and R_F. Summing moments about the front axle and equating to zero:

$$2R_R(a+b+c+d) - 2F(a+b) = 0$$

Summing forces in the vertical direction and equating to zero:

$$2R_F + 2R_R - 2F = 0$$

Solving these for the reaction forces gives:

$$R_F = F\frac{(c+d)}{(a+b+c+d)}, \qquad R_R = F\frac{(a+b)}{(a+b+c+d)} \tag{4.4}$$

Next we will put each element, *a* through *f*, into static equilibrium beginning with an element, *a*, which is loaded by one of the reaction forces.

Looking at the exploded structure we can begin by putting the structural surface *a* into equilibrium. Remembering that we can only have shear loads along edges, this panel has the known reaction force R_F along the front edge (and $-R_F$ along the rear edge by static equilibrium in the vertical direction). An unknown load Q acts on the top and bottom edges, as shown in Figure 4.18. Summing moments about point o and equating to zero:

$$Qh - R_F a = 0$$

$$Q = R_F \frac{a}{h}$$

(4.5)

Now taking each element, *a* through *f*, Figure 4.17, we can use equilibrium to determine all internal loads.

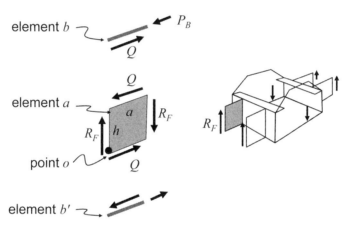

Figure 4.18 Internal loads on motor compartment panel.

For the upper bar element, *b*, Figure 4.18, the applied shearing load is Q from element a, and we can have an end force P_B. Equilibrium along the horizontal direction gives $P_B = Q$. Similarly, for element *b'*, we find $P_B' = Q$.

For the dash panel, *c*, Figure 4.19a, the applied loads are R_F from element *a*, with reactions at the outer edges of P_C, so $P_C = R_F$. The cowl panel, *d*, Figure 4.19b, is loaded by the upper bar, *b*, and has reaction loads, P_D, at its edges, so $P_D = Q$. The front portion of the floor pan, *e*, Figure 4.19c, is loaded on each side by the lower bar, *b'*, with load Q and by the side panel, *f*, by an equal and opposite load, $P_E = Q$.

Finally, looking at the side frame, structural surface *f*, Figure 4.20, we have the vertical load R_F from the dash, Q rearward at the belt line from the cowl panel, Q forward at the base of the front hinge pillar from the floor, and similar loads from the panels at the rear of the vehicle.

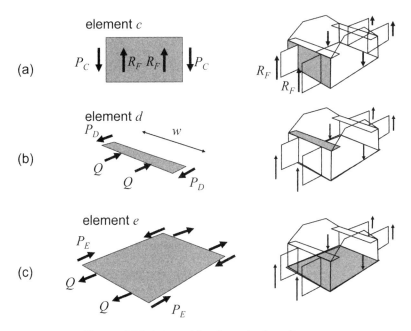

Figure 4.19 Internal loads under bending.

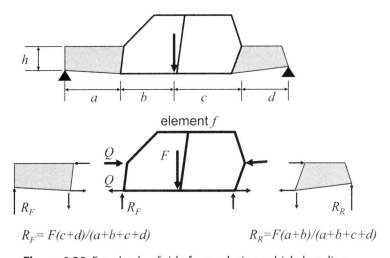

$$R_F = F(c+d)/(a+b+c+d) \qquad R_R = F(a+b)/(a+b+c+d)$$

Figure 4.20 Free-body of side frame during vehicle bending.

Figure 4.21 summarizes the results for these equilibrium analyses. We have identified the load paths and loads resulting from the bending strength requirement for each element of the body. Each element must be capable of reacting the loads shown without excessive permanent deformation for the overall body structure to meet the bending strength requirement. Let us look at three of these structural subsystems in more detail to see how this information can be used in design.

134

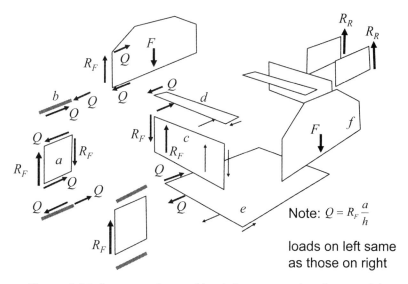

Figure 4.21 Summary: Internal loads for structural surface model.

The motor compartment side panel, element a, is loaded by the shearing loads shown in Figure 4.22a. This panel must react these loads in a bending strength test. However, this high, thin panel can be prone to elastic plate buckling under this loading condition, Figure 4.22b. Using the principles developed in Chapter 3, we can look at ways to increase the shear buckling critical stress. One mass-effective method is to add ribs to reduce the width of the buckling plate, Figure 4.22c.

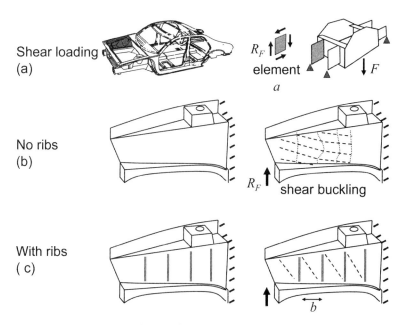

Figure 4.22 Buckling of motor compartment side panel.

The cowl, element d, is loaded as shown in Figure 4.19b. Under these loads we can identify the bending moments along this element in the plan view, Figure 4.23. Given this moment diagram, we can now design a section to react this moment without failing using yield of the effective section as the failure criterion. This will ensure that the cowl will be adequate under global bending.

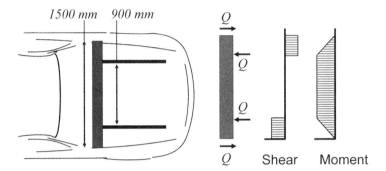

Figure 4.23 Loads on cowl surface during bending.

Finally, consider the side frame element f, loaded as shown in the bending test, Figure 4.20. This element is made of several beams: front hinge pillar, A pillar, roof rail, B pillar, C pillar, and rocker, Figure 4.24. Our objective is now to see what moments are applied to each beam by these loads so that we may design the beam sections. As a framework of beams, this structural subsystem is statically indeterminate; that is, we cannot use equilibrium equations only to determine the moments in each beam. The moments also depend on the relative stiffness of each of the beams. Using the loading shown in Figure 4.25, a small finite element model of the side frame may be applied to identify internal bending moments for each beam, Figure 4.26. Once the moments are known for a beam, the sections can be designed which react the moments without yield of the effective section.

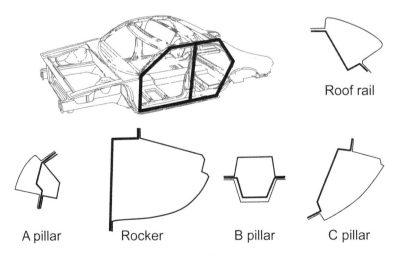

Figure 4.24 Side frame beams.

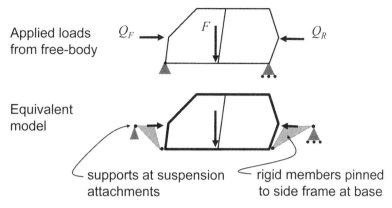

Applied loads from free-body

$Q_F \longrightarrow$ F $\longleftarrow Q_R$

Equivalent model

supports at suspension attachments   rigid members pinned to side frame at base

Figure 4.25 Side frame planar beam model.

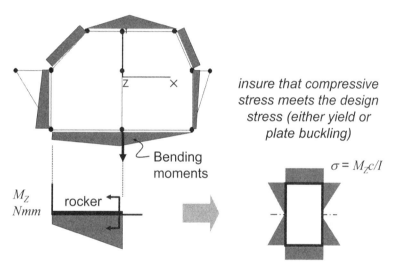

insure that compressive stress meets the design stress (either yield or plate buckling)

Bending moments

$\sigma = M_Z c / I$

M_Z
Nmm rocker $\longleftarrow$

Figure 4.26 Finite element analysis for internal loads.

4.3.1 Summary: Bending strength

In this section, we have shown how the applied loads in the global bending strength test can be flowed down to the loads on each individual structural subsystem of the body. To accomplish this flow down of load requirements, we used a highly idealized body structure of simple structural surfaces and bar elements, Figure 4.21. We set each element into static equilibrium to find the internal element loads. Then, knowing the subsystem loads, we showed how the section design principles developed in the previous chapter can be used to find the appropriate sections to react these loads. We have thus gone from a strength requirement on the body system to ensuring the strength of each structural element in the body under bending.

4.4 Analysis of Body Bending Stiffness

In the previous section we examined how the bending strength requirement implies a strength requirement for each structural element, and how we can use these requirements to design the elements. We will continue to focus on the global bending stiffness requirement and flow down that requirement to stiffness requirements for a primary structural subsystem—the side frame.

We focus on the side frame due to its effect on bending stiffness, and model the body structure as shown in Figure 4.27. We consider the structure from the front suspension attachment rearward to the front of the side frame as rigid and connected with two pinned joints to the side frame [6]. One pinned joint is at the base of the A pillar, and one is at the base of the hinge pillar. Similarly, we use a rigid element from the rear suspension attachment forward to the side frame and attached in the same way as shown in Figure 4.27. With this configuration, the side frame is loaded as in our earlier free-body shown in Figure 4.20.

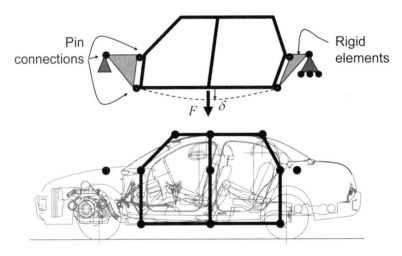

Figure 4.27 Very basic side frame finite element model. (Courtesy of the American Iron and Steel Institute, UltraLight Steel Auto Body)

We will follow the design procedure shown in Figure 4.28. Beginning with the global bending stiffness requirement, we will create a side frame design concept by making an initial estimate for the section size of each of the beam elements in the side frame. We will then predict the bending stiffness performance using a finite element model based on Figure 4.27. We will then compare the estimated performance to the required stiffness and adjust the beam sizes until the requirement is met at acceptable mass.

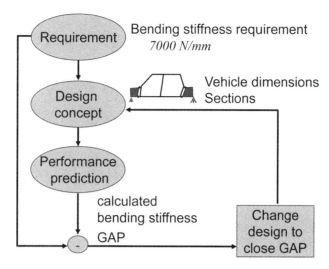

Figure 4.28 Systems design procedure.

Consider example data for this model. Figure 4.29 shows the initial guess for beam section size—height, width, and thickness for the rectangular sections. These beams are used in the simple finite element model [7] shown in Figure 4.30, in which each beam is rigidly connected to adjacent beams at a node point. The model is restrained at the front suspension attachment point by restraining deflection in all three directions (but allowing rotation about all three axes), and restrained at the rear suspension attachment point in the vertical direction and out-of-plane direction. A downward load is applied at the node where the B pillar attaches to the rocker, simulating the H point bending load, Figure 4.30.

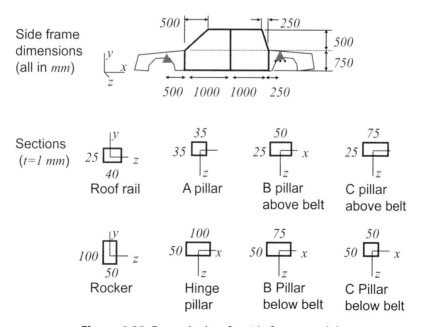

Figure 4.29 Example data for side frame model.

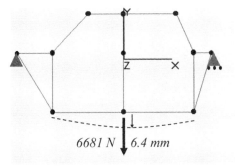

Body bending stiffness:

$K=6681\ N\ /\ 6.4\ mm$ $=1044\ N/mm$ per side
$=2088\ N/mm$ body bending stiffness
(30% of the *7000 N/mm* target*)*

Figure 4.30 First-order analysis of side frame with rigid joints.

The result of this analysis is the deflected shape of the side frame. By taking the ratio of the applied load to the deflection at the node of load application, we can calculate the bending stiffness for the side frame, Figure 4.30.

4.4.1 Importance of joint flexibility

If we were to compare the actual stiffness to what we have predicted with this model, we would find the predicted stiffness is approximately twice the actual. We have neglected a very important physical behavior of the thin-walled beam sections: whenever two or more thin-walled beams are joined, there is considerable localized deformation, Figure 4.31. This localized deflection has the effect of a flexible joint between the beams [8]. Thus, our assumption that the beams are rigidly connected to each other is in error. By adjusting the finite element analysis (FEA) to contain flexible joints, we can achieve much better correlation with actual stiffness. Rather than attaching a beam end rigidly to a node, we will instead attach the beam to a node through a rotational stiffness representing the joint, Figure 4.32.

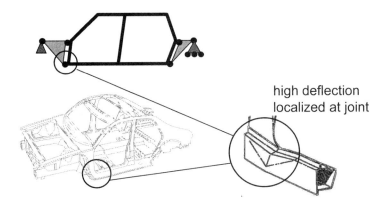

high deflection
localized at joint

Figure 4.31 Observed localized deformation at joint.

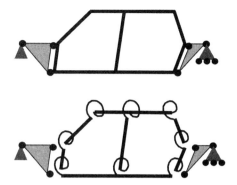

Figure 4.32 Modified model with joint flexibility.

The rotational stiffness value is determined by extracting the physical joint from the body structure, restraining all beam connections except one, and applying a moment at that beam connection, Figure 4.33. The resulting rotational deflection is measured and a moment-rotational angle curve plotted. The slope of this curve is the joint stiffness. Benchmark data for side frame joints of four vehicles are shown in Figure 4.34, and joint stiffnesses for five different joints are compared.

The absolute value for joint stiffness is somewhat difficult to interpret. For example, is a joint stiffness of $0.2x10^6$ Nm/rad $(1.77x10^6$ in $lb/rad)$ a very stiff or very flexible joint? To answer this question, it is helpful to compare the joint stiffness to the bending stiffness of the beam to which it is attached, Figure 4.35a. We can then define joint efficiency, f, as the ratio of the combined stiffness of the beam with joint to the stiffness of the beam alone (assuming a rigid joint).

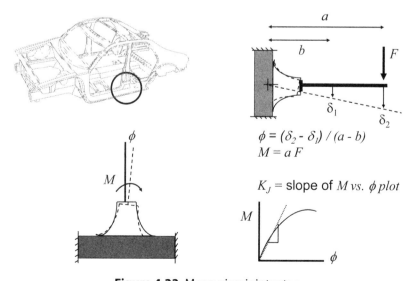

$$\phi = (\delta_2 - \delta_1) / (a - b)$$
$$M = a F$$

K_J = slope of M vs. ϕ plot

Figure 4.33 Measuring joint rates.

141

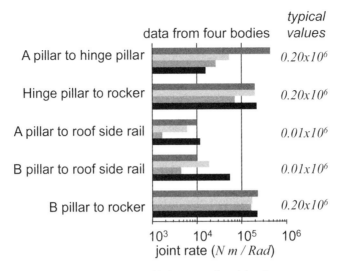

Figure 4.34 Typical joint rates for side view.

$$f = \frac{K_{BEAM \& JOINT\ SYSTEM}}{K_{BEAM}} \tag{4.6}$$

Since the beam stiffness and joint stiffness are in series,

$$K_{BEAM \& JOINT\ SYSTEM} = \frac{K_{JOINT} K_{BEAM}}{K_{JOINT} + K_{BEAM}} \tag{4.7}$$

Substituting Equation 4.7 into 4.6 gives,

$$f = \frac{K_{JOINT}}{K_{JOINT} + K_{BEAM}}$$

Note that for a beam loaded by an end moment, Figure 35b,

$$K_{BEAM} = \frac{M}{\theta} = \frac{2EI}{L} \tag{4.8}$$

Then the joint efficiency, f, is given by

$$f = \frac{K_{JOINT}}{K_{JOINT} + \dfrac{2EI}{L}}$$

$$f = \frac{1}{1 + \dfrac{2EI}{LK_{JOINT}}} \tag{4.9}$$

where:

   L = Beam length

   EI = Beam section stiffness

   K_{JOINT} = Joint stiffness

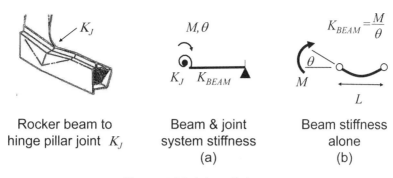

| Rocker beam to hinge pillar joint K_J | Beam & joint system stiffness (a) | Beam stiffness alone (b) |

Figure 4.35 Joint efficiency.

When the joint stiffness, K_{JOINT}, is much larger than the beam stiffness, (2EI/L), the denominator of Equation 4.9 is approximately one, and the efficiency approaches *100%*. Thus a *100%* efficient joint would be rigid with all of the beam stiffness being utilized. A very low efficiency indicates most of the deformation of the beam-joint system is caused by the joint deformation.

Example: Joint efficiency

Consider the steel rocker beam from the side-frame model of Figure 4.29. To the rocker, we will place a typical joint for the Hinge Pillar-to-rocker joint, Figure 4.34. With these assumptions, we have

$I=4.15 \times 10^5 \, mm^4$ (computed from the values $h=100 \, mm$, $w=50 \, mm$, $t=1 \, mm$)

$K_{JOINT}=0.2 \times 10^6 \, Nm/rad=0.2 \times 10^9 \, Nmm/rad$ from Figure 4.34

$L=1000 \, mm$, the length of the rocker beam from Figure 4.29

Inserting these values into Equation 4.9 gives

$$f = \frac{1}{1+\dfrac{2(207000 N / mm^2)(4.15 \times 10^5 \, mm^4)}{(1000 mm)(0.2 \times 10^9 \, Nmm / rad)}}$$

$$f = 0.537$$

So for this beam-joint system, the presence of the joint reduces the stiffness to approximately one half that of the beam.

Consider again our side-frame model of Figure 4.29. We now apply reasonable joint stiffnesses to three of the joints, Figure 4.36 and re-run the FEA. The resulting side frame stiffness, Figure 4.37, is *83%* of that with rigid joints, Figure 4.30, and is in better agreement with physically measured values. We are now content that we have a valid model to predict side frame stiffness given an initial guess at beam section sizes. What if our initial guess does not give a stiffness which meets the

143

required bending stiffness? Intuitively, we would begin to increase section sizes or joint stiffnesses until we attain adequate bending stiffness following the iterative loop of Figure 4.28. But which beams to adjust first, and once we achieve the requirement, how can we be sure that it is the minimum mass solution? In the next section, we describe a means to efficiently improve the structure's stiffness.

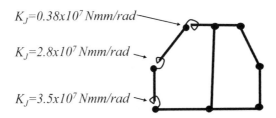

Figure 4.36 Model with flexible joints.

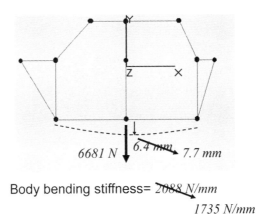

Figure 4.37 Effect of joint stiffness on bending stiffness.

4.4.2 Strain energy and stiffness

As the bending load is applied to the body structure, the structure deflects at the point of load application. As the applied force moves through this displacement, it does work. This external work is stored as strain energy in each of the structural elements as they deform under the load. The strain energy in a structural element may be calculated if we know the applied moments.

For example, in Figure 4.38 the strain energy of an end-loaded beam is developed as a function of the end moments on the beam, resulting in the relationship:

$$e_{BEAM} = \frac{L}{6EI}(M_1^2 + M_1M_2 + M_2^2) \tag{4.10}$$

where:

L = Beam length

EI = Section stiffness

$M_{1,2}$ = Applied moment at each end

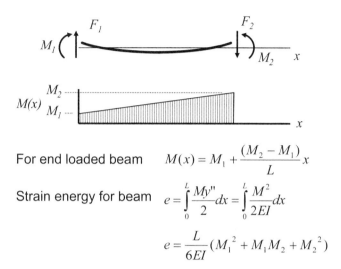

For end loaded beam $\quad M(x) = M_1 + \dfrac{(M_2 - M_1)}{L}x$

Strain energy for beam $\quad e = \displaystyle\int_0^L \frac{My''}{2}dx = \int_0^L \frac{M^2}{2EI}dx$

$$e = \frac{L}{6EI}(M_1{}^2 + M_1M_2 + M_2{}^2)$$

Figure 4.38 Strain energy of a beam loaded at ends.

The strain energy for a joint of stiffness K_{JOINT}, Figure 4.39, with applied moment, M, and resulting angle of rotation, ϕ, is:

$$e = \frac{M\phi}{2} = \frac{M^2}{2K_{JOINT}}$$

$$e = \frac{\phi^2 K_{JOINT}}{2}$$

(4.11)

where:

K_{JOINT} = Joint stiffness

M = Applied moment at each end

ϕ = Deflection angle (rad)

Thus, knowing the moment applied to a structural element—either a beam or joint—we can identify the strain energy stored in it by using Equation 4.10 or 4.11.

We can now answer the question of which beam or joint to improve first to meet the body bending stiffness requirement most efficiently:

To improve the stiffness of a structural system, increase the performance of the structural element with the highest fraction of strain energy.

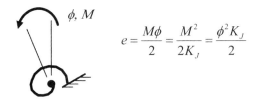

$$e = \frac{M\phi}{2} = \frac{M^2}{2K_J} = \frac{\phi^2 K_J}{2}$$

Figure 4.39 Strain energy of a joint.

Example: Automobile seat mount system

Consider the automobile seat mount system, Figure 4.40, consisting of a beam connected by a flexible joint to a rocker which we consider as rigid. The seat applies a downward load and we are interested in the vertical stiffness. Assume that the current system design does not meet the stiffness requirement, and we wish to know which element to change—the beam or the joint—to increase the stiffness of the system. Using the values given in Figure 4.40 we can calculate the strain energy in each element.

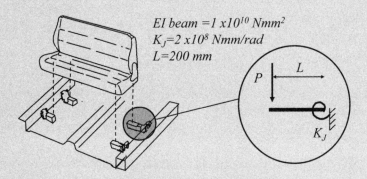

EI beam =1 x10¹⁰ Nmm²
Kⱼ=2 x10⁸ Nmm/rad
L=200 mm

Figure 4.40 Seat mount system.

The strain energy in the beam using Equation 4.10 is

$$e_{BEAM} = \frac{200mm}{6(1 \times 10^{10}\ Nmm^2)}(M_2^2) = 3.33 \times 10^{-9}\ M_2^2$$

where:
M_2 = Moment at the connection with the rocker

The strain energy in the joint using Equation 4.11 is

$$e_{JOINT} = \frac{M_2^2}{2(2 \times 10^8\ Nmm\ /\ rad)} = 2.5 \times 10^{-9}\ M_2^2$$

Since the strain energy in the beam is greater than that in the joint, an incremental change in beam stiffness will have a larger effect on overall system stiffness than an incremental change in joint stiffness.

While the use of strain energy in this very simple example is not essential, in more complex systems such as the side frame it is much more difficult to identify which beam or joint to improve. Strain energy for these complex structures becomes a critical tool for iterative improvement of the structure. For more complex structural systems, strain energy can be calculated, using the FEA model which evaluates Equations 4.10 and 4.11, for each beam and joint within the software.

Example: Side frame design to meet the bending requirement

We continue with the side frame model presented first in Figure 4.27. The initial guess at beam section sizes is shown in Figure 4.29, and the predicted bending stiffness for these beams with joints is shown in Figure 4.37. This predicted stiffness does not meet the bending requirement of $K \geq 7000\ N/mm$, Figure 14. Now consider another result of the FEA, Figure 4.41, which is the strain energy percent in each element. To improve side-frame bending stiffness most efficiently, we increase the beam section with the highest strain energy—in this case, the front rocker beam. We increase the beam height, the most mass-effective way to increase moment of inertia, and re-run the FEA. This will result in a new estimate for stiffness and also a different strain energy distribution. We continue this process in an iterative fashion until we improve the stiffness performance to meet the requirement. This process requires several iterations and is often automated within the FEA software.

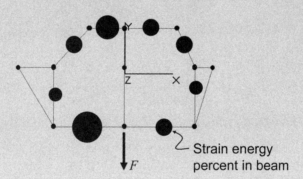

Figure 4.41 Strain energy in side frame.

Once we have sized the sections to meet the bending stiffness requirement, we consider the bending strength requirement. We follow the procedure shown in Figure 4.42 and look at the maximum stress in each beam to ensure the stress is within the design stress level. With thin-wall design, careful attention is given to the compressive stresses to ensure they are within the plate buckling stress limits. If stress is greater than the design stress we have a choice of:

1. Increasing the buckling design stress by inhibiting elastic plate buckling using the methods described in Chapter 3

2. Choosing a material with increased yield when buckling is not the limiting failure condition

3. Reducing the stress by increasing the section properties.

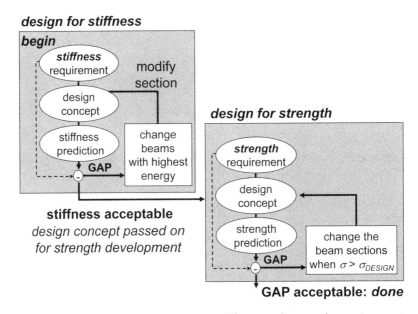

Figure 4.42 Iterative process to meet stiffness and strength requirements.

4.4.3 Note on the bending stiffness changes due to side doors

The model we have used for bending performance has neglected side doors. It has been shown experimentally that the static bending stiffness is unchanged with or without doors attached. To understand this non-obvious behavior, consider the deflected shape of the side frame with an idealized door in place, Figure 4.43. To add to the global bending stiffness, the door system must generate the loads shown at the hinges and latch. However, current designs for the hinge and latch do not have sufficient stiffness to generate these loads. Alternative designs for door attachment can provide this stiffness. For example, Figure 4.44 shows a highly rigid hinge attachment design [9].

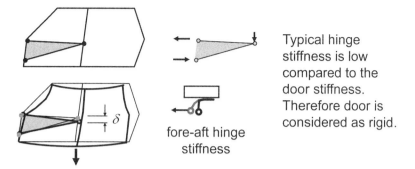

Typical hinge stiffness is low compared to the door stiffness. Therefore door is considered as rigid.

fore-aft hinge stiffness

Figure 4.43 Effect of door on body bending stiffness.

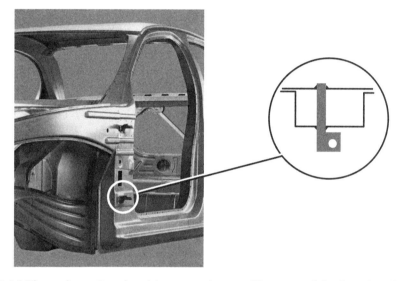

Figure 4.44 Through-section door hinge attachment. (Courtesy of the American Iron and Steel Institute, UltraLight Steel Auto Body)

4.4.4 Summary: Bending stiffness

In this section, we have shown how the global bending stiffness requirement can be flowed down to identify the required properties of individual beams and joints. We used a simple Finite Element Analysis of the side frame to predict deflections and also element strain energy. Strain energy informed us regarding which beams have the most influence on global bending stiffness. We have also shown how joint stiffness plays an important role in global bending stiffness performance.

Often the structural element which most influences global bending stiffness is a joint rather than a beam element. In the following section, we look at principles of joint design for stiffness.

4.5 Principles of Good Joint Design

In the previous section we showed the importance of the joints between beams to body stiffness. In this section we will discuss design principles which result in joints with high stiffness.

First, we classify two cases for joint bending stiffness: in-plane bending and out-of-plane bending. Consider the intersection of the rocker beam and B pillar beam, Figure 4.45. The two beams define a plane, and when the applied moment causes beam rotation within that plane we have the case of in-plane bending. Now consider the same joint but loaded in a way that causes the B pillar to rotate out of the plane containing the two beams, Figure 4.46. This is the out-of-plane bending case. In most instances, in-plane bending joint stiffness is of interest under global body bending, while the out-of-plane bending joint stiffness is of interest under global body torsion. In the following, we will focus on design principles for in-plane bending but realize that the general conclusions follow for the out-of-plane case.

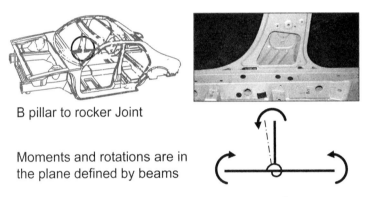

B pillar to rocker Joint

Moments and rotations are in the plane defined by beams

Figure 4.45 In-plane bending. (Photo courtesy of A2Mac1.com Automotive Benchmarking)

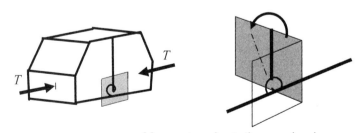

Moment and rotation are in plane perpendicular to the plane defined by beams

Figure 4.46 Out-of-plane bending.

We will develop principles for stiff joints by first looking at a case of an undesirable flexible joint, that is one with much local deformation where the beams join. In this simplified joint, two beams with rectangular section join in a planar T shape similar to the B pillar to rocker joint, Figure 4.47. The section width for the B pillar is smaller than the width of the rocker section. The rocker is restrained at either end, and a rearward load is applied at the top of the B pillar beam, creating the stress distribution shown in Figure 4.48. Because we are dealing with thin-walled sections, the corners of the section have a relatively higher stress than the center of the walls. We can idealize this stress distribution by assuming that all of the stress is taken by the corners of the section, as shown in Figure 4.49. Now isolating the top surface of the rocker section and considering it as a simply supported plate, we can see that the corner loads from the B pillar are applied to the central portion of this plate, Figure 4.50. As a thin-walled plate, it has little bending rigidity to react these centrally applied normal loads, and there is considerable deformation of the plate as shown in Figure 4.51.

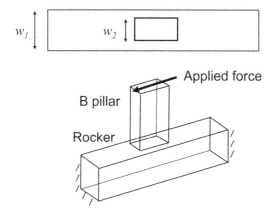

Figure 4.47 Simplified joint of thin-walled sections under in-plane bending.

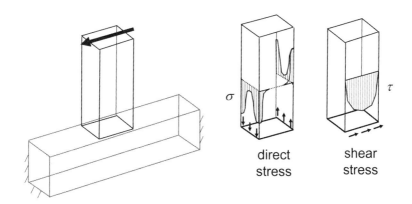

Figure 4.48 Stress distribution for loaded joint.

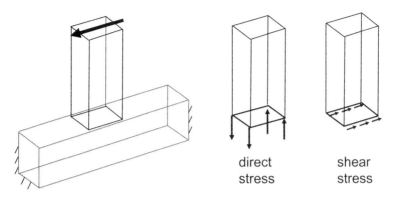

Figure 4.49 Idealized stress distribution for thin-walled section.

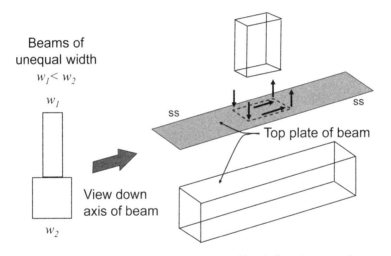

Figure 4.50 Reaction stresses generated by deforming top plate.

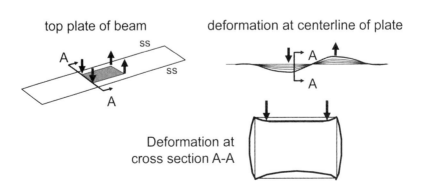

Figure 4.51 Deformation of top plate for joint with beams of unequal width.

This condition, in which a corner load from one beam normally loads a plate surface of the connecting beam, is the basis for the high joint flexibility we are trying to avoid. As an improvement to this joint configuration, consider that both beams are now of the same width, Figure 4.52. Now the corner loads from the B pillar are reacted by shearing in the side surface of the rocker. Considerably less deformation occurs under this shear action, and the joint flexibility is greatly reduced. This provides a guideline for stiff joint construction:

For high joint stiffness, the shear walls of the connected beams should be aligned at the joint and flow smoothly from one beam to another.

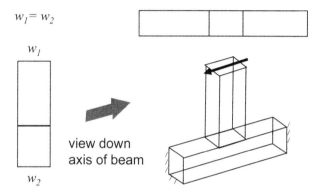

Figure 4.52 Joint with beam side walls coincident.

To better understand this improved joint configuration, consider the shear sides of the two beams. Let us model these shear walls with shear resistant members and bars and load them as shown, Figure 4.53. To understand how the applied in-plane load flows through the joint, each element can be placed into static equilibrium, Figure 4.54. This analysis shows that the applied load can be reacted efficiently by shear and compression. However, two areas must be treated carefully; the shear panel at the intersection of the beams, panel B, is relatively highly loaded in shear and can be prone to shear buckling. Also, the bar elements at the front and rear of the B pillar, bar S_3 and S_4, are relatively highly loaded in compression and are also prone to compressive buckling. One means to solve both of these concerns is by using the rib pattern shown in Figure 4.55a. The V rib pattern increases the shear buckling stress of the panel, while the vertical ribs provide a path for the compressive load. Examining Figure 4.54, note that the corner force, X_D, acting on bar S_3 is reduced as the span, b, is increased; $X_D = hF/b$. One means to increase this span is to provide a filleted transition, as shown in Figure 4.55b.

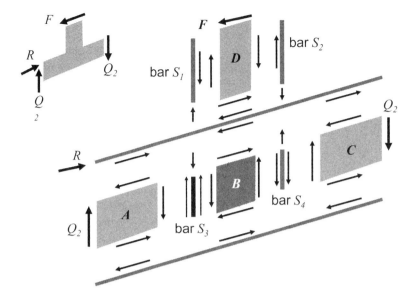

Figure 4.53 Internal loads: In plane bending.

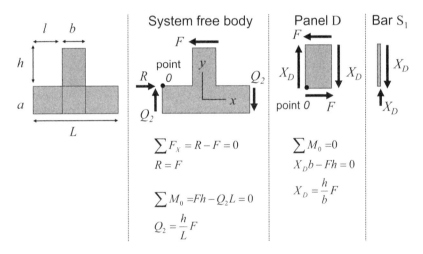

Figure 4.54a Free body diagram: In plane bending.

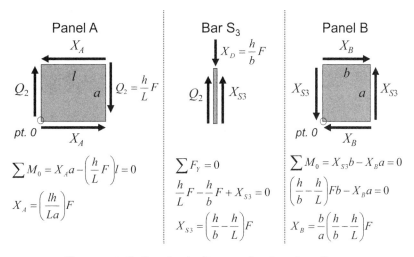

Panel A

X_A

l

a

Q_2 $Q_2 = \dfrac{h}{L}F$

pt. 0 X_A

$$\sum M_0 = X_A a - \left(\frac{h}{L}F\right)l = 0$$

$$X_A = \left(\frac{lh}{La}\right)F$$

Bar S₃

$X_D = \dfrac{h}{b}F$

Q_2 X_{S3}

$$\sum F_Y = 0$$

$$\frac{h}{L}F - \frac{h}{b}F + X_{S3} = 0$$

$$X_{S3} = \left(\frac{h}{b} - \frac{h}{L}\right)F$$

Panel B

X_B

b

a

X_{S3} X_{S3}

pt. 0 X_B

$$\sum M_0 = X_{S3}b - X_B a = 0$$

$$\left(\frac{h}{b} - \frac{h}{L}\right)Fb - X_B a = 0$$

$$X_B = \frac{b}{a}\left(\frac{h}{b} - \frac{h}{L}\right)F$$

Figure 4.54b Free body diagram: In plane bending.

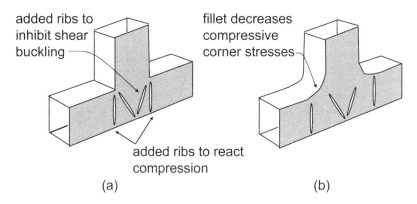

added ribs to inhibit shear buckling

fillet decreases compressive corner stresses

added ribs to react compression

(a)   (b)

Figure 4.55 Modification to joint shear wall.

Figure 4.56 shows experimental data for in-plane bending stiffness of three alternatives to the untreated planar T joint we have been analyzing [10]. A doubler plate over the shear wall, Figure 4.56a, inhibits shear buckling, although with some mass penalty. A filleted transition, Figure 4.56b, reduces the forces applied by the corners by increasing the distance between the corner forces. Finally, by adding a bulkhead under the upper section corners, Figure 4.56c, an additional load path is provided to react the corner loads but at some mass penalty. Similar data for out-of-plane bending alternatives [11] are provided in Figure 4.57.

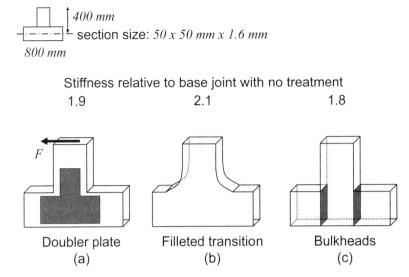

400 mm
section size: *50 x 50 mm x 1.6 mm*
800 mm

Stiffness relative to base joint with no treatment

| 1.9 | 2.1 | 1.8 |

Doubler plate
(a)

Filleted transition
(b)

Bulkheads
(c)

Figure 4.56 Increasing joint rigidity for in plane bending.

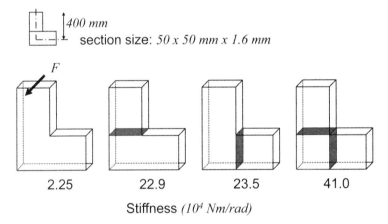

400 mm
section size: *50 x 50 mm x 1.6 mm*

| 2.25 | 22.9 | 23.5 | 41.0 |

Stiffness *(10⁴ Nm/rad)*

Figure 4.57 Effect of added bulkheads on out of plane joint rigidity.

4.5.1 Examples of body joint design

We have discussed joint design principles using an idealized set of rectangular beams. Now let us look at the application of these principles to more representative joints.

A-pillar-to-hinge-pillar joint

Figure 4.58a shows an A pillar and hinge pillar from a typical vehicle. The A-pillar surface is just behind the windshield edge, while the hinge pillar outer surface is considerably inboard due to positioning of the hinge attachment. This creates an undesirable lack of continuity for the shear surfaces of these two beams at the belt line. With careful coordination of exterior styling and door hinging, it is possible to achieve the alignment of the shear surface for both beams and a much stiffer joint, Figure 4.58b.

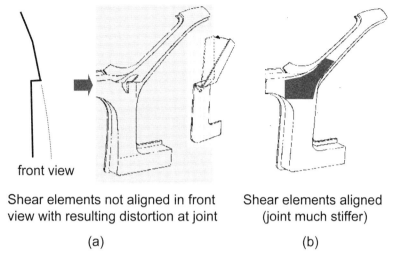

front view

Shear elements not aligned in front
view with resulting distortion at joint

(a)

Shear elements aligned
(joint much stiffer)

(b)

Figure 4.58 Hinge-pillar-to-A-pillar joint.

Hinge-pillar-to-rocker joint

Typical construction for the hinge-pillar-to-rocker joint is shown in Figure 4.59a. In
many cases, there is an offset, s, between the inner shear plane of the rocker and the
inner shear plane of the hinge pillar, Figure 4.59b. While we would ideally like these
planes to be aligned with $s=0$ for the stiffest joint, other constraints may prevent this
from occurring. When $s>0$, the rocker section distorts under in-plane loading, as shown
in Figure 4.59c. In this case, the rocker distortion behaves as a series of springs reacting
the rotational motion of the hinge pillar, Figure 4.60a. (This is similar to the behavior of
thin-walled sections under a point load which we looked at in Chapter 3.) The larger
the offset, s, the more flexible is each slice of the rocker section. Figure 4.60b plots the
joint efficiency vs. the rocker section stiffness for various offset dimensions. It can be
seen that for relatively small offsets ($s>5$ mm), the joint efficiency is greatly reduced.

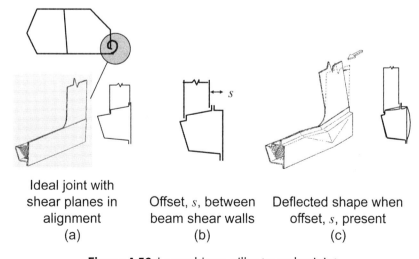

Ideal joint with
shear planes in
alignment

(a)

Offset, s, between
beam shear walls

(b)

Deflected shape when
offset, s, present

(c)

Figure 4.59 Lower-hinge-pillar-to-rocker joint.

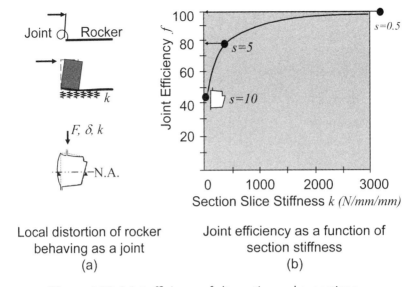

Local distortion of rocker
behaving as a joint
(a)

Joint efficiency as a function of
section stiffness
(b)

Figure 4.60 Joint efficiency of alternative rocker sections.

Floor-cross-member-to-rocker joint

The floor-cross-member-to-rocker joint shown in Figure 4.61 is loaded in out-of-plane bending. If the spot-weld flanges are perpendicular to the tensile and compressive loads in the top and bottom of the cross-member section, Figure 4.61a, the welds will be in peel and have considerable local distortion. By forming the weld flange as shown in Figure 4.61b, the loads are more nearly transferred in shear and a much stiffer joint results.

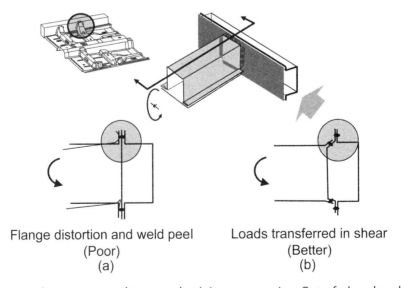

Flange distortion and weld peel
(Poor)
(a)

Loads transferred in shear
(Better)
(b)

Figure 4.61 Seat-cross-member-to-rocker joint construction: Out-of-plane bending.

4.5.2 Joint behavior at abrupt geometric transitions

Thus far we have been considering the localized deformation at the physical attachment between two or more beams. Often we see a similar large localized defection within a single beam, which contains an abrupt geometric transition, Figure 4.62. Even though no physical joint is present, we can treat the transition area as a joint stiffness with a beam on either side. A notable example is a beam with two relatively straight portions and a central curved portion, Figure 4.63. If we look at in-plane bending for this beam, we would see that the curved portion has high localized angular deformation and acts as a flexible joint connecting the two straight portions of the beam. Isolating the curved portion and applying moments to either side gives rise to compressive stresses along the top surface and tensile stresses along the bottom surface, Figure 4.63a.

Now consider a side view of the curved section of the beam, Figure 4.63b. The compressive load, P, acts on the top surface of the section, and we can see that an upward force, w, is required to place the surface into static equilibrium, Figure 4.63b. The magnitude of this upward force is given by:

$$w = \frac{P}{R}$$

(4.12)

where:

w = Compressive force per unit length acting on web

P = Compressive force acting on the strip

R = Radius of curvature of the strip.

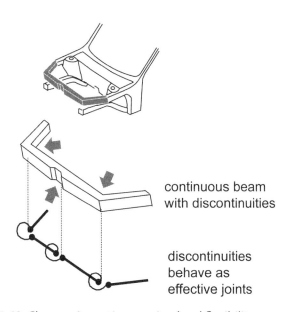

continuous beam with discontinuities

discontinuities behave as effective joints

Figure 4.62 Changes in section causing local flexibility.

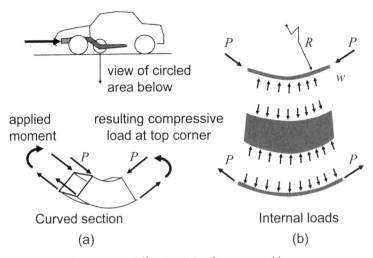

Figure 4.63 Effective joint for a curved beam.

This force, w, is a compressive force provided by the web of the beam pushing the upper and lower beam surfaces apart. Because the magnitude of this force is inversely proportional to the beam radius, beams with more abrupt curvature (smaller R) have a higher compressive force in the web. This compressive web force distorts the section, as shown in Figure 4.64a, and results in joint-like behavior. To improve this joint stiffness, the radius of curvature should be increased, or radial ribs included which react the compressive web loads, Figure 4.64b.

Often, the geometric transition is in a straight beam for the purpose of clearing some component of the vehicle, Figure 4.65. Again, a smooth transition will result in a stiffer effective joint, even at the expense of reducing beam section well before the component, Figure 4.66.

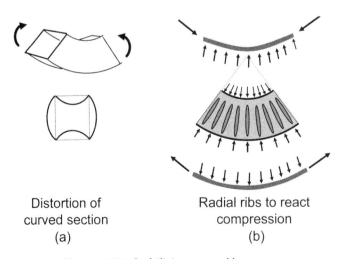

Figure 4.64 Stabilizing curved beam.

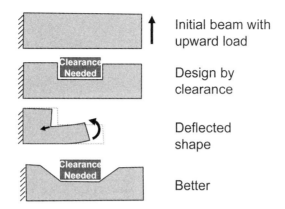

Figure 4.65 Section transition as a joint.

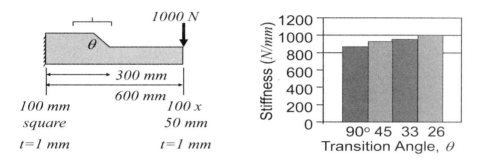

Figure 4.66 Smoothness of section transition.

4.5.3 Summary-Joint design

Two types of joint loading were defined—in-plane and out-of-plane, Figures 4.45 and 46. For in-plane loading, several design principles for stiff joint construction were presented. Shear walls of the connected beams should be aligned and flow smoothly from one beam to another, Figure 4.58. Even with the shear walls aligned in this way, however, the side wall at the beam intersection is relatively highly loaded in shear, and can be prone to shear buckling. Also the front and rear corners of the loaded beam place high compression stresses in the side wall, which is prone to compressive buckling, Figure 4.53. Careful placement of ribs or bulkheads can provide load paths for these conditions, Figures 4.55, 56, and 57. A severe geometric transition in a single beam can act as a flexible joint. To make this transition as stiff as possible, it should be gradual and smooth, Figures 4.64 and 66.

References

1. Pawlowski, J., *Vehicle Body Engineering*, Business Books, London, 1969, p. 124.

2. "Lincoln LS and the Parameter Development Vehicle," *Automotive Engineering International*, SAE International, Warrendale, PA, April, 1999, p. 32.

3. Thomson, W., *Vibration Theory and Applications*, Prentice-Hall, NJ, 1965, p. 275.

4. Fenton, J., *Vehicle Body Layout and Analysis*, Mechanical Engineering Publications, Ltd, London, 1980, pp. 55–57.

5. Brown, J.C., Robertson, A.J., Serpento, S.T., *Motor Vehicle Structures*, SAE International, Warrendale, PA, 2001, pp. 68–75.

6. Kikuchi, N. & Malen, D., Course notes for ME513 Fundamentals of Body Engineering, University of Michigan, Ann Arbor, MI, 2007.

7. Zienkiewicz, O. & Cheung, Y., *The Finite Element Method in Structural and Continuum Mechanics*, McGraw-Hill, NY, 1967.

8. Chang, D., "Effects of Flexible Connections on Body Structural Performance," SAE Paper No. 740041, SAE International, Warrendale, PA, 1974.

9. AISI, *Ultra Light Steel Auto Body-Final Report*, American Iron and Steel Institute, Southfield, MI, 1998.

10. Sunami, Y., et al., "Analysis of Joint Rigidity In-Plane Bending of Plane-Joint Structures," *Japan Society of Automotive Engineers,* April 1988.

11. Sunami, Y., et al., "Out-of-Plane Bending of Plane-Joint Structures," *Japan Society of Automotive Engineers*, July 1990.

Chapter 5
Design for Body Torsion

In this chapter we consider the overall body structure being twisted. As we did with the bending in Chapter 4, we will consider two types of body torsion requirements: strength and stiffness.

5.1 Body Torsion Strength Requirement

In defining the torsion strength requirement, we are seeking a vehicle-use condition which applies a maximum torque to the body, and yet a condition where the user would expect the body to recover its shape with little to no permanent deformation upon removal of the torque. The *twist ditch* maneuver is such a condition. Here, one wheel falls into a ditch and becomes unsupported, Figure 5.1. Putting the vehicle into static equilibrium for this condition, we see the twist ditch torque, T_{MAX}, is given by

$$T_{MAX} = W_{AXLE} \frac{t}{2} \tag{5.1}$$

where:

W_{AXLE} = Weight for the axle with the highest static load,

t = Track for that axle.

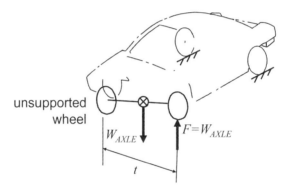

Figure 5.1 Body loaded in twist ditch condition.

A test set-up for this strength requirement is shown in Figure 5.2. The vehicle is supported at the suspension attachment points at one end, and loaded at the suspension attachment points at the other end. At the loaded end an upward deflection, δ, is imposed on one side and an equal downward deflection imposed on the other side, producing a twisting couple. Load cells at the loaded end measure the magnitude of the twisting couple. The angle of twist of the body, ϕ, is also measured as

$$\phi = \frac{2\delta}{w} \tag{5.2}$$

where:

δ = Deflection at each loaded suspension attachment

w = Width at the loaded points

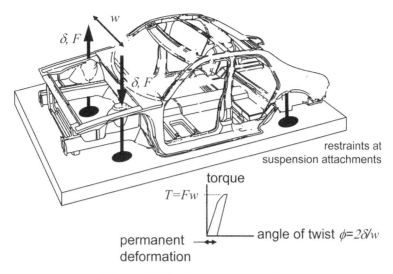

Figure 5.2 Torsion test convention.

For this requirement, we expect the body to suffer little permanent deformation after the twist ditch torque is removed. The chart in Figure 5.3 shows the twist ditch torques for a range of vehicles.

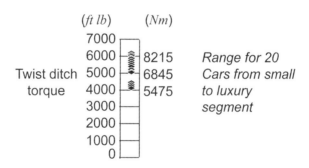

Figure 5.3 Torsion strength benchmarking.

5.2 Body Torsion Stiffness Requirement

From the body torsion test, we can measure a torsion stiffness—the slope in the linear region of the applied couple vs. angular rotation, ϕ. We will now develop the rationale for a body torsion stiffness requirement. Two important functions require high torsional stiffness:

1. To ensure good handling properties, the body should be torsionally stiff relative to the suspension stiffness [1].

2. To ensure a solid structural feel and minimize relative deformations which result in squeaks and rattles.

5.2.1 Ensure good handling

As the vehicle turns a corner, it rolls on the suspension ride springs, Figure 5.4. This rolling causes a weight transfer from the inside wheels to the outside wheels and can affect the steering characteristics of the vehicle. During suspension design, the body is assumed to be rigid, and suspension parameters are set with this assumption. We wish to set a body torsional stiffness requirement high enough such that this assumption of body rigidity is approximately correct. We can do this by making the torsional stiffness of the body many times stiffer than the roll rate of the suspension system.

Roll gain: Degrees of vehicle roll per g of lateral acceleration: θ/n

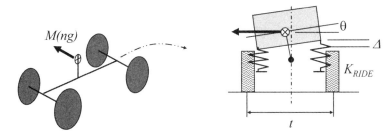

Figure 5.4 Vehicle roll stiffness.

Figure 5.5 illustrates the vehicle in a rolled condition. Assuming for now that the body is infinitely rigid, we can derive the roll stiffness based on the ride spring rates [2]. Typical values are shown with the resulting roll stiffness, K_{ROLL}, equal to approximately *1000 Nm/° (738 ft lb/°)*. Now view the system of roll springs of the front and rear suspension and the torsional stiffness of the flexible body as a series

torque applied to vehicle by lateral acceleration

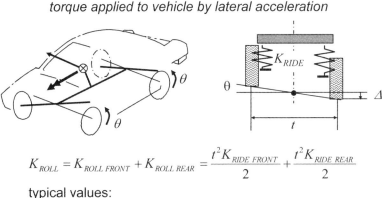

$$K_{ROLL} = K_{ROLL\ FRONT} + K_{ROLL\ REAR} = \frac{t^2 K_{RIDE\ FRONT}}{2} + \frac{t^2 K_{RIDE\ REAR}}{2}$$

typical values:

$t = 1560\ mm$
$K_{RIDE} = 23.4\ N/mm$
$K_{ROLL} = 57000\ Nm/rad = 1000\ Nm/deg$

Figure 5.5 First-order estimate of vehicle roll stiffness.

connection of springs, Figure 5.6. We wish to have the stiffness of this spring system, which includes the torsionally flexible body, Figure 5.6b, approximately equal to the ideal system consisting of only the suspension roll rates, Figure 5.6a. The graph of Figure 5.7 plots the ratio of the stiffness with a torsionally flexible body, K_{EFF}, to the suspension stiffness with a rigid body, K_{ROLL}, against the ratio of body torsional stiffness to suspension roll stiffness. We wish to have K_{EFF}/K_{ROLL} to approach one. Therefore we need the body torsional stiffness, K_{BODY}, to be *10* times the suspension roll stiffness, K_{ROLL}, to achieve K_{EFF}/K_{ROLL}=0.9. For typical passenger cars, this places a torsional stiffness requirement at K_{BODY}=*10000 Nm/° (7375 ft lb/°)* for suspension handling reasons.

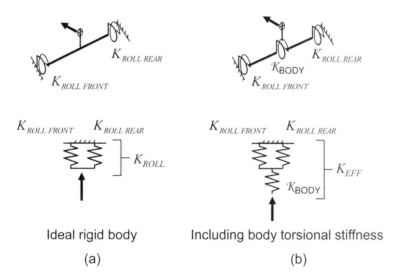

Figure 5.6 Vehicle roll stiffness with torsionally flexible body.

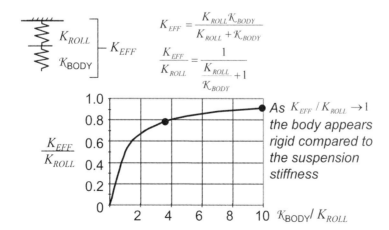

Figure 5.7 Effective torsional rate of flexible body and suspension.

5.2.2 Ensure solid structural feel

The second vehicle function demanding high torsional stiffness is to ensure the feel of solidness over road irregularities. This is related to the fundamental natural frequency of the body twisting mode; in general, higher natural frequency yields a more desirable solid feel. Figure 5.8 shows benchmark torsional resonant frequency data for several vehicles. As with bending vibration, sufficiently solid feel results when the vehicle torsional frequency is in the *22 to 25 Hz* range.

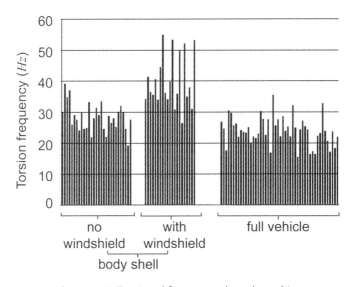

Figure 5.8 Torsional frequency benchmarking.

Consider the torsional resonance as a single-degree-of-freedom mass-spring oscillator where the point mass is the effective inertia of the rigidly attached vehicle subsystems and the spring is the torsional rigidity of the body. Thus, the body torsional stiffness requirement is related to the vehicle torsional frequency target. Figure 5.9 shows the measured torsional stiffness for several vehicles meeting this desirable vehicle torsional frequency range. A value of *12,000 Nm/° (8850 ft lb/°)* is seen as a good level of performance to meet both structural feel and also minimize relative deformations which cause squeaks and rattles.

Earlier we noted that the torsional stiffness requirement must satisfy two functions: 1) ensure good handling properties, and 2) ensure solid structural feel. Note that the suggested torsional stiffness based on structural feel, *12000 Nm/° (8850 ft lb/°)*, exceeds the stiffness based on suspension roll stiffness, *10000 Nm/° (7400 ft lb/°)*, so we select the greater of these as the requirement. (However, in the case of performance vehicles and race cars, the increased suspension roll stiffness of these vehicles makes the suspension stiffness function the dominant requirement for torsion.)

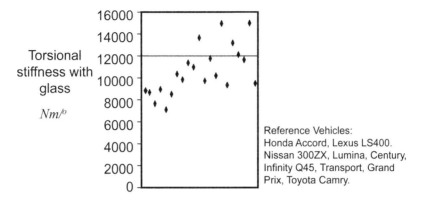

Figure 5.9 Torsional stiffness test and benchmarking. (Courtesy of the American Iron and Steel Institute, UltraLight Steel Auto Body)

To summarize, we have shown how both strength and stiffness requirements may be established for body torsion. Typical values for these requirements are shown in Figure 5.10 for a midsize vehicle. We will now look at the design of body structure to meet these requirements. The next section treats design for the strength requirement, followed by a section on design for the stiffness requirement.

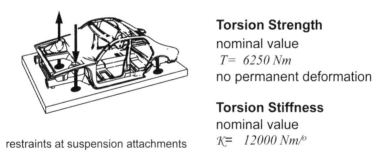

Torsion Strength
nominal value
$T = 6250\ Nm$
no permanent deformation

Torsion Stiffness
nominal value
$K = 12000\ Nm/^o$

restraints at suspension attachments

Figure 5.10 Typical torsional requirement for a mid-size vehicle.

5.3 Internal Loads During Global Torsion: Load Path Analysis

First we will consider the design of structure to meet the body strength requirement. As with the bending case, our objective is to understand how this global body requirement flows down to loads on structure elements. Once we have the loads on individual structural elements, we can then design those elements using the principles of Chapter 3.

5.3.1 Shear-resistant members

Early computer modeling of body torsional behavior during the 1970s idealized the structure as a framework of beams, Figure 5.11, very similar to our side-frame model for bending. However, these early analyses consistently predicted a more torsionally flexible structure with stiffness at *10–30%* of experimental values. When panels were added to the model, the predictions were much more accurate. These early studies showed that surfaces—shear-resistant members—are the dominant structure in reacting torsion loading, and that they can explain the behavior of the body loaded in torsion. Therefore, in the models of this chapter we will idealize the body structure as a set of shear-resistant members.

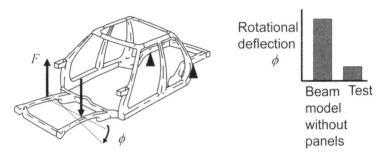

Figure 5.11 Beam model without panels.

We will begin by looking at a very simple model of a body: a *closed box*, Figure 5.12. The structural elements making up this box are structural shear surfaces first introduced in our analysis of the body strength in bending, Figure 5.13. We imagine the box loaded at the front corners with a twisting couple and an equal and opposite couple acting on the rear corners. The magnitude of this couple is the torsional strength requirement developed in the previous section. We are interested in how these applied loads flow into the individual surfaces [3].

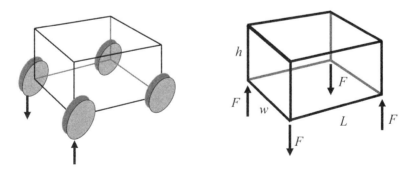

Figure 5.12 Box model.

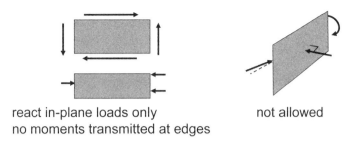

react in-plane loads only
no moments transmitted at edges

not allowed

Figure 5.13 Idealized surface element.

Now let us explode the box and look at the internal loads on each surface, Figure 5.14. These loads can be determined, as we did with the bending strength analysis, by setting each surface into static equilibrium [4]. To begin, look the front left applied force. It must be reacted by both the front surface left edge and the side surface front edge, as shown in Figure 5.14b. Let α be the fraction of the applied load taken by the front surface, and $(\alpha–1)$ the fraction taken by the side surface where $0<\alpha<1$. Now set the front surface and side surfaces into equilibrium by taking moments at a corner, Figure 5.15a and b. Finally, consider the bottom surface equilibrium. The forces on this surface must be equal and opposite to those applied by the front surface (along the front edge), and by the side surface (along the side edge), Figure 5.15c. Equilibrium of the bottom then tells us that $\alpha=1/2$ or the applied force is shared equally by the front and side surfaces, regardless of the dimensions of the box.

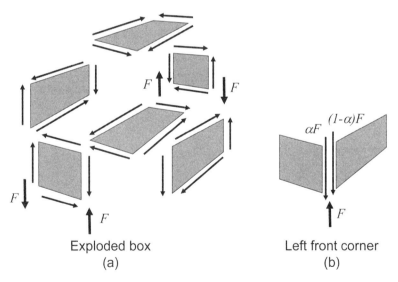

Exploded box
(a)

Left front corner
(b)

Figure 5.14 Box model internal loads.

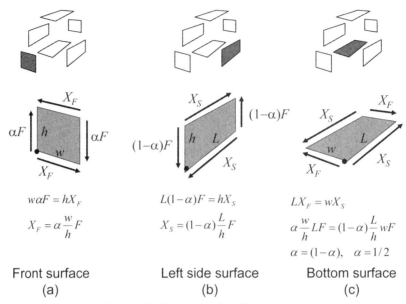

$$w\alpha F = hX_F$$

$$X_F = \alpha \frac{w}{h} F$$

Front surface
(a)

$$L(1-\alpha)F = hX_S$$

$$X_S = (1-\alpha)\frac{L}{h}F$$

Left side surface
(b)

$$LX_F = wX_S$$

$$\alpha \frac{w}{h} LF = (1-\alpha)\frac{L}{h} wF$$

$$\alpha = (1-\alpha), \quad \alpha = 1/2$$

Bottom surface
(c)

Figure 5.15 Free body of box model.

From the analysis of this simple structure under torsional loading, we see that 1) *all* surfaces are loaded, 2) that the internal loads are independent of material properties, and 3) that each surface is necessary to react the applied torsional couple: removal of any single surface will not allow the required equilibrium and the box will collapse. Also note that the *shear flow*—the shear force per unit of length of the edge it acts upon—is equal for all edges, Figure 5.16.

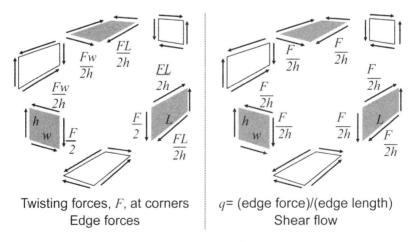

Twisting forces, F, at corners
Edge forces

$q=$ (edge force)/(edge length)
Shear flow

Figure 5.16 Monocoque box in torsion.

Example: Shear loads on rear hatch opening

Even with such a simple model, we can determine internal shear loads within a body structure. Consider the van shown in Figure 5.17 and loaded by the twist ditch torque. We are interested in the shearing loads which need to be reacted by the rear hatch structure, Figure 5.17a. First we must convert the torque requirement to a force couple using $F=T/w$. Using the equations developed for our box model, we immediately find that the shearing loads on the back structural shear surface are as shown in Figure 5.17b. With this information, we could begin to develop a rear hatch structure which will react these shear loads without excessive permanent deformation.

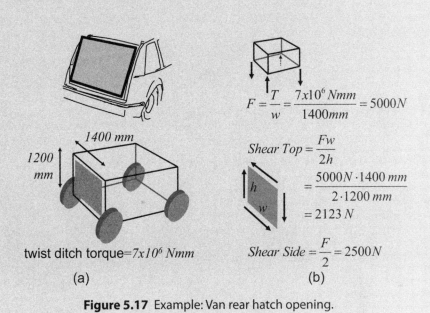

$$F = \frac{T}{w} = \frac{7x10^6\, Nmm}{1400mm} = 5000N$$

$$Shear\ Top = \frac{Fw}{2h}$$
$$= \frac{5000N \cdot 1400\, mm}{2 \cdot 1200\, mm}$$
$$= 2123\, N$$

$$Shear\ Side = \frac{F}{2} = 2500N$$

1400 mm

1200 mm

twist ditch torque=$7x10^6\, Nmm$

(a)   (b)

Figure 5.17 Example: Van rear hatch opening.

Note that the shear-resistant structural elements representing each face of the box do not have to be flat surfaces, as we have shown in the illustration of Figure 5.13. A shear-resistant member is any element which can resist the match-boxing deformation shown in Figure 5.18 top. This could be a flat panel as we have shown, a crowned panel, a panel with a hole or ribs, or a framework of beams, Figure 5.18 bottom. Examples of shear-resistant beam frameworks include the side-frame, rear hatch opening, and windshield ring, among others. Less-conventional shear-resistant members include diagonal straps in which only the strap under tension reacts the shear loads, and panels which have been loaded beyond their shear buckling load. Such buckled panels will still resist shearing deformation, with the buckled waves acting as tension members—this state is referred to as *diagonal tension*.

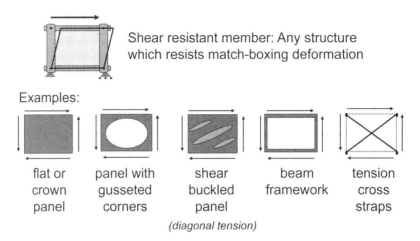

Shear resistant member: Any structure which resists match-boxing deformation

Examples:

| flat or crown panel | panel with gusseted corners | shear buckled panel | beam framework | tension cross straps |

(diagonal tension)

Figure 5.18 Shear-resistant members.

In preparation to look at a more realistic model for the body subjected to torsion loading, let us consider a box-shaped passenger cabin loaded by a torque at the front face and an equal and opposite torque at the rear face, Figure 5.19a.

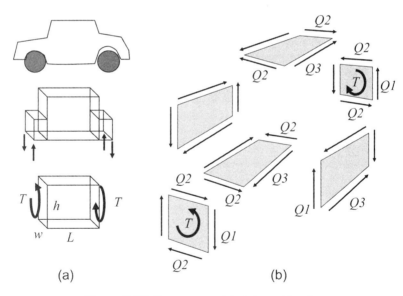

Figure 5.19 Passenger cabin internal loads.

We again place each surface into static equilibrium, Figure 5.19b, but this time we will use a matrix formulation [5]. The three independent equilibrium equations can be written as

$$+wQ_1 + hQ_2 = T \qquad \text{equilibrium of front face}$$

$$-LQ_2 + wQ_3 = 0 \qquad \text{equilibrium of top face}$$

$$-LQ_1 + hQ_3 = 0 \qquad \text{equilibrium of side face}$$

We can now define a column matrix of applied torques

$$\mathbf{T} = \begin{bmatrix} T \\ 0 \\ 0 \end{bmatrix}, \tag{5.3}$$

a column matrix of shear loads

$$\mathbf{Q} = \begin{bmatrix} Q_1 \\ Q_2 \\ Q_3 \end{bmatrix}, \tag{5.4}$$

and a coefficient matrix

$$\mathbf{A} = \begin{bmatrix} w & h & 0 \\ 0 & -L & w \\ -L & 0 & h \end{bmatrix}. \tag{5.5}$$

With these definitions, the equilibrium equations can be written as

$$\mathbf{AQ=T} \tag{5.6}$$

and as we are interested in the internal shear loads, this can be solved as

$$\mathbf{Q=A^{-1}T} \tag{5.7}$$

which gives for this case:

$$\begin{bmatrix} Q_1 \\ Q_2 \\ Q_3 \end{bmatrix} = \frac{1}{2whL} \begin{bmatrix} hL & h^2 & -hw \\ wL & -wh & w^2 \\ L^2 & Lh & Lw \end{bmatrix} \begin{bmatrix} T \\ 0 \\ 0 \end{bmatrix}$$

$$\text{or } Q_1 = \frac{T}{2w}, \quad Q_2 = \frac{T}{2h}, \quad Q_3 = \frac{TL}{2wh}$$

We now have the tools to allow us to look at the more realistic model for the body subjected to torsion loading, Figure 5.20. This is the same shear surface and bar model we used in the analysis of bending loads. Now we will apply the twist ditch torque as force couple, $R_F = T_{MAX}/w'$, acting on the motor compartment side surfaces. An equal and opposite couple is applied to the rear surfaces to place the body system into equilibrium. The model is now exploded, and the unknown internal loads drawn with the assumed positive directions, Figure 5.21. Note that the loads are symmetrically opposite about the plane dividing the body into right and left sides. As in the bending case, we begin by examining a surface which has an external load applied: the right motor compartment side surface *a*.

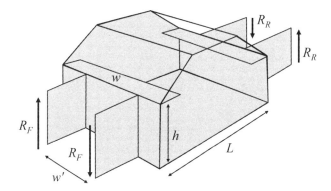

Figure 5.20 Structural surface and bar model.

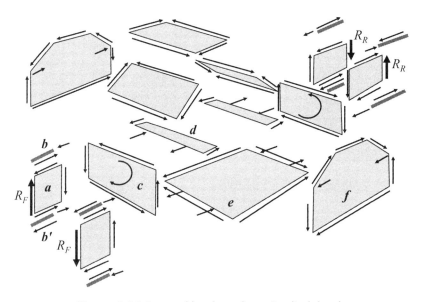

Figure 5.21 Internal loads under twist ditch load.

Proceeding by placing surface *a* into equilibrium, we find the internal shearing load, Q, on the top and bottom edges, Figure 5.22. The load on the upper edge is reacted by the bar, *b*, and we see a load, P_{B}, is required at its end to place the bar into equilibrium. Similarly, the load on the bottom edge of the surface is reacted by bar, *b´*.

As in the bending case, we see the motor compartment side, surface *a*, is loaded in shear, and we can use the principles developed in Chapter 3 to design this panel to meet these shear loads. As the magnitude of shear loads differs from the bending case, the worst case—bending or torsion—is used to size the panel.

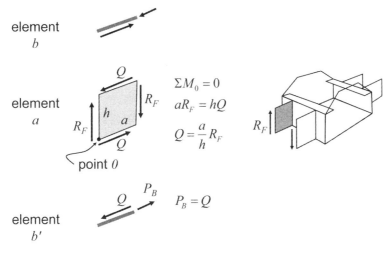

Figure 5.22 Internal loads on motor compartment panel.

Isolating the cowl surface, surface **d** in Figure 5.23, we see a rearward load applied by the left bar **b** and, due to asymmetry, an equal forward load applied by the right bar. To place the cowl surface into equilibrium, equal and opposite loads, P_D, are applied at the ends of the cowl. We can now identify all the loads applied to the cowl during twist ditch torsion loading, and can determine the moment diagram shown in Figure 5.24. We can now design a cowl section to react these moments without failing. With a thin-wall section, we must account for buckling, and determine if the elements of the section in compression are buckled. But we must note that the twist ditch torsion load may be in either the clockwise or counterclockwise direction. Therefore the moments on the cowl may be reversed in sign, and the compressed side of the section reversed. The designed section must accommodate both conditions.

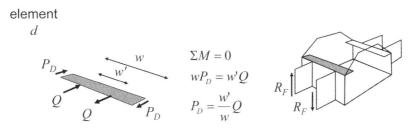

Figure 5.23 Internal loads on cowl and package shelf.

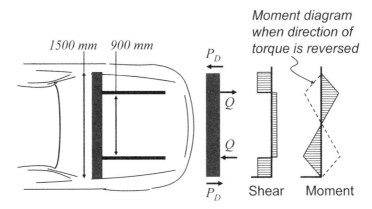

Figure 5.24 Internal loads on cowl bar.

We can now turn our attention to the cabin surfaces, Figure 5.25. Note that, like our earlier simple model of a six-sided box, the cabin forms a space totally enclosed by surfaces, although now we have eight sides. The cabin is loaded by a twisting couple applied to the dash by the motor compartment sides, *a* and *a´*. The couple has the same magnitude as the twist ditch torque. A similar, but opposite, arrangement occurs at the rear of the cabin structure with the couple applied to the rear surface.

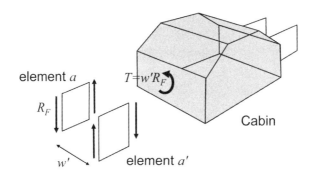

Figure 5.25 Twist ditch loads transmitted to cabin as a torque.

Again we explode the cabin surfaces and label the seven unknown shear loads, Q_1 to Q_7, as shown, Figure 5.26. As with the box model, we can set each surface into static equilibrium and express the set of equilibrium equations in matrix form. Beginning with the dash surface with an external torque applied, we sum moments about a corner point and set to zero,

$$h_1Q_1 + wQ_2 - T = 0 \qquad \text{equilibrium of dash surface}$$

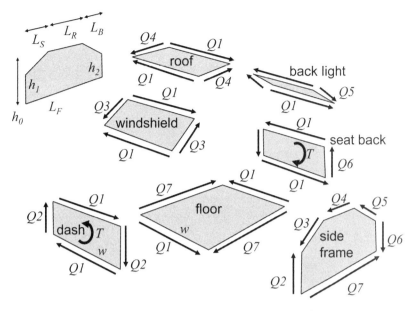

Figure 5.26 Shear loads on cabin panels.

Proceeding with equilibrium equations for all seven surfaces (only one side surfaces gives an independent equation), we end up with seven linear equations with seven unknown shear loads. These equations may be expressed in matrix form as we did with the simple six-sided box case in Equations 5.3 though 5.6:

$$
\begin{bmatrix}
h_1 & w & 0 & 0 & 0 & 0 & 0 \\
L_S & 0 & -w & 0 & 0 & 0 & 0 \\
L_R & 0 & 0 & -w & 0 & 0 & 0 \\
L_B & 0 & 0 & 0 & -w & 0 & 0 \\
h_2 & 0 & 0 & 0 & 0 & w & 0 \\
L_F & 0 & 0 & 0 & 0 & 0 & -w \\
0 & 0 & 0 & (h_o - h_1) & [L_F(h_o - h_2) + L_2(h_2 - h_1)]/L_B & -L_F & h_1
\end{bmatrix}
\begin{bmatrix}
Q_1 \\ Q_2 \\ Q_3 \\ Q_4 \\ Q_5 \\ Q_6 \\ Q_7
\end{bmatrix}
=
\begin{bmatrix}
T \\ 0 \\ 0 \\ 0 \\ T \\ 0 \\ 0
\end{bmatrix}
\begin{matrix}
\text{dash} \\ \text{windshield} \\ \text{roof} \\ \text{backlight} \\ \text{rear seat panel} \\ \text{floor} \\ \text{side frame}
\end{matrix}
\quad (5.8)
$$

In this set of equations, **AQ=T**, the applied torque is known and we wish to find the internal shearing loads Q_1 to Q_7. With the cabin dimensions, we can substitute numerical values for the coefficient matrix, **A**, and solve for the shearing loads, **Q=A⁻¹T**.

Several important points should be made on determining internal shear loads:

1. The loads were identified independently of material, so whether steel, aluminum, plastic or other, the structural elements must react these loads under body torsion

2. Each surface must be present to react its loads, otherwise the body will collapse

3. Each structural surface has the ability to react shear loads, and the actual structural element can be a beam framework, a curved panel, or others as discussed previously and shown in Figure 5.18.

Example: Cabin shear loads under torsion

The dimensions and twist ditch torque for a typical mid-size sedan, Figure 5.27, were used in the above matrix equation to determine the internal shear loads. Figure 5.28 shows the resulting matrix formulation (top) and the solution for shear loads on each surface (bottom). Note in particular the loading on the windshield surface, Q_1, and side frame, Q_7, as we will examine these more closely in the following section on torsional stiffness.

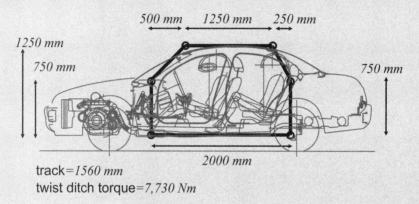

Figure 5.27 Example sedan data. (Courtesy of the American Iron and Steel Institute, UltraLight Steel Auto Body)

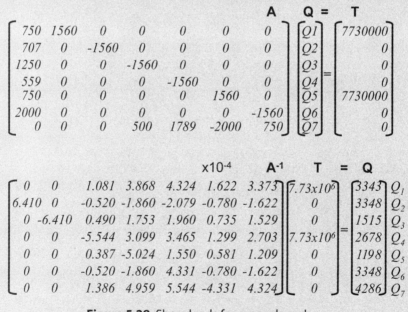

Figure 5.28 Shear loads for example sedan.

5.3.2 Summary: Torsion strength

In this section, we have shown how the applied torque in the global torsion strength test can be *flowed down* to the shear loads on each individual structural element. To accomplish this flow down of requirements, we used a model consisting of simple structural surfaces and bar elements, Figure 5.26. We set each element into static equilibrium to find a set of equations for the shear loads, set them into matrix form, and solved by inverting the coefficient matrix, Equation 5.8. Then, knowing the subsystem loads, the section design principles developed in the Chapter 3 can be used to find the appropriate structural elements to react these loads. We have thus gone from a torsion strength requirement on the body system to ensuring the strength of each structural element under torsion.

5.4 Analysis of Body Torsional Stiffness

In the previous section we examined how the torsion strength requirement implies loading on each structural element, and how we can ensure that each element is sufficiently strong. We will now focus on the global torsion stiffness requirement and flow down the requirement to shear stiffness requirements of the individual structural elements.

We begin by looking again at the six-sided box model, but now we look at the elastic angular deflection of the box under a torsional load, T, Figure 5.29. The angular deflection, θ, is the relative rotation between the front and rear surfaces. The torsional stiffness of the box, K, is the ratio T/θ. We wish to develop an equation which will predict torsional stiffness given the box dimensions, surface thicknesses, and material properties. To do this we will use energy methods, and we first develop an equation for the shear strain energy of a surface [6].

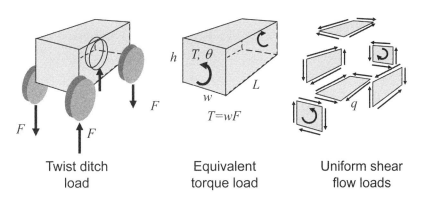

Twist ditch load   Equivalent torque load   Uniform shear flow loads

Figure 5.29 Van box model for torsional deformation.

5.4.1 Shear strain energy of a surface

When a surface is loaded in shear, the surface distorts into the diamond shape shown in Figure 5.30, and energy is stored as in an elastic spring. The shear strain energy, e, for a surface in uniform shear is given by:

$$e = \int_{VOLUME} \frac{\tau\gamma}{2} dV \qquad (5.9)$$

where:

    τ = Shear stress

    γ = Shear strain

and the integral is over the panel volume.

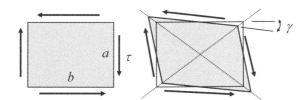

Figure 5.30 Panel under uniform shear.

We assume that the panel is in uniform shear, and both τ and γ are constant over the panel. This equation can be expressed in a useful form in terms of the panel dimensions, a, b, t, the material shear modulus, G, and the shear flow, q:

$$\tau = \frac{q}{t}, \quad V = abt, \quad \text{and} \quad G = \frac{\tau}{\gamma}, \quad \text{then} \quad e = \int_{VOLUME} \frac{\tau^2}{2G} dV = \frac{\tau^2}{2G} abt$$

$$e = q^2 \frac{ab}{2(Gt)} \qquad (5.10)$$

Note that the stored energy is inversely proportional to shear rigidity, (Gt). That is, surfaces with higher shear rigidity have lower stored strain energy for a given loading. We will now use this result for shear strain energy to determine the torsional stiffness of a closed box structure.

5.4.2 Energy balance for torque-loaded box

Look again at the closed box of Figure 5.29. As the applied torque, T, rotates through the angle, θ, it does an amount of work, $W = \frac{1}{2}T\theta$. This elastic work is stored within the panels as shear strain energy. By equating external work to the total stored energy, we can determine an expression for rotational deflection:

Work done by external torque = Total shear strain energy in all surfaces

From Equation 5.19,

$$\frac{1}{2}T\theta = \sum_{\text{ALL SURFACES}} \frac{1}{2}q^2 \left[\frac{ab}{(Gt)}\right]_{\text{SURFACE i}}$$

Recall that shear flow for a closed box, Figure 5.16, is

$$q = \frac{F}{2h} = \frac{\left(\dfrac{T}{w}\right)}{2h} = \frac{T}{2wh}$$

Substituting into the above,

$$\frac{1}{2}T\theta = \frac{1}{2}\left(\frac{T}{2wh}\right)^2 \sum_{\text{ALL SURFACES}} \left[\frac{ab}{(Gt)}\right]_{\text{SURFACE i}}$$

$$\theta = T\left(\frac{1}{2wh}\right)^2 \sum_{\text{ALL SURFACES}} \left[\frac{ab}{(Gt)}\right]_{\text{SURFACE i}}$$

$$K = \frac{T}{\theta} = (2wh)^2 \frac{1}{\displaystyle\sum_{\text{ALL SURFACES}} \left[\frac{ab}{(Gt)}\right]_{\text{SURFACE i}}}$$

The important result below gives the torsional stiffness of a closed six-sided box in terms of the properties for each surface.

$$K = (2wh)^2 \frac{1}{\left[\dfrac{ab}{(Gt)}\right]_{\text{SURFACE 1}} + \left[\dfrac{ab}{(Gt)}\right]_{\text{SURFACE2}} + \left[\dfrac{ab}{(Gt)}\right]_{\text{SURFACE 3}} + \left[\dfrac{ab}{(Gt)}\right]_{\text{SURFACE 4}} + \left[\dfrac{ab}{(Gt)}\right]_{\text{SURFACE 5}} + \left[\dfrac{ab}{(Gt)}\right]_{\text{SURFACE 6}}}$$

$$(5.11)$$

where:

K = Torsional stiffness of box

G = Shear modulus

w = Width of box

h = Height of box

a, b = Dimension of a side surface

t = Thickness of side surface

5.4.3 Series spring analogy

To gain a physical understanding of the way the rigidity, (Gt), for each of the six surfaces combine to give the torsional stiffness of the box, consider a set of six linear springs in series, Figure 5.31. Now look at the mathematical formulation for such springs in series which relates the combined stiffness to the stiffness of each spring:

$$\frac{1}{K_{EQ}} = \frac{1}{K_1} + \frac{1}{K_2} + \frac{1}{K_3} + \frac{1}{K_4} + \frac{1}{K_5} + \frac{1}{K_6}$$

$$K_{EQ} = \frac{1}{\left[\dfrac{1}{K_1}\right] + \left[\dfrac{1}{K_2}\right] + \left[\dfrac{1}{K_3}\right] + \left[\dfrac{1}{K_4}\right] + \left[\dfrac{1}{K_5}\right] + \left[\dfrac{1}{K_6}\right]} \qquad (5.12)$$

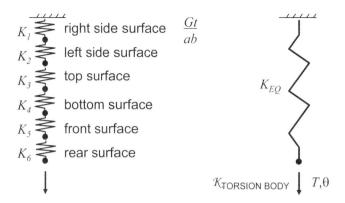

Figure 5.31 Spring analogy for torsional stiffness.

Rearranging the box torsional stiffness equation, Equation 5.11 can be rewritten as:

$$K = (2wh)^2 \frac{1}{\left[\frac{1}{\left(\frac{(Gt)}{ab}\right)}\right]_{SURF\ 1} + \left[\frac{1}{\left(\frac{(Gt)}{ab}\right)}\right]_{SURF\ 2} + \left[\frac{1}{\left(\frac{(Gt)}{ab}\right)}\right]_{SURF\ 3} + \left[\frac{1}{\left(\frac{(Gt)}{ab}\right)}\right]_{SURF\ 4} + \left[\frac{1}{\left(\frac{(Gt)}{ab}\right)}\right]_{SURF\ 5} + \left[\frac{1}{\left(\frac{(Gt)}{ab}\right)}\right]_{SURF\ 6}}$$

$$(5.13)$$

Compare the form of Equation 5.12 for linear springs in series with Equation 5.13 for our box in torsion. View the quantity $\left(\frac{(Gt)}{ab}\right)$ for each surface in Equation 5.13 as analogous to the stiffness K_i in Equation 5.12, and the forms are the same except for the constant, $(2wh)^2$. It can be seen that each surface of a box in torsion contributes to the torsional stiffness in the same way that each spring contributes to the equivalent stiffness in a series of springs.

As we pull at the end of a series of springs, Figure 5.31, the deflection is dominated by the most flexible spring in the group. If our objective is to increase stiffness, the most efficient way is by increasing the stiffness of the least stiff spring—stiffening the more-stiff springs will do little. This suggests a strategy to meet torsional stiffness requirements:

To increase torsional stiffness, identify which surface is the most flexible—lowest $\left(\frac{(Gt)}{ab}\right)$ —and improve it.

Example: Application of torsional stiffness equation

As an example, let us model a van with the dimensions shown in Figure 5.32 as a six-sided box with ideally flat, non-buckling, steel surfaces, *1 mm* thick. Substituting values into Equation 5.13 we find $K=6.95 \times 10^{10}$ Nm/rad $=1,200,000$ Nm/°. Comparing this value with the benchmark torsional stiffness data of Figure 5.9, we can see this answer is about *100* times stiffer than measured data! Clearly there is a problem.

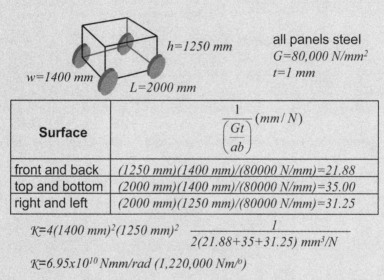

Surface	$\dfrac{1}{\left(\dfrac{Gt}{ab}\right)}\,(mm/N)$
front and back	(1250 mm)(1400 mm)/(80000 N/mm)=21.88
top and bottom	(2000 mm)(1400 mm)/(80000 N/mm)=35.00
right and left	(2000 mm)(1250 mm)/(80000 N/mm)=31.25

$$K=4(1400\ mm)^2(1250\ mm)^2\ \frac{1}{2(21.88+35+31.25)\ mm^3/N}$$

$$K=6.95 \times 10^{10}\ Nmm/rad\ (1,220,000\ Nm/°)$$

Figure 5.32 Example: Torsional stiffness of a box van.

In the above example, the estimated torsional stiffness was *100* times stiffer than anticipated. The problem is not with Equation 5.13 but with our use of ideal flat panels for each surface. In doing so, we have assumed that the surfaces remain perfectly flat during loading. In this unlikely case, the stiffness is very high as anyone who has twisted a closed shoe box knows. In reality, the surfaces of a vehicle body differ considerably from an ideal flat plate; they often have considerable out-of-plane shape such as crown, they have holes and cut-outs, often they are a framework of beams with flexible joints. Despite these realities, Equation 5.13 is still valid if we use the *effective shear rigidity, (Gt)EFF*, for these real surfaces.

5.4.4 Effective shear rigidity for structural elements

Using the *effective shear rigidity, (Gt)$_{EFF}$,* for each real-world panel in Equation 5.13, we can predict torsional stiffness for a body structure. To determine $(Gt)_{EFF}$, we consider the behavior of a test panel in a pinned frame fixture, Figure 5.33. This test fixture is made of four rigid bars connected at their ends by pin joints. Two of the joints are connected to ground, a shearing load is applied at the opposite side, and deflection is measured in line with the load. With no panel in the fixture, it offers no resistance to deformation, and deforms in a diamond shearing shape as shown in Figure 5.33.

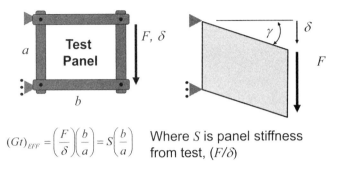

$$(Gt)_{EFF} = \left(\frac{F}{\delta}\right)\left(\frac{b}{a}\right) = S\left(\frac{b}{a}\right)$$ Where S is panel stiffness from test, (F/δ)

Figure 5.33 Test for effective (Gt).

Now consider a flat panel to be connected along the bars, and under load the panel will be deformed in a state of uniform shear, Figure 5.33. We wish to relate the applied load and measured deflection to the shear rigidity, (Gt), of the flat panel. Using the definition of shear modulus, we can arrive at the simple relationship:

$$G = \frac{\tau}{\gamma}, \quad \tau = \frac{F}{at}, \quad \gamma = \frac{\delta}{b}$$

$$(Gt) = \left(\frac{F}{\delta}\right)\left(\frac{b}{a}\right) = S\left(\frac{b}{a}\right), \quad \text{where } S = \left(\frac{F}{\delta}\right)$$

(5.14)

Where:

(Gt) = Inferred shear rigidity for the panel

S = Measured stiffness, (slope of the load F vs. deflection δ curve)

a = Panel dimension of the side to which the load is applied

b = Adjacent side dimension

This test suggests a means to determine the effective shear rigidity for any real world-panel: Imagine the panel held in the fixture of Figure 5.33, apply a shearing load, F, and measure the resulting deflection, δ. From the plot of F *vs.* δ, find the stiffness, S (slope). The effective shear rigidity is then given by Equation 5.14. The generation of the stiffness, S, may be determined by an actual physical test, by a strength-of-material type analysis, or by finite element analysis.

Now we have a strategy for calculation of vehicle torsional stiffness: use the surface model of Figure 5.29 and the resulting Equation 5.13 as before, but substitute *effective shear rigidity* for each real-world shear-resistant element as found from Equation 5.14

Examples using effective shear rigidity

To illustrate the use of Equations 5.13 and 5.14 we will look at four shear-resistant members: a van rear hatch perimeter, a general crown panel, the windshield-adhesive system, and the body side frame.

Example: Van rear hatch perimeter

We consider the perimeter of a van rear hatch opening as made of four rigid bars connected by flexible joints, Figure 5.34. If we imagine this model placed in the shear test fixture described above, we can apply basic energy analysis to determine the stiffness. Under a load, F, the frame deflects an amount, δ, and the work done by the external force is

$$w = \frac{1}{2}F\delta$$

This work is stored in each of the corner joints, and equating the external work with the energy stored in the joints, we have

$$\frac{1}{2}F\delta = 4\left(\frac{1}{2}K_J\,\theta^2\right)$$

Since $\theta = \dfrac{\delta}{b}$ we have for the stiffness of the framework

$$S = \frac{F}{\delta} = \frac{4k_J}{b^2}$$

Using Equation 5.14, we determine that the effective shear rigidity for the frame is

$$(Gt)_{EFF} = \frac{4k_J}{ab} \tag{5.15}$$

where:

  $(Gt)_{EFF}$ = Effective shear rigidity of a rigid frame with flexible joints

  k_J = Joint rate

  a, b = Dimensions of frame

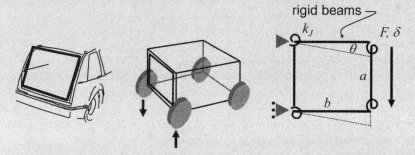

Figure 5.34 Effective shear rigidity of a frame with flexible joints

Let us now apply this result to the torsional stiffness of the box van of Figure 5.32. In that earlier example, each surface was an ideal flat, non-buckling panel. Let us replace the rear panel with an open frame of rigid links with a typical joint stiffness value, $k_J = 0.1 \times 10^8$ Nmm/rad

Using Equation 5.15 we find that the effective shear rigidity for the rear opening is

$$(Gt)_{EFF} = \frac{4(0.1 \times 10^8\ Nmm\,/\,rad)}{(1250mm)(1400mm)} = 22.86\,N\,/\,mm$$

Compare this value to our previous value for the flat, 1-mm-thick panel: $(Gt)=(80{,}000\ N/mm)(1\ mm)=80{,}000\ N/mm$. So the frame is much more flexible than the original assumption of a flat panel.

Now we can use Equation 5.13 to determine the resulting torsional stiffness for this van made of flat, *1-mm*-thick steel panels *except* for the frame substituted for the rear surface, Figure 5.35. The resulting value for torsional stiffness, Figure 5.35, is much lower than that computed previously for the van with all flat surfaces, Figure 5.32, and more in line with reality. Note that only one surface of the closed box need be flexible to reduce the stiffness for the whole box. This is an illustration of the series spring view of torsional stiffness from Figure 5.31.

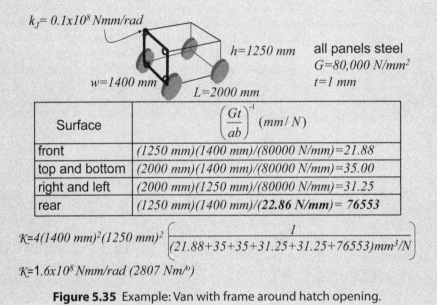

$k_J = 0.1 \times 10^8 \ Nmm/rad$

$h=1250 \ mm$ all panels steel
$G=80,000 \ N/mm^2$
$t=1 \ mm$

$w=1400 \ mm$

$L=2000 \ mm$

Surface	$\left(\dfrac{Gt}{ab}\right)^{-1} (mm/N)$
front	(1250 mm)(1400 mm)/(80000 N/mm)=21.88
top and bottom	(2000 mm)(1400 mm)/(80000 N/mm)=35.00
right and left	(2000 mm)(1250 mm)/(80000 N/mm)=31.25
rear	(1250 mm)(1400 mm)/(22.86 N/mm)= 76553

$$K=4(1400 \ mm)^2(1250 \ mm)^2 \left[\frac{1}{(21.88+35+35+31.25+31.25+76553)mm^3/N}\right]$$

$$K=1.6 \times 10^8 \ Nmm/rad \ (2807 \ Nm/°)$$

Figure 5.35 Example: Van with frame around hatch opening.

The above example shows the influence of the hatch opening on torsional stiffness of a van. In practice, efforts are made to use the hatch door to increase the shear rigidity of the rear surface. Typically, the hinge and latch are not sufficiently stiff to do this, and mechanisms to wedge the door into the opening are used, Figure 5.36 & 5.37.

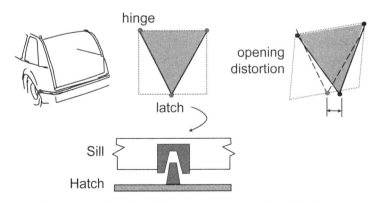

hinge

opening
distortion

latch

Sill

Hatch

Figure 5.36 Van hatch latch effect on torsional stiffness.

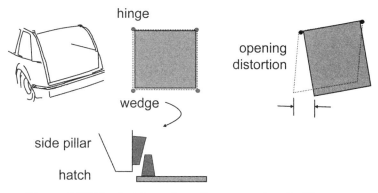

Figure 5.37 Van hatch wedge effect on torsional stiffness.

Example: Crown panel

Many panels on a vehicle are crowned in shape. The reason for this is that the crown shape improves panel stiffness for normal loading such as dent resistance and panel vibration. To determine the effective shear rigidity for panels of various crown heights, an FEA model of the shear test fixture was constructed for typical panel dimensions, Figure 5.38a. The resulting $(Gt)_{EFF}$ Figure 5.38b, shows that for crowns from *10–40 mm*, the $(Gt)_{EFF}$ is much smaller than for a flat panel. This explains why our initial calculation of torsional stiffness using flat panel values gave us unrealistically high values.

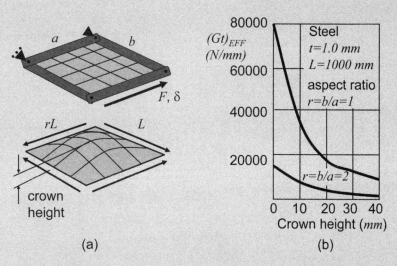

(a)   (b)

Figure 5.38 Effective shear rigidity of crown panels based on FEA.

Example: Windshield-adhesive system

We now understand that all surfaces enclosing the cabin must act as shear-resistant members. The windshield is one of these surfaces. Of the alternative means to retain the windshield, Figure 5.39, the most effective for shear resistance is adhesive bonding. In this retention system, the windshield is bonded to the body flange at its perimeter. As the body is torsionally loaded, the windshield opening distorts in shear and loads the adhesive in a wiping motion, Figure 5.40.

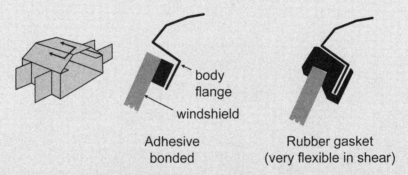

Figure 5.39 Windshield retention alternatives.

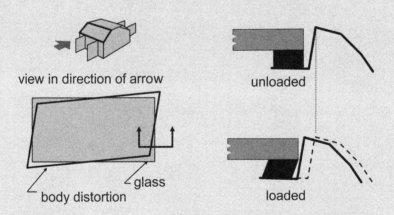

Figure 5.40 Effective shear rigidity of windshield-adhesive system.

A spring system is formed by the three members: the windshield opening frame, the windshield, and the adhesive, as shown in Figure 5.41a. In Figure 5.41b we show the relative shear stiffness for each of these members by itself. As can be seen, the frame is relatively flexible, K_3, and the parallel path of windshield with adhesive, K_7 & K_8, provides the main path for shearing stiffness. As a measure of this additional stiffness, the torsional resonances are compared with and without windshield glass, Figure 5.42. For many vehicles there is a marked increase in torsional frequency with glass installed, indicating a torsionally stiffer body. For those vehicles which do not show this increase, the body windshield opening perimeter is very stiff.

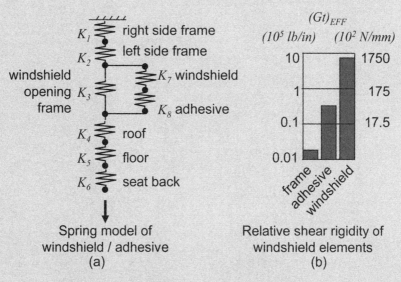

Figure 5.41 Windshield model for torsional stiffness.

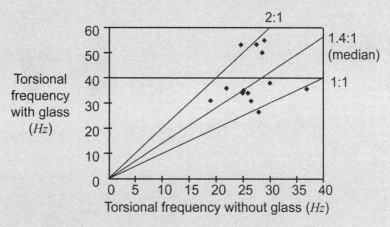

Figure 5.42 Effect of windshield on torsion frequency.

Example: Side frame

In the chapter on bending, we examined the sedan side-frame under bending loads. Now we can look at the influence of side-frame performance on torsional stiffness. We replace the side surfaces in our torsion model with the beam framework of the side-frame, Figure 5.43. We can determine the effective shear rigidity of the side frame by imagining it in the shear test fixture, Figure 5.44. As the side frame is a redundant structure, we use FEA to simulate its behavior in this fixture. We restrain the side-frame at the hinge-pillar-to-rocker joint, and at the rocker-to-C-pillar joint with a pin to ground. We apply a shearing load across the top of the side-frame along the roof rail, Figure 5.45, and solve for the resulting horizontal deflection. Using Equation 5.14 we can then identify the effective shear rigidity for this side-frame as shown at the bottom of Figure 5.45.

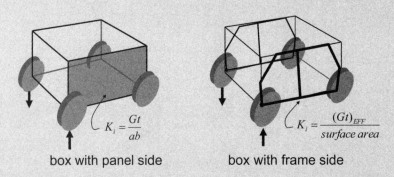

box with panel side $\qquad$ box with frame side

$$K_i = \frac{Gt}{ab} \qquad K_i = \frac{(Gt)_{EFF}}{surface\ area}$$

Figure 5.43 Side-frame contribution to torsional stiffness.

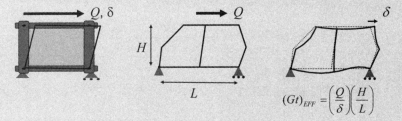

$$(Gt)_{EFF} = \left(\frac{Q}{\delta}\right)\left(\frac{H}{L}\right)$$

Figure 5.44 Effective shear rigidity of side frame.

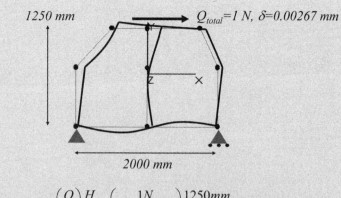

$$(Gt)_{EFF} = \left(\frac{Q}{\delta}\right)\frac{H}{L} = \left(\frac{1N}{.00267mm}\right)\frac{1250mm}{2000mm} = 234N/mm$$

Figure 5.45 Side-frame FEA under shear loading.

Note that this is the same side-frame FEA model used earlier in the bending analysis of Chapter 4. However, the restraints and loading for the torsion case are different, resulting in a different deflected shape and also a different strain energy distribution.

5.4.5 Torsional stiffness of a vehicle cabin

Earlier we looked at the internal loads of a sedan cabin, Figure 5.46. The torsional stiffness of that cabin can be found by generalizing Equation 5.13 from the six-sided box to a cabin with more surfaces enclosing the volume, Figure 5.47. Again, the work done by the external torque will equal the shear strain energy in all surfaces:

$$\frac{1}{2}T\theta = \sum_{\text{ALL SURFACES}} \frac{1}{2}q^2\left[\frac{ab}{(Gt)}\right]_{\text{SURF } i}$$

Dividing both sides by T^2,

$$\frac{\theta}{T} = \left(\frac{q}{T}\right)^2 \sum_{\text{ALL SURFACES}} \left[\frac{\text{area of surface } i}{(Gt)_{EFF}}\right]_{\text{SURF } i}$$

$$K = \frac{1}{\left(\frac{q}{T}\right)^2 \sum_{\text{ALL SURFACES}} \left[\frac{\text{area of surface } i}{(Gt)_{EFF}}\right]_{\text{SURF } i}} \qquad (5.16)$$

where (q/T) is found by solving the matrix equation $\mathbf{Q=A^{-1}T}$, Equation 5.7, and q is the resulting shear flow on any non-loaded surface.

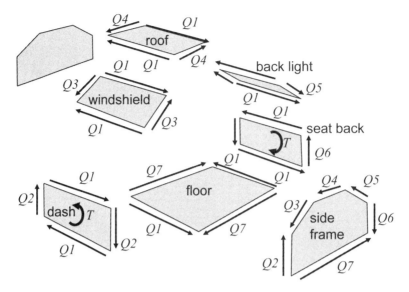

Figure 5.46 Cabin surfaces.

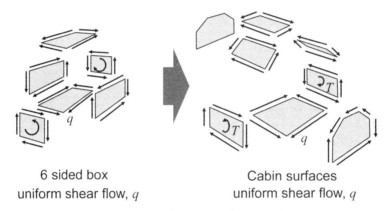

6 sided box
uniform shear flow, q

Cabin surfaces
uniform shear flow, q

Figure 5.47 Uniform shear flow in cabin.

These relationships provide a calculation procedure for the torsional stiffness of a realistic cabin structure:

1. Apply a torque, T, to the structure and solve for the internal shear loads, Q_i, using $\mathbf{Q=A^{-1}T}$ (An example of this was done in earlier in Figures 5.27 & 5.28 and Equation 5.8.)

2. Find the shear flow, q, by taking any of the surfaces not loaded by the external torque and dividing that shear load by the length of the side over which the load acts. Then form the ratio (q/T) with the torque being the value used in step one

3. Determine the effective shear rigidity, $(Gt)_{EFF}$, for each panel using Equation 5.14

4. Substitute the values for (q/T), $(Gt)_{EFF}$, and surface area into Equation 5.16 to determine the torsional stiffness of the cabin.

194

Example: Sedan torsional stiffness with side-frame

Let us use the procedure outlined above to determine the cabin torsional stiffness for the sedan cabin dimensions from Figure 5.27. For the two side surfaces, we will use the side-frame described above in "Example: Side frame" and, for simplicity, let all the other surfaces remain flat, 1-mm-thick steel panels. Proceeding through the four steps outlined above:

1. We can use the results from the torsional strength analysis done earlier for this sedan cabin, Figure 5.28. There we found, for $T=7,730,000$ Nmm, the shear loads, Q_1 through Q_7, which are given in Figure 5.28 with the convention of Figure 5.26. For example, the shear load on the roof panel side is $Q_4=2678$ N.

2. With the roof panel side length of 1250 mm, Figure 5.27, we find

$$q=(2678N)/(1250mm)=2.1414N/mm$$

and

$$(q/T)=(2.1414N/mm)/(7,730,000Nmm)=2.77 \times 10^{-7} mm^{-2}.$$

3. The effective shear rigidity for each side-frame from the previous "Example: Side frame," Figure 5.45, is

$$(Gt)_{EFF}=234 \text{ N/mm}$$

and for the other surfaces

$$(Gt)=80,000 \text{ N/mm}^2(1 \text{ mm})$$

4. Now using Equation 5.16, Figure 5.48, we find that the cabin torsional stiffness is $K=11,423$ Nm/°.

Panel	Area of panel (mm^2)	Effective shear rigidity $(Gt)_{EFF}$ (N/mm)	$\left[\dfrac{\text{area of surface i}}{(Gt)_{EFF\,i}} \right]$ (mm^3/N)
dash	1170000	80000	14.6
windshield	1103087	80000	13.8
roof	1950000	80000	24.4
back light	872067	80000	10.9
seat back	1170000	80000	14.6
floor	3120000	80000	39.0
side frame-left	2312500	234	9882.5
side frame-right	2312500	234	9882.5
		sum	19882.3

$$\mathcal{K} = \cfrac{1}{\left(\dfrac{q}{T}\right)^2 \sum_{ALL\,SURFACES}\left[\dfrac{\text{area of surface i}}{(Gt)_{EFF\,i}}\right]} = \begin{array}{l} 6.55 \times 10^8 \text{ Nmm/rad} \\ (11491 \text{ Nm/o}) \end{array}$$

Figure 5.48 Cabin structure torsional stiffness.

If this cabin system stiffness is not adequate with respect to the requirement, we can return to the side-frame FEA of Figure 5.45 and improve the shearing stiffness. Again, we can use strain energy to direct us to the beams and joints which will have the largest impact on improving stiffness.

5.4.6 Summary: Torsion stiffness

We have investigated the torsional behavior of enclosed structures constructed of shear-resistant surfaces. Using energy principles, we developed Equation 5.16, which relates the shear stiffness of the individual surfaces making up the cabin structure to the body torsional stiffness, Figure 5.26. For perfectly flat panels, this shear rigidity is *(Gt)*—the product of shear modulus and thickness. However, for the real surfaces which make up the cabin surfaces, we need to identify the *effective shear rigidity, (Gt)$_{EFF}$.* This is done by visualizing the real surface loaded in a shear test fixture, Figure 5.33. By evaluating the load and deflection in this test fixture, the effective shear rigidity for the panel can be determined using Equation 5.14. As with the strength load case, we found that all surfaces must be present to provide torsional stiffness, and that the least stiff surface dominates the stiffness of the body. A convenient way to visualize the cabin torsional stiffness is as a series combination of springs, with each spring being one panel's effective shear rigidity, Figure 5.31.

5.5 Torsional Stiffness of Convertibles and Framed Vehicles

Our focus in this chapter on torsion has been on the body as a monocoque structure, that is, enclosed by shear-resistant surfaces. This focus is motivated by the efficiency of this type of structure in reacting torsional loading. However, other alternatives exist, Figure 5.49, although they seldom approach the torsional stiffness efficiency achieved by the monocoque structure, Figure 5.50. In this section we will discuss some of the most notable alternative structures.

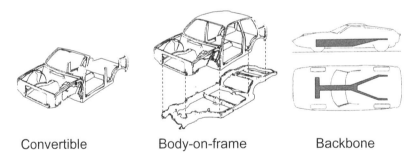

Convertible   Body-on-frame   Backbone

Figure 5.49 Alternatives to monocoque body structure.

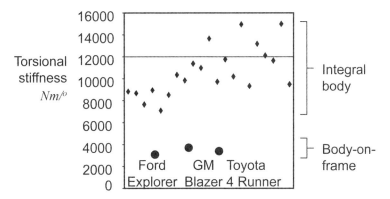

Figure 5.50 Body-on-frame torsional stiffness benchmarking.

5.5.1 Torsional stiffness of convertibles

We have stressed the need for all surfaces of the vehicle body 'box' to be present to react torsional loads. In a convertible, we have removed the top of the box, and lost the ability to react loads via shear resistant surfaces. This can be visualized with the series spring analogy, Figure 5.51, where the series chain is 'broken' with the absence of the top surface. In order to provide adequate torsional rigidity, we add another spring in parallel—a lower load path—to resist torsional loads. A common means to provide this load path is to use *differential bending* of the rocker beams.

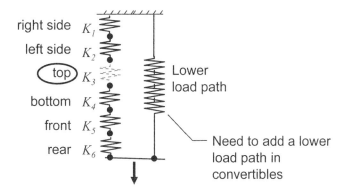

Figure 5.51 Convertibles with top shear-resistant member removed.

Consider the lower structure of a convertible to be idealized as two cross members and two side rail beams, Figure 5.52a. The front cross member shown is located at the dash, and the rear cross member is located at the rear seat back. In this model, both cross members are viewed to be infinitely stiff in torsion about their longitudinal axis, but very flexible in bending. Now let a torque be applied to the front cross member, and a reacting torque applied to the rear cross member. The deformed shape will be as shown in Figure 5.52a. Looking at the side rail beam in the side view, Figure 5.52b, we see it has zero slope at its ends. This is because the cross members are infinitely rigid in torsion and will have zero twist along their cross-car axes. Also, because of the cross members' flexibility in bending, the side rails are not twisted down their axis. Thus the side rail is in a state of pure bending with zero slope at either end, and it is this *differential bending* which gives this framework its torsional rigidity.

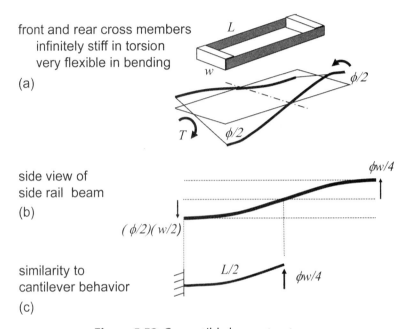

Figure 5.52 Convertible lower structure.

We now quantify the effect of differential bending on torsion stiffness. The structural elements, Figure 5.52, are loaded by a torque, T, and rotate through a total angle, ϕ an angle of $(\phi/2)$ at one end and $(-\phi/2)$ at the other. The downward motion at the front of the side rail is $(w/2)(\phi/2)$, with an identical upward motion at the rear of the side rail. Looking at the rail in the side view, Figure 5.52b, we can see that the behavior for the front half is identical to a cantilever beam of length $(L/2)$, Figure 5.52c. Using the deflection equation for a cantilever

$$Tip\ Defection = \frac{FL^3}{3EI}$$

we can calculate the force, F, required for a deflection of $(w/2)(\phi/2)$.

$$F = 3EI\left(\frac{w\phi}{4}\right)\left(\frac{2}{L}\right)^3$$

This end force appears at both ends of the front cross member, so we have the applied torque needed to achieve the deflection ϕ. This yields a first-order equation for torsional stiffness, K, of a frame with differential bending:

$$\frac{T}{\phi} = \frac{wF}{\phi} = \frac{w}{\phi}\left[3EI\left(\frac{w\phi}{4}\right)\left(\frac{2}{L}\right)^3\right]$$

$$K = \frac{6w^2EI}{L^3} \qquad\qquad (5.17)$$

where:

   EI = Side rail bending stiffness

   L = Length between front and rear cross members

   w = Frame width

Example: Torsional stiffness of a frame under differential bending

Considering a frame with a side rail section of *100-mm*-square and *2-mm*-thick steel (I=*1.33x10⁶ mm⁴*), with frame width of *1500 mm* and length *3000 mm*, Equation 5.17 yields

$$K = \frac{6(1500mm)^2(207000N/mm^2)(1.33x10^6\,mm^4)}{(3000mm)^3} = 137,655,000\,Nmm/rad$$

or K=*2415 Nm/º* –a low but not unreasonable stiffness.

Differential bending is frequently applied to structure for convertibles. In that application, the torsionally rigid cross members are realized in practice by a large closed box section at the dash and at the rear seat back, Figure 5.53. This approximates the very rigid six-sided box model we have used earlier but on the scale of a cross member rather than the whole cabin. A difficulty with realizing differential bending in practice is the cross-member-to-side-rail joint. The zero-slope end condition in the side rail requires a very large bending moment to be transferred by the cross member, Figure 5.53 bottom. This large moment can cause stress concentrations at the joint, which can lead to durability problems if not treated during design.

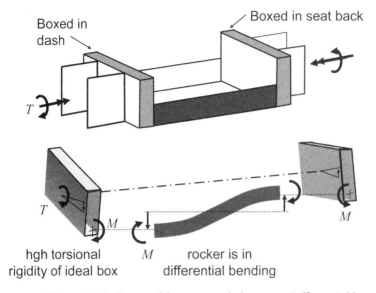

Figure 5.53 Convertible structural elements: Differential bending.

5.5.2 Torsional stiffness of body-on-frame vehicles

A common arrangement for both passenger and utility vehicles is the *body-on-frame* configuration. In this configuration, a body shell is attached to a ladder frame with several elastomeric body mounts, Figure 5.54. These body mounts allow relative motion between the frame and body, both in the vertical direction (compression), and in lateral direction (shear). The primary function of the body mounts is the isolation of structure-borne noise and vibration from the frame into the body.

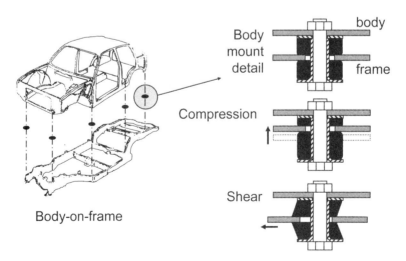

Figure 5.54 Body mounts.

If we consider a torque being applied to the frame though the suspension, the frame and body will tend to twist about different longitudinal axes, Figure 5.55. This type of twisting action causes shearing deformation in the body mounts, which reduces the stiffness of this system; the torsional stiffness of the vehicle can be less than the sum of the torsional stiffnesses of the body and the frame.

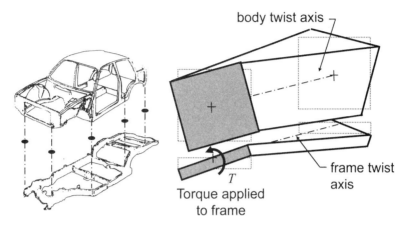

Figure 5.55 Body-on-frame idealized torsion model.

To understand how the frame, body, and body mounts combine to give a vehicle torsional stiffness, consider Figure 5.56a. Here we have the system of the body, with torsional stiffness K_1, the frame, with torsional stiffness K_2, and four body mounts at each corner, spaced by width, w, and length, L. The mount is viewed as two linear springs: one in the horizontal direction (shear stiffness k_X), and one in the vertical direction (compressive stiffness k_Y), Figure 5.56b.

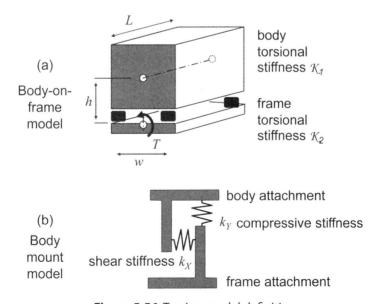

Figure 5.56 Torsion model definitions.

201

A torque, T, is applied at the front of the frame, Figure 5.57, with an equal and opposite torque applied at the rear. The frame then twists through ϕ_2 about its own twist axis in the plane of the frame, and the body twists through an angle ϕ_1 about its twist axis at height, h, above the frame. Note that ϕ_1 and ϕ_2 are very small angles so $\sin\phi=0$ and $\cos\phi=1$. The vehicle system can be exploded to show the internal loads, F_X and F_Y, and we can put both the body and frame into static equilibrium, Figure 5.57. Now, examining the kinematic relationships, Figure 5.58, we can identify the deflection across the mount:

$$\delta_{VERTICAL} = \frac{w}{2}\phi_2 - \frac{w}{2}\phi_1$$

$$\delta_{LATERAL} = h\phi_2$$

Finally, we write the constitutive relationship (load-deflection behavior) for the mount. A set of four equations result with the unknowns Fx, Fy, ϕ_1, ϕ_1:

$$wF_Y - 2hF_X - K_1\phi_1 = 0$$

$$wF_Y + K_2\phi_2 = T$$

$$F_X - K_Xh\phi_1 = 0$$

$$F_Y + K_Y\frac{w}{2}\phi_1 - K_Y\frac{w}{2}\phi_2 = 0$$

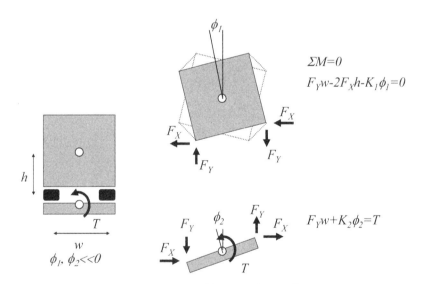

$\Sigma M=0$

$F_Yw-2F_Xh-K_1\phi_1=0$

$F_Yw+K_2\phi_2=T$

Figure 5.57 Equilibrium relationships.

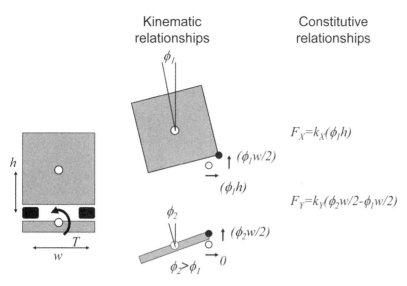

ϕ_1

$F_X=k_X(\phi_1 h)$

$(\phi_1 w/2)$

$(\phi_1 h)$

$F_Y=k_Y(\phi_2 w/2 - \phi_1 w/2)$

ϕ_2

$(\phi_2 w/2)$

$\phi_2 > \phi_1$ 0

h

T

w

Figure 5.58 Kinematic and constitutive relationships.

We may solve this set and then take the ratio of T/ϕ_2 to arrive at the vehicle torsional stiffness, $K_{VEH}=T/\phi_2$:

$$K_{VEH} = K_2 + K_1\psi + 2h^2 k_X \psi$$

$$\psi = \cfrac{1}{1 + \cfrac{2h^2 k_X}{\left(\dfrac{w^2}{2} k_Y\right)} + \cfrac{K_1}{\left(\dfrac{w^2}{2} k_Y\right)}} \tag{5.18}$$

where:

 K_1 and K_2 = Torsional stiffness of the body and frame, respectively,

 k_X and k_Y = Mount stiffnesses in the horizontal and vertical directions

 h = Height of the body twist axis above the plane of the frame,

 w = Width between body mounts

 ψ = Body-frame coupling term which indicates how tightly coupled are the twisting actions of the frame and body; larger ψ is greater coupling.

Equation 5.18 shows that the vehicle torsional stiffness consists of all of the frame stiffness, K_2, plus a portion of the body stiffness, $K_1\psi$, plus the unexpected term, $2h^2 K_X \psi$. This last term depends on the shear stiffness, k_X, of the mounts. To understand this behavior, we take typical values for a body-on-frame sedan, and for the body mount select an extremely soft shear stiffness and vary the mount's compressive stiffness, Figure 5.59. Note that the frame torsional stiffness is typically very low in comparison to the body.

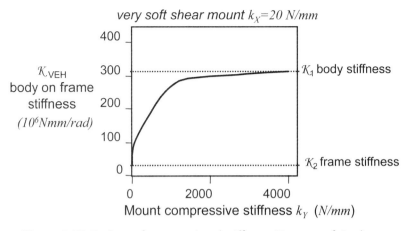

Figure 5.59 Body-on-frame torsional stiffness: Mounts soft in shear.

The resulting behavior, Figure 5.59, shows that for compressively soft mounts, the body is not coupled to the twisting motion of the frame and the vehicle stiffness approaches that of the frame alone. For compressively stiff mounts, the body and frame are highly coupled, and the vehicle stiffness approaches the sum of the body and frame stiffnesses.

Now using the same vehicle data, let us additionally vary the shear stiffness of the body mount. The result, Figure 5.60, shows that by increasing the shear stiffness of the body mounts, we can increase vehicle stiffness *beyond* the sum of the body and frame torsional stiffness. Physically, the reason for this is that the body and frame have different twist axes. By increasing the mount shear stiffness, the body and frame fight against one another for the axis to twist about. The combined twist axis is above the frame. As the combined twist axis moves above the frame, the frame itself becomes a shear-resistant member, Figure 5.61. This shearing of the frame contributes to a greater torsional stiffness for the vehicle system. Thus, the shear stiffness of the frame becomes an important design consideration.

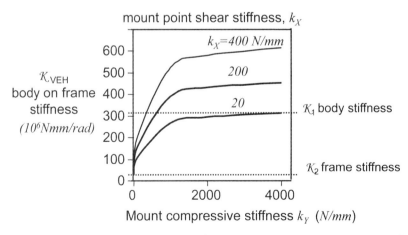

Figure 5.60 Body-on-frame torsional stiffness: Varying mount point shear stiffness.

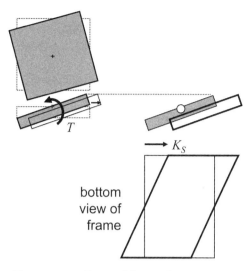

Figure 5.61 Shear of frame during torsion.

5.5.3 Torsional stiffness of a ladder frame

As discussed above, the ladder frame influences vehicle torsional rigidity not only by its *torsional stiffness* but also by its *shear stiffness*. In this section, we will look briefly at both of these conditions.

The evolution of the automobile frame, Figure 5.62, has tended towards closed sections for both side rails and cross members to improve torsional stiffness. Also, improved joints at the crossmember-to-siderail have improved both torsional and shear stiffness. To understand this evolution, let us look at a simple frame of two side rails and two cross members loaded at both ends by equal and opposite torques, Figure 5.63. Loaded in this way, at any section along the perimeter, there will be a moment in the cross vehicle direction, M_{zz}, as shown in Figure 5.64a. This moment acts as a uniform torque for the cross member, and a linearly varying bending

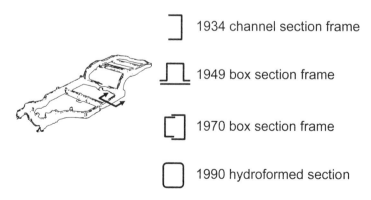

Figure 5.62 Typical frame configurations.

205

moment for the side rail. There will also be a fore-aft moment, M_{XX}, which will be a linearly varying bending moment for the cross member and a uniform torque for the side rail, Figure 5.64b. Thus, all members are loaded both in bending and in torsion.

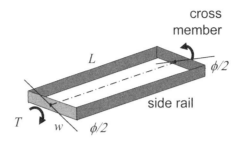

Figure 5.63 Simple frame.

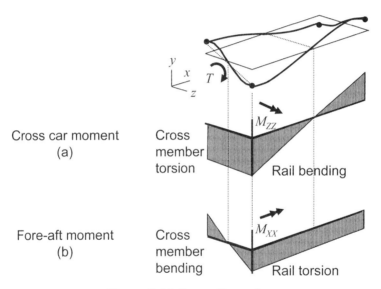

Figure 5.64 Frame: General case.

It is helpful to consider two limiting cases for this simple frame. The first we have already looked at under differential bending—the case of cross members being very rigid in torsion. In that case the torsional rigidity of the frame is due only to *bending* of the side rails, Equation 5.17. Now consider another limiting case where the cross member is infinitely rigid in bending and very flexible in torsion, Figure 5.65. For this case, each side rail is twisted through the same angle, ϕ, with no bending occurring. The torsional stiffness of the frame is due only to *twisting* of the side rails, and the frame stiffness is the sum of the torsional stiffness of the rails,

$$K = 2\left(\frac{GJ_{EFF}}{L}\right) \tag{5.19}$$

where:

    G = Shear modulus

    J_{EFF} = Torsion constant of side rail

    L = Length of side rail

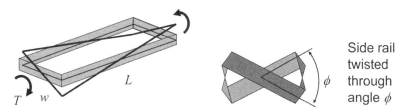

Figure 5.65 Cross members infinitely rigid in bending and very flexible in torsion.

In these two limiting cases, we see that both twisting as well as bending of the side rails can add to the frame torsional stiffness. (We could have considered these limiting cases with the side rails as infinitely stiff, and drawn parallel conclusions about bending and twisting of the cross members.)

For the general case of frame torsion, both the cross members and side rails are under both bending and torsion, as in Figure 5.64. A Finite Element Analysis may be used to understand the relative contribution of each of these conditions. Figure 5.66 shows data for a generic planar frame of steel with section size *100 mm (4 in.)* square and *1 mm (0.04 in.)* thick. A torsional stiffness of *2030 Nm/° (1500 ft lb/°)* results with the strain energy distribution shown. The torsion strain energy is larger than the bending strain energy for both the cross members and the side rails. This implies that increasing the torsional properties would have the largest influence on frame stiffness. This sensitivity to torsional section properties is typical for most common automotive frames. Note also the relatively low torsional stiffness value compared with the typical monocoque body values of *10,000–12000 Nm/o (7375–8850 ft lb/°)*.

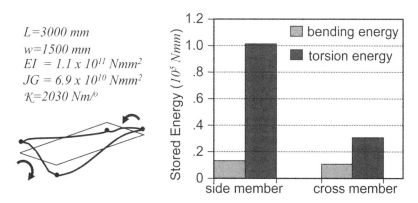

Figure 5.66 Strain energy in frame under torsion.

Finally, in discussing body-on-frame torsional behavior, the importance of frame shear stiffness, k_S, was pointed out, Figure 5.61. This plan-view shearing stiffness depends greatly on the crossmember-to-siderail joints, Figure 5.67a. As an approximation for describing this stiffness, consider a frame consisting of rigid beam elements connected by flexible joints with stiffness, k_J. Loading the frame in shear with a load, F, a deflection, δ, will result. Using energy principles, we can write the shear stiffness of the frame, k_S, as seen at each body mount attachment:

$$k_S = \frac{4k_J}{L^2} \qquad (5.20)$$

where:

k_S = Frame shear stiffness in plan view

k_J = Joint stiffness

L = Length of side rail

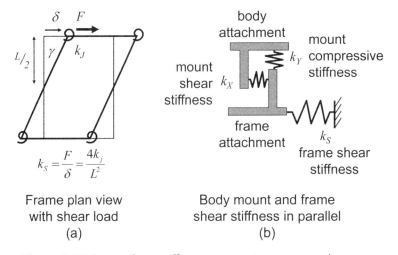

Frame plan view
with shear load
(a)

Body mount and frame
shear stiffness in parallel
(b)

Figure 5.67 Frame shear stiffness as seen at mount attachment.

For a typical joint stiffness of k_J=1x10⁹ Nmm/rad (8.85x10⁶ in lb/rad) and a 4500 mm (177 in.) long frame, using Equation 5.20 yields k_S≈200 N/mm (1142 lb/in). This value is very near the shear stiffness for a body mount. In our analysis of the body-on-frame, Figure 5.60, we found that a body mount with sufficient shear stiffness can increase the vehicle torsional rigidity beyond the sum of the body and frame stiffnesses. However, the body mount acts as a spring in series with the frame shear stiffness, and the frame shear flexibility puts an upper limit on the stiffness of the combined body mount-frame, Figure 5.67b. In efforts to increase the frame shearing stiffness, we can increase the joint rate with gussets added to the cross-member-to-rail joints, or in the extreme case, with an X configuration of rails added to the frame, Figure 5.68.

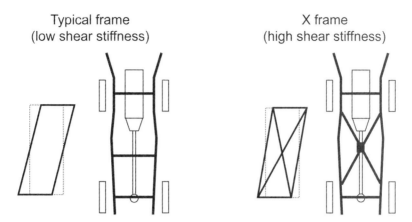

Figure 5.68 Shear stiffness of frame.

5.5.4 Torsional stiffness of backbone frame vehicles

For cases where vehicle packaging allows, a large central closed section can be a very effective structure for torsional stiffness. Usually this structure arrangement is limited to seating arrangements where a large high tunnel can be tolerated, such as open two-seat sport cars. For preliminary sizing of a backbone structure, the equations for closed thin-walled sections may be used for both strength and stiffness estimates, Figure 5.69. As these sections tend to have large width-to-thickness ratios, careful attention must be paid to elastic shear buckling of the walls. Frequently, diagonal rib patterns on the backbone sides are used to inhibit shear buckling.

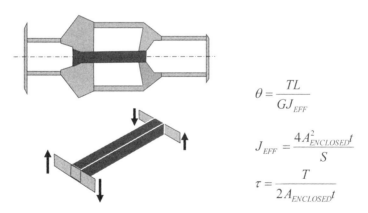

$$\theta = \frac{TL}{GJ_{EFF}}$$

$$J_{EFF} = \frac{4A_{ENCLOSED}^2 t}{S}$$

$$\tau = \frac{T}{2A_{ENCLOSED}t}$$

Figure 5.69 Backbone frame.

5.5.5 Torsional resistance of sandwich plates

An infrequently used but effective means to resist torsional loads is with a thick plate in the space between the occupant's foot and the ground clearance plane. For mass efficiency, this plate is constructed as a laminate with thin outer faces of a stiff material, and a shear resisting core of low density, Figure 5.70a. Under torsional loading, the plate deforms as shown in Figure 5.70b and the means of reacting the external torsion loads is by plate bending.

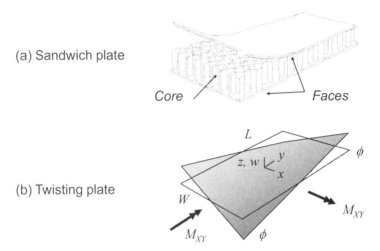

(a) Sandwich plate

(b) Twisting plate

Figure 5.70 Under-floor thick plate.

To develop a first-order model of this condition, consider the out-of-plane deflection at a point, (x,y), to be given by

$$w = Cxy$$

where C = a constant

This describes the twisted shape shown in Figure 5.70b. For this shape, the angle of rotation at each end of the plate, $(\pm L/2)$, is given by

$$\frac{\partial w}{\partial x} = \phi = Cy, \quad \phi = \pm C\frac{L}{2}$$

or the total angle of rotation, θ, for the plate from front to rear is

$$\theta = 2\phi = CL$$

Now consider the applied moments required to distort the plate into this shape. Using the plate equations from Equation 3.21, the moments depend on the second partial derivatives of the deflection function above:

$$\frac{\partial^2 w}{\partial x^2} = 0, \quad \frac{\partial^2 w}{\partial y^2} = 0, \quad \frac{\partial^2 w}{\partial x \partial y} = C$$

$$M_X = -D\left(\frac{\partial^2 w}{\partial x^2} + \mu \frac{\partial^2 w}{\partial y^2}\right) = 0$$

$$M_Y = -D\left(\mu \frac{\partial^2 w}{\partial x^2} + \frac{\partial^2 w}{\partial y^2}\right) = 0$$

$$M_{XY} = -D(1-\mu)\left(\frac{\partial^2 w}{\partial x \partial y}\right) = -DC(1-\mu)$$

So, not surprisingly, the applied moment along all edges is the twisting moment, M_{XY}, per unit of edge length, Figure 5.70b. A means to interpret this moment is to divide the plate along each side into small, equal increments of width and length, Δx and Δy, Figure 5.71a. Now consider a series of couples applied along each edge of the plate of magnitude $F\Delta x$ along the width and $F\Delta y$ along the length. These couples provide the twisting moment, M_{XY}, according to:

$M_{XY}\Delta x = F\Delta x$ along the width, and $M_{XY}\Delta y = F\Delta y$ along the length.

The magnitude of these forces is therefore

$$F = M_{XY}$$

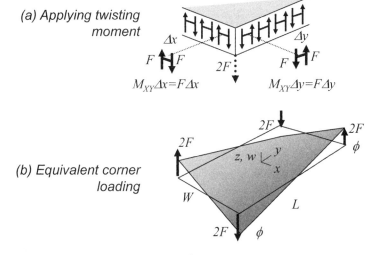

(a) Applying twisting moment

(b) Equivalent corner loading

Figure 5.71 Plate twisting.

Now consider this series of up and down forces along the edges of the plate, Figure 71a. The forces cancel in up-down pairs except at the corners of the plate where they sum to a corner force of 2F. Thus an equivalent loading for the plate is by corner forces, Figure 5.71b, with magnitude

$$2F = 2M_{XY} = 2DC(1-\mu)$$

So the twisting couple acting on the plate is

$$T = (2F)W = 2DC(1-\mu)W$$

The torsional stiffness for a general plate is then

$$K = \frac{T}{\theta} = \frac{2DC(1-\mu)W}{CL} = 2D(1-\mu)\frac{W}{L} \qquad (5.21)$$

where:

   K = Torsion stiffness of a plate

$$D = \frac{EI}{\left(1-\mu^2\right)}$$

   W = Plate width

   L = Plate length

In the example below, we will consider the specific case of a sandwich plate with very stiff material (high E) for the faces, and a much more flexible material (low E) but shear resistant for the core.

Example: Torsional stiffness of a sandwich floor pan

Consider a flat floor pan constructed of a laminate of steel faces, *0.5 mm* thick with a shear-resistant core. The total sandwich height is *100 mm*. The dimensions of the floor pan are L=*3500 mm*, W=*1500 mm*.

The moment of inertia for a unit of width for this plate, Figure 5.72, is

$$EI = E_{FACE}\left[2(t\cdot1)\left(\frac{h}{2}\right)^2\right] = \frac{E_{FACE}th^2}{2}$$

$$D = \frac{EI}{\left(1-\mu^2\right)} = \frac{E_{FACE}th^2}{2\left(1-\mu^2\right)}$$

$$K = \frac{2D(1-\mu)W}{L} = 2\left[\frac{E_{FACE}th^2}{2(1-\mu)(1+\mu)}\right](1-\mu)\frac{W}{L}$$

$$K = \frac{E_{FACE}th^2}{(1+\mu)}\frac{W}{L} \qquad \qquad (5.22)$$

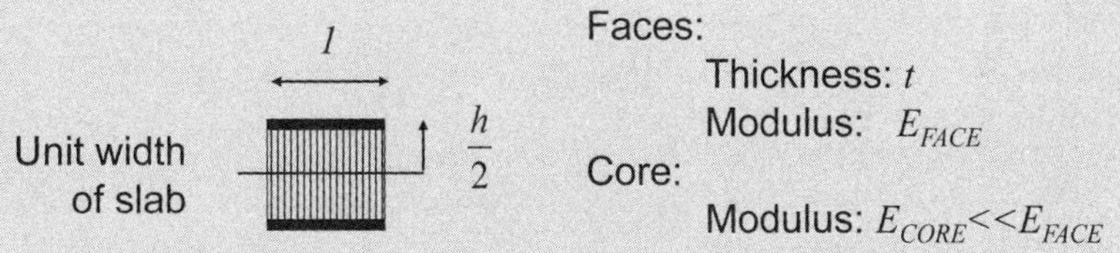

Figure 5.72 Moment of inertia for sandwich.

Where Equation 5.22 is the stiffness of a sandwich laminate. Substituting values for this example,

$$K = \frac{(207000N/mm^2)(0.5mm)(100mm)^2}{(1+.3)}\frac{1500mm}{3500mm}$$

$$K = 341,200,000 Nmm/rad(5986 Nm/o)$$

This value is a large fraction of our nominal vehicle torsional stiffness of *12000 Nm/º*, and shows the effectiveness of this structural system. Of course, a load path from the suspension attachments to this plate must be provided to realize this stiffness.

References

1. Milliken, W. & Milliken, D., *Chassis Design, Principles and Analysis,* SAE International, Warrendale, PA, 2002, pp. 93–101.

2. Gillespie, T. D., *Fundamentals of Vehicle Dynamics,* SAE International, Warrendale, PA, 1992, pp. 210–214.

3. Fenton, J., *Vehicle Body Layout and Analysis*, Mechanical Engineering Publications, Ltd, London, 1980, p. 93.

4. Pawlowski, J., *Vehicle Body Engineering*, Business Books, London, 1969, pp. 149–150.

5. Brown, J.C., Robertson, A. J., Serpento, S.T., *Motor Vehicle Structures*, SAE International, Warrendale, PA, 2001, pp. 75–84.

6. Kikuchi, N. & Malen, D., Course notes for ME513 Fundamentals of Body Engineering, University of Michigan, Ann Arbor, MI, 2007.

Chapter 6

Design for Crashworthiness

One of the primary functions of the automobile body is to protect occupants in a collision. Because of the variability of field impacts, many governments have defined standard crash tests and minimum performance levels. For the USA, these standards are contained in the Federal Motor Vehicle Safety Standards [1], with similar standards for the European Union, Japan, Korea, Australia, and others.

6.1 Standardized Safety Test Conditions and Requirements

For impact tests which influence the overall vehicle structure, these standard tests may be categorized into four major groups: front impact, side impact, rear impact, and roll-over resistance. Figure 6.1 and Figure 6.2 illustrate the test conditions for these four groups.

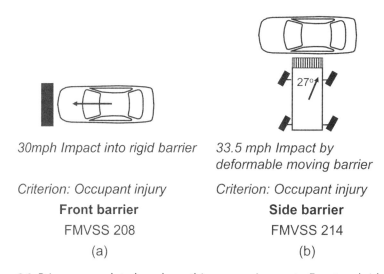

30mph Impact into rigid barrier

Criterion: Occupant injury
Front barrier
FMVSS 208
(a)

33.5 mph Impact by deformable moving barrier

Criterion: Occupant injury
Side barrier
FMVSS 214
(b)

Figure 6.1 Primary mandated crashworthiness requirements: Front and side impact.

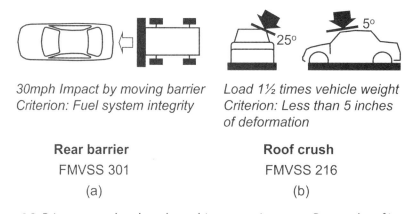

30mph Impact by moving barrier
Criterion: Fuel system integrity

Rear barrier
FMVSS 301
(a)

Load 1½ times vehicle weight
Criterion: Less than 5 inches of deformation

Roof crush
FMVSS 216
(b)

Figure 6.2 Primary mandated crashworthiness requirements: Rear and roof impacts.

These government standards establish a *minimum* performance level for the vehicles to be sold in the respective country. The insurance industry and consumer groups have developed tests which evaluate vehicles *beyond* the minimum government standards. The New Car Assessment Program (NCAP) is one such set of tests. The NCAP evaluation is based on the probability of injury for a specific test, measured with a star scale ranging from one star—higher probability of injury, to five stars— lower probability of injury, Figure 6.3.

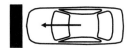

US-NCAP Front barrier:
35 mph full face rigid barrier

Criterion: Star rating based on combination of head injury criteria (HIC) and chest acceleration for driver and passenger

% chance of serious injury		
	Frontal	Side
	35 mph	33 mph
★★★★★	<10%	<5%
★★★★	11-20	6-10
★★★	21-35	11-20
★★	36-45	21-25
★	>46	>26

Star rating system

Figure 6.3 Customer-based crashworthiness: New car assessment program.

While these standards are extensive and cover many test configurations, we shall focus on two important conditions: full front barrier and side impact. By examining these two conditions in detail, we will develop general approaches and tools which can be used to examine other conditions.

6.2 Front Barrier

In this section, we will look at the condition of a moving vehicle impacting a rigid barrier. Although this is but one of several conditions for front impacts, the physical understanding and analysis we develop for this test case may be expanded for other conditions. The test conditions are shown in Figure 6.4. A vehicle of mass M is moving at a constant speed, V. It will just touch the rigid, unmovable barrier at time $t=0$. As the vehicle deforms, the speed of the vehicle center-of-mass will gradually reduce until it reaches $V=0$ at which time a maximum deformation, Δ, occurs. Figure 6.4b illustrates the vehicle before and after the test.

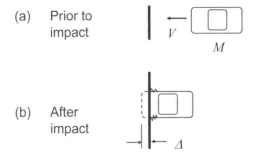

(a) Prior to impact

(b) After impact

Figure 6.4 Full front impact.

Let us look at the sequence of events for a typical midsized vehicle as it undergoes such an impact, Figure 6.5 [2]. At time $t=0$ the vehicle is moving at velocity $V=V_O$ and the front bumper is just touching the barrier face. At time $t=15$ *msec*, the bumper has collapsed and the motor compartment midrails and side rails are being loaded. At $t=30$ *msec* the mid rails have begun to crumple in an accordion fashion, and the powertrain has just touched the barrier and begins to decelerate. At $t=45$ *msec*, the midrails and upper rails continue to crumple, the powertrain has decelerated to zero velocity and the wheels have impacted the barrier. Finally at $t=90$ *msec*, the vehicle has decelerated to $V=0$ with the motor compartment crumpled by some deformation, Δ.

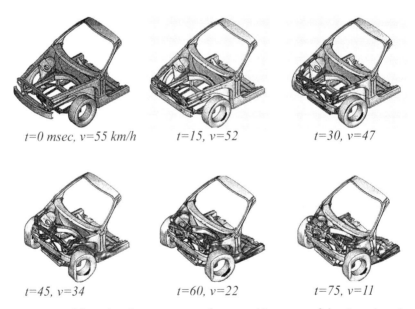

t=0 msec, v=55 km/h   *t=15, v=52*   *t=30, v=47*

t=45, v=34   *t=60, v=22*   *t=75, v=11*

Figure 6.5 Typical front barrier sequence of events. (Courtesy of the American Iron and Steel Institute, UltraLight Steel Auto Body)

For this test, primary data are collected on the load applied to the barrier face and on the acceleration of the vehicle mass center. The acceleration is then integrated to determine velocity. Figure 6.6a [2] shows a typical graph for barrier face load vs. time. Early in the crash event, we see relatively low loads generated by the bumper collapsing, followed by a spike when the midrails are initially loaded. Another spike occurs at $t=35$ *msec* when the rigid powertrain impacts the barrier and is suddenly decelerated. This is followed by a relatively constant load for the rest of the event as the motor compartment structure continues to crumple.

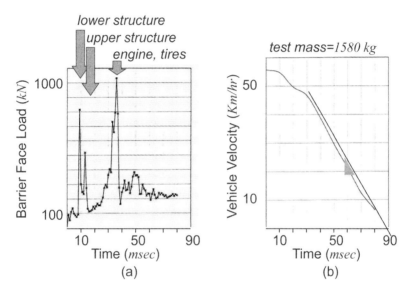

Figure 6.6 Typical front barrier time histories. (Courtesy of the American Iron and Steel Institute, UltraLight Steel Auto Body)

The corresponding graph of vehicle mass center velocity vs. time is shown in Figure 6.6b. Note that the slope of this curve at any time, t, is the acceleration of the vehicle mass center at that time. Velocity initially reduces gradually up to $t=30$ *msec* followed by a steeper linear reduction through the end of the event. The slope during this linear velocity reduction is the cabin deceleration.

For such a test we are interested in minimizing occupant injury. A first-order indication of this injury is given by the acceleration of the vehicle center of mass—a lower level of acceleration is less injurious. Let us consider the most basic analytical model to describe the impact and the resulting acceleration.

6.2.1 Basic kinematic model of front impact

Let us model this event with a point mass, M, representing the mass of the vehicle [3]. Attached to this mass is an element which generates a constant force, F_0, as it collapses, representing the crumpling front structure, Figure 6.7. We wish to develop graphs similar to Figure 6.6b for acceleration, velocity, and deformation for this simple model.

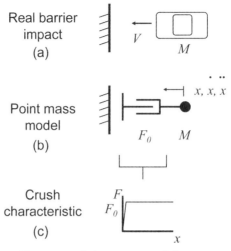

Figure 6.7 Point mass model.

Consider the force element just touching the barrier at time $t=0$. Summation of forces acting on the mass gives:

$$M\ddot{x} = -F_0$$

$$\frac{d^2x}{dt^2} = -\frac{F_0}{M} \qquad \text{acceleration of point mass}$$

$$\frac{d\,x}{dt} = -\frac{F_0}{M}t + C_1$$

The initial conditions are at $t=0$, $dx/dt=Vo$, the impact speed. Therefore,

$$\frac{d\,x}{dt} = -\frac{F_0}{M}t + V_0 \qquad \text{velocity of point mass}$$

Integrating a second time:

$$x = -\frac{F_0}{2M}t^2 + V_0 t + C_2$$

The initial conditions are at $t=0$, the deformation of the vehicle front end is zero or $x=0$:

$$x = -\frac{F_0}{2M}t^2 + V_0 t \qquad \text{deformation for point mass}$$

Finally, we can ask at what time, t_{FINAL}, is the crash event completed. This occurs when $dx/dt=0$, or:

$$\frac{d\,x}{dt} = -\frac{F_0}{M}t_{FINAL} + V_0 = 0$$

final time for point mass: $t_{FINAL} = \dfrac{MV_0}{F_0}$

The resulting behavior for this point mass model is shown in Figure 6.8.

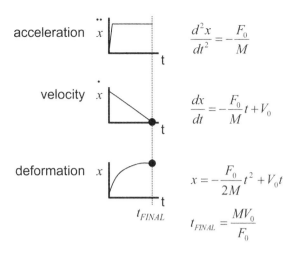

acceleration $\ddot{x}$

$$\frac{d^2x}{dt^2} = -\frac{F_0}{M}$$

velocity $\dot{x}$

$$\frac{dx}{dt} = -\frac{F_0}{M}t + V_0$$

deformation x

$$x = -\frac{F_0}{2M}t^2 + V_0 t$$

t_{FINAL}

$$t_{FINAL} = \frac{MV_0}{F_0}$$

Figure 6.8 Behavior of point mass model.

Example: Point mass impact behavior

Consider a vehicle of mass *1580 kg*, impacting a rigid barrier at *55 kph*, and an average motor compartment crush load of *300 KN*. Applying the above equations, the acceleration is:

$$\frac{d^2x}{dt^2} = -\frac{F_0}{M} = \frac{300,000N}{1580kg} = 189.87 \, m/\sec^2 = 19.37g$$

The time at the end of the crash event is:

$$t_{FINAL} = \frac{MV_0}{F_0} = \frac{1580kg(55kph)(1000m/km)/(3600\sec/hr)}{300000N} = 0.0805 \, \sec$$

The total deformation at t_{FINAL} is:

$$x = -\frac{F_0}{2M}t^2 + V_0 t =$$

$$= -\frac{300000N}{2(1580kg)}(.0805\sec)^2 + (55kph)(1000m/km)/(3600\sec/hr)(.0805\sec) =$$

$$x = 0.614m$$

We can see by comparison of the velocity history with Figure 6.6b that this very simple model can reasonably predict impact behavior.

In this initial model we have assumed an ideal case of uniform load, F_0, as the structure collapses, Figure 6.7c. Let us refine the model by allowing crush force properties other than uniform. In Figure 6.9 we show a load-deformation curve with the same area under the curve as the square wave used before. (Thus each curve will result in the same work during deformation.) We can characterize this curve by a crush efficiency factor,

$$\eta = \frac{F_{AVG}}{F_{MAX}} \quad (0 < \eta < 1) \tag{6.1}$$

As we are applying these forces to a point mass, M, the resulting acceleration for the mass will be $a=F/M$. Therefore we can also express the crush efficiency factor as

$$\eta = \frac{F_{AVG}}{F_{MAX}} = \frac{Ma_{AVG}}{Ma_{MAX}} = \frac{a_{AVG}}{a_{MAX}} \qquad (6.2)$$

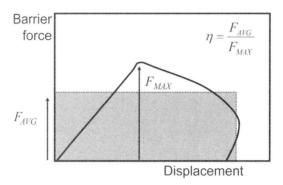

Figure 6.9 Characterizing barrier force.

We can now ask which of the four deformation curves of Figure 6.10a is preferable in minimizing occupant injury. Each of these deformation curves was applied to an occupant injury model with the resulting head acceleration shown in Figure 6.10b. The more square-shaped the curve, η approaching one, the lower the head injury. Thus, as we design the collapsing structure of the motor compartment, we will attempt to approach a square wave shape.

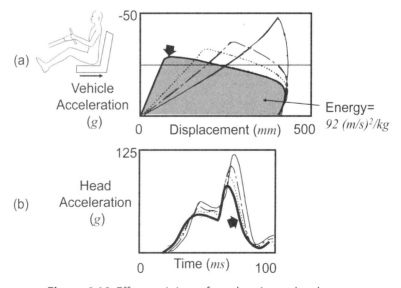

Figure 6.10 Effect on injury of acceleration pulse shape.

Now that we have established the general shape for the cabin acceleration, let us consider the magnitude of the peak acceleration. Figure 6.11 illustrates the velocity-time histories for several vehicles in a *30 mph (48 km/h)* front barrier test. Looking at the slope of these curves, we see a range of *20 g* to *30 g* for peak acceleration. In general, a lower peak cabin acceleration is less injurious, and we take the lower end of this benchmark data, *20 g*, as a target.

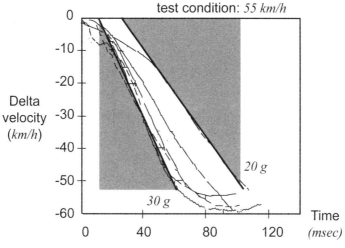

Figure 6.11 Typical velocity-time histories.

To summarize, we have established desirable characteristics for cabin acceleration during impact. First, limit maximum acceleration to approximately *20 g*, and second, make this acceleration as uniform as possible, that is make $\eta = a_{AVG}/a_{MAX} \sim 1$. These characteristics for cabin acceleration are directly related to front structure requirements since the inertia loads are those to be reacted by the structure.

The crushable space in the motor compartment, Figure 6.12, is an important variable in body design, as we will show in the next section. To measure this space, we first identify the cabin zone which we desire to keep from deforming. Ideally this zone wraps completely around the passengers. In practice, we can accept some amount of intrusion into the toe pan area on the order of *50 mm* to *120 mm (2 in. to 4.75 in.)* without increasing injury. The crush zone extends forward from the cabin zone to the front of the vehicle. The crush space, *Δ*, is the fore-aft dimension of the crush zone less the stack-up of rigid elements within that zone. Some subsystems within the crush zone are completely rigid, such as the engine block, and their total fore-aft dimension must be subtracted. Others can be crushed to some degree, and we make some assumption as to their final crush dimensions for inclusion in the crush space. For example, the radiator is assumed to crush to *50%* of its original thickness.

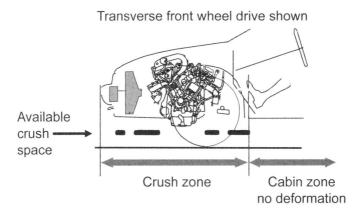

Transverse front wheel drive shown

Available crush space →

Crush zone

Cabin zone no deformation

Figure 6.12 Available crush space.

6.2.2 Structural requirements for front barrier

During preliminary vehicle design, we are interested in establishing body structural requirements:

1. The maximum cabin acceleration, a_{MAX}

2. The necessary crushable space, Δ

3. The average crush force, F_{AVG}

These requirements cannot be set independently. We will show this by looking at a work-energy balance of the vehicle before and after impact.

Consider the vehicle immediately before impact, Figure 6.13a. The kinetic energy is $\frac{1}{2}MV_0^2$ where V_0 is the impact speed for the test. After impact we have zero kinetic energy, but work has been done to the system during collapse equal to the average crush force over the crush distance, $F_{AVG}\Delta$, Figure 6.13b. Equating kinetic energy to work done

$$\frac{1}{2}MV_0^2 = F_{AVG}\Delta$$

and substituting $F_{AVG}=\eta F_{MAX}$ and $F_{MAX}=a_{MAX}$ yields

$$a_{MAX} = \frac{V_0^2}{2\eta\Delta} \tag{6.3}$$

which is a relationship between maximum cabin acceleration, a_{MAX}, (which is related to occupant injury), crush space, Δ, (which is related to vehicle styling and packaging), crush efficiency, η, (which is related to body structure performance), and test speed, V_0. If we look at benchmark test data, Figure 6.14, we can see the inverse relationship between maximum cabin acceleration during impact and crush space. This illustration also shows the practical maximum value for crush efficiency, $\eta{\sim}0.8$, when using crumpling of thin-walled structures as the energy-absorbing mechanism.

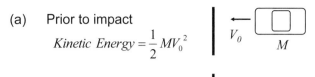

(a) Prior to impact

$$Kinetic\ Energy = \frac{1}{2}MV_0^2$$

(b) After impact

$$Kinetic\ Energy = 0$$

$$Work = F_{AVG}\Delta$$

Figure 6.13 Full front impact.

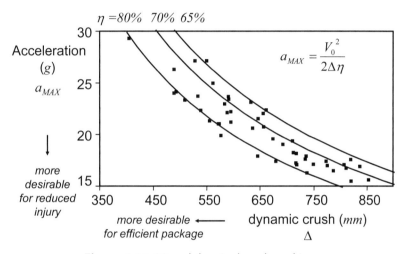

$\eta = 80\%$ 70% 65%

Acceleration (g)

a_{MAX}

$$a_{MAX} = \frac{V_0^2}{2\Delta\eta}$$

↓

more desirable for reduced injury

more desirable ⟵ for efficient package

dynamic crush (mm)

Δ

Figure 6.14 30 mph barrier benchmarking.

The relationship of Equation 6.3 suggests a procedure for establishing front body structural requirements for crashworthiness:

1. Determine the maximum allowable cabin decelerations based on occupant injury, a_{MAX}

2. Determine a consistent structural efficiency and crush space, (η, Δ) using Equation 6.3

3. Compute the average and maximum allowable crush forces which the vehicle must generate during impact, F_{AVG}, F_{MAX}, using the efficiency, η, from step 2

4. Allocate these total forces from step 3 to the structural elements within the vehicle front end.

Some rules of thumb exist for this allocation of step 4: 20% of the force is generated by the upper structure load path just under the top of fender, 50% of the force is generated by the mid-rail structure, 20% by the lower cradle, and 10% by the hood and fenders, Figure 6.15.

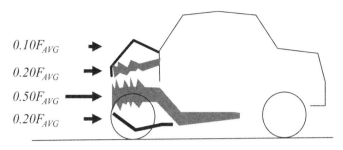

Figure 6.15 Typical barrier load partition.

To summarize, we have clearly defined the cabin area where we want minimal deformation which may injure the occupants. It is the cabin area which we want to decelerate within the accelerations levels we have identified earlier. We also defined a crush zone which we expect to deform upon impact and which must generate the forces which decelerate the cabin zone in the desired manner. Now that we have targets for average and maximum allowable crush force, we can now size structural elements. Some of the elements in the crush zone are sized to provide forces as they crush, and other structural elements around the cabin zone are sized to ensure that area is not intruded upon, Figure 6.16. We now look at tools to aid in sizing both these classes of structural elements.

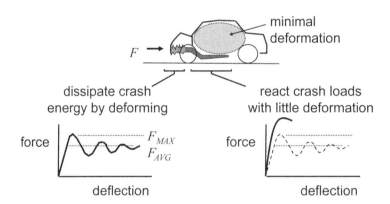

Figure 6.16 Structure requirements for front barrier.

6.2.3 Beam sizing for energy absorption

First let us consider the structural elements in the crush zone and focus on the major load path—the mid rail. Using the procedure of the previous section, we have identified the required average crush force, and also the maximum allowable force. An efficient means to generate an approximately square wave force over a large distance is through progressive column crush of a thin-walled section, Figure 6.17 [4]. In this figure we see a thin-walled square section undergoing an axial compressive load, F. For small loads, the column is under simple compressive stress, $\sigma = FA$, A being the cross-section area, Figure 6.17a. As the compressive load is gradually increased, the elastic buckling load for the section walls is reached and the walls buckle, Figure 6.17b. As the load continues to increase past the ultimate load of the buckled walls, a load is reached where the corners of the section cripple and the load drops, Figure 6.17c. Once the crippled corner bottoms out, the load begins to increase again, Figure 6.17d. This process repeats itself, forming an accordion pattern and a load-deformation curve which oscillates about an average crush force, Figure 6.17e & f.

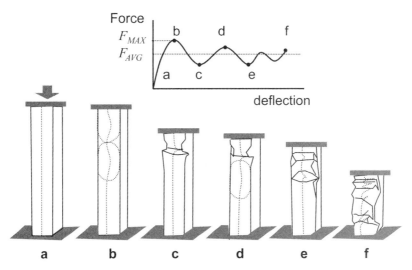

Figure 6.17 Square section under axial compressive load. (Courtesy of SAE International)

This physical behavior is very useful for energy absorption as it generates a high average crush force with a square wave character. An empirical relationship for predicting forces during crush, Figure 6.18, is given in Equation 6.4 [4]. Note that this relationship was developed for a square steel section loaded by static (very slowly applied) forces. The units are *N, mm, N/mm²*.

$$P_M = 386t^{1.86}b^{0.14}\sigma_Y^{0.57}$$
$$P_{MAX} = 2.87P_M$$
$$P_1 = 1.42P_M \tag{6.4}$$
$$P_2 = 0.57P_M$$

where:

P_M = Static mean crush force (*N*)

P_{MAX} = Maximum crush force (*N*)

P_1 and P_2 = crush loads shown in Figure 6.18 (*N*)

t = Material thickness (*mm*)

b = Section width and height (*mm*)

σ_Y = Material yield stress (*N/mm₂*)

Using Equation 6.4, we can determine the thickness, t, section width, w, and yield stress, σ_Y, to generate the required average static crush force.

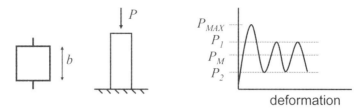

Figure 6.18 Axial crush of thin-walled steel square sections. (Courtesy of SAE International)

Motor compartment packaging or manufacturing may require flange locations and section shapes other than that shown in Figure 6.18. Guidance for alternative flange positions is given in Figures 6.19, and guidance for sections other than square in Figure 6.20 [5].

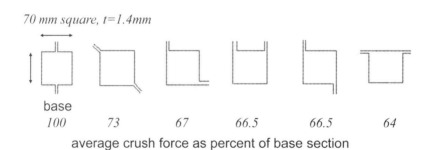

Figure 6.19 Flange position effect on average crush force. (Courtesy of the American Iron and Steel Institute, Automotive Steel Design Manual))

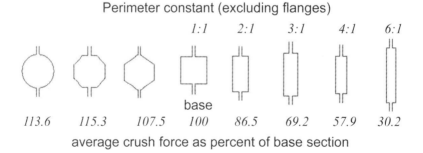

Figure 6.20 Section shape effect on average crush force. (Courtesy of the American Iron and Steel Institute, Automotive Steel Design Manual)

A final adjustment to Equation 6.4 is made for dynamic loading. Crush load requirements are load levels which must be generated during a *dynamic* impact in which the loading is done over a short time. Many materials are strain rate sensitive and generate higher stress when loaded rapidly (steel is one such material). Figure 6.21 compares the loads generated in both a static test along with the same section tested by impacting at *48 km/h (30 mph)* [5]. These data may be used to adjust the values predicted by Equation 6.4 to arrive at dynamic crush force. From Figures 6.20, the effectiveness of a polygon (Shapes B and C) is evident. Many contemporary designs apply this concept, Figure 6.22, and also use shape optimization to improve energy absorbed [6].

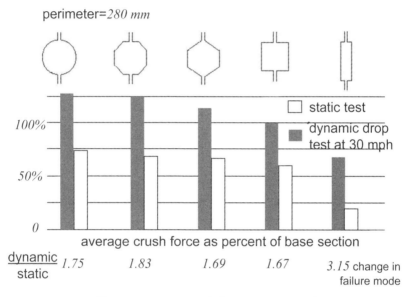

Figure 6.21 Dynamic effect on average crush force. (Courtesy of the American Iron and Steel Institute, Automotive Steel Design Manual)

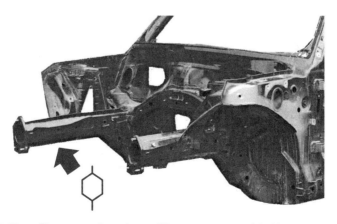

Figure 6.22 Use of hexagonal sections. (Photo courtesy of A2Mac1.com, Automotive Benchmarking)

Note that along with the mean crush force, P_M, Equation 6.4 predicts the maximum load, P_{MAX}, at *2.87* times the mean load. This is the load needed to initiate the first crippled corner. With this factor, the crush efficiency for an unmodified square section is:

$$\eta = F_{AVG}/F_{MAX} = P_M/(2.87P_M) = 0.35$$

which is rather poor compared with the benchmark data from Figure 6.14. To improve this efficiency, we need to reduce the maximum load. Understanding that this load is caused by the initial corner crippling, darts or beads are often added to initiate the crippling, Figure 6.23. The precise placement and geometry of these *crush initiators* is found by simulation or experimentation. Once the initial crippling load has been reduced, we are left with the subsequent crippling cycles at load P_1, Figure 6.16. Equation 6.4 tells us these will occur at 1.42 times the mean crush force for a crush efficiency of $\eta = P_M/(1.42P_M) = 0.7$. This is very much in line with the benchmark data of Figure 6.14.

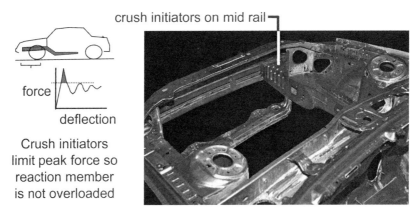

Figure 6.23 Crush initiators. (Photo courtesy of A2Mac1.com, Automotive Benchmarking)

6.2.4 Beam sizing for cabin reaction structure

Now that we have sized the structural element in the crush zone, we can turn our attention to the structure whose function is to ensure the cabin zone does not excessively deform, Figure 6.24. In typical body construction, an under-floor structure reacts the crush loads, Figure 6.25. This structure must react the maximum loads generated by the midrail, F_{MAX}, without excessive deformation into the cabin area. In this case *excessive* deformation is a level which does not influence occupant trajectory during impact and does not increase injury: approximately *50–120 mm (2–4.75 in.)* of deformation into the cabin zone. As we expect the reaction structure to permanently deform a small amount during impact, using yield of the outer fiber as failure criterion would result in an inefficient design. Instead we use the *limit analysis* to size the reaction structure.

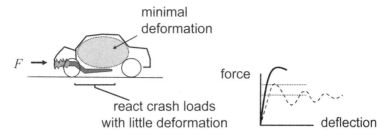

Figure 6.24 Reaction member requirements.

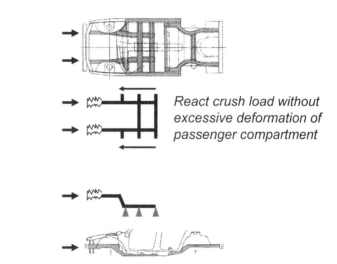

React crush load without excessive deformation of passenger compartment

Figure 6.25 Maintaining integrity of the passenger compartment.

6.2.5 Limit analysis design

In *limit analysis* we look at a structure as reaching its load capacity just as it behaves as a mechanism with *plastic hinges,* Figure 6.26 [7, 8, 9, 10]. For example, in this structure we consider the horizontally applied load to be gradually increasing. Initially the reaction structure will react this load elastically; as the load is increased we will begin to see yielding in the areas shown. If we were designing a structure in which initiation of permanent deformation was the failure criterion, this state would indicate the failure load. However, for a reaction structure, we can continue to increase the applied load until the yield zones extend across the section at the two key locations shown in Figure 6.26. We define these locations to be plastic hinges. Continuing to increase the applied load beyond this point, load A in Figure 6.26, will cause the reaction structure to behave as a mechanism with rigid links connected at the plastic hinge locations, Figure 6.27. At this load, the structure will distort without bound. We define the limit load as the ultimate load-carrying ability for the structure. We will analyze this behavior using the model shown in Figure 6.27. Here we view the structure as consisting of rigid links and ideally plastic hinges. Before we do an analysis to predict the limit load for a particular structure, let us look closer at the behavior of an individual plastic hinge.

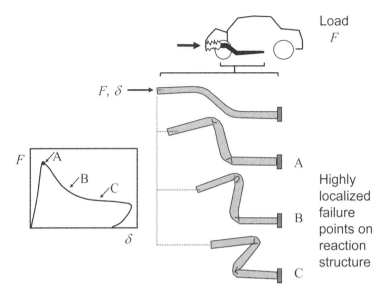

Figure 6.26 Reaction member: Plastic hinge location. (Courtesy of the American Iron and Steel Institute)

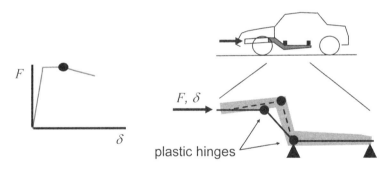

Figure 6.27 Limit analysis of structures.

6.2.6 Plastic hinge behavior

Consider a simple plastic hinge, Figure 6.28 [11]. Here a cantilever beam has a vertical tip load. Focus on the stresses at the section next to the wall. With s mall applied loads, the stress distribution is linear and elastic, condition *a* in Figure 6.28. Releasing the load in this range will result in a return to the original shape. This behavior is shown in the load-deflection curve, Figure 6.28a. As the load is increased, a load is reached where the outer fiber of the section reaches yield, condition *b*. If we continue loading a small amount above this load and then release the load, the beam will show some permanent deformation. However, we have not reached the maximum or ultimate load which the beam can react. Continuing to increase the tip load, the yielded (plastic) region will continue to increase toward the middle of the section with the stress distribution being shown in Figure 6.28c. The ultimate load-carrying capacity of the beam has even now not yet been met.

Eventually the whole section is in a state of yield with compressive yield stress on the lower part of the section and tensile yield stress on the upper part of the section, Figure 6.28d. At this load the yielded section acts like a pinned joint with a resisting moment—like a rusty gate hinge. Once we have reached this load, the beam will continue to deflect without bound in the manner of a rigid bar pinned at the wall. This defines the limit load—the ultimate load-carrying capacity of this structure.

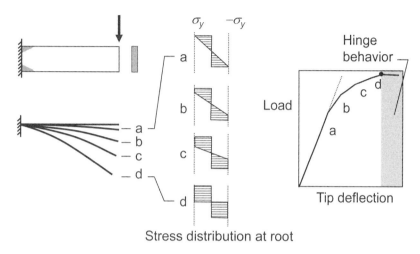

Figure 6.28 Limit analysis example: Cantilever beam.

We can model a plastic hinge as shown in Figure 6.29 as a pinned joint with a uniform resisting moment, M_p throughout the range of angular deformation. The value for M_p depends on the section dimensions and material yield stress [12]. To arrive at a value for M_p, consider the physical behavior of a square thin-walled section shown in Figure 6.30.

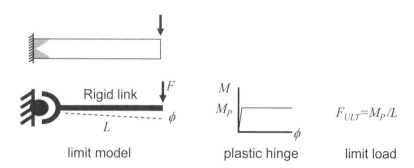

Figure 6.29 Plastic hinge first-order model.

233

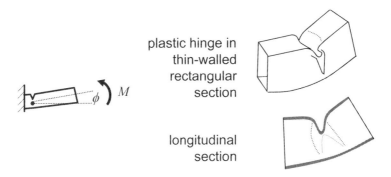

Figure 6.30 Plastic hinge formation in thin-walled sections.

For this loading, we place the section in uniform tensile yield below the plastic neutral axis and uniform compressive yield above the neutral axis. In the example shown in Figure 6.31 we have assumed the top plate of the section, which is in compression, has buckled and carries no load; in the calculation below we therefore ignore it. The plastic neutral axis location is found by equating horizontal force on the section to zero, Figure 6.31.

$$(-\sigma_Y)A_U + (+\sigma_Y)A_L = 0$$
$$A_U = A_L$$

where:

A_U = Cross section area above the plastic neutral axis

A_L = Cross section area below the plastic neutral axis

σ_Y = Material yield stress

The plastic neutral axis location is then defined as the axis where the area above and below are equal. (Note that the elastic and plastic neutral axes are not necessarily the same.)

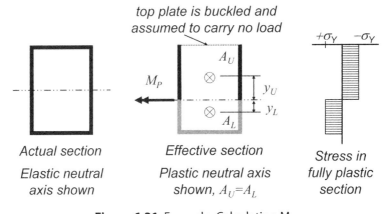

Figure 6.31 Example: Calculating M_p.

The plastic moment, M_p, is now found by summing moments about the plastic neutral axis, Figure 6.31. This result is given by

$$M_P = \sigma_Y(A_U y_U + A_L y_L) \tag{6.5}$$

where:

M_p = Plastic moment

A_U = Cross section area above the plastic neutral axis

A_L = Cross section area below the plastic neutral axis

y_U = Distance from the plastic neutral axis to the center of area for the upper area, A_U

y_L is the distance from the plastic neutral axis to the center of area for the lower area, A_L

σ_Y = Material yield stress

To better understand design principles for an effective plastic hinge, consider a square section shown in Figure 6.32. We calculate the moment for three cases: a) the moment which causes the yield of the outermost fiber, b) the moment for the fully plastic state when all elements of the section are fully effective, and c) the moment in the fully plastic state when the compression element of the section has buckled and carries no load. Figure 6.33 shows the values for these moments identified on the moment-angular deflection curve for the beam. To provide the highest load reaction capability, we would like the plastic hinge to follow *Curve A*. However, if elements of the section buckle before yield, the path will be *Curve C*. If buckling occurs after yield, the path will be *Curve B*—both *B* and *C* being less effective than *Curve A*. To obtain *Curve A*, we must inhibit buckling of the compression element even beyond the yield stress. Figure 6.34 illustrates an approach to achieve this using section reinforcement.

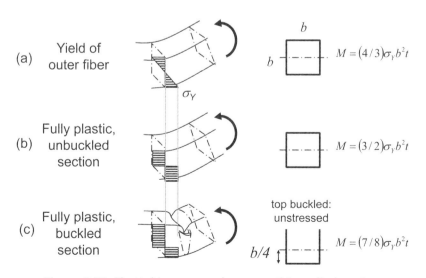

(a) Yield of outer fiber

$M = (4/3)\sigma_y b^2 t$

(b) Fully plastic, unbuckled section

$M = (3/2)\sigma_y b^2 t$

(c) Fully plastic, buckled section

top buckled: unstressed

$M = (7/8)\sigma_y b^2 t$

Figure 6.32 Plastic hinge example: square thin-walled section.

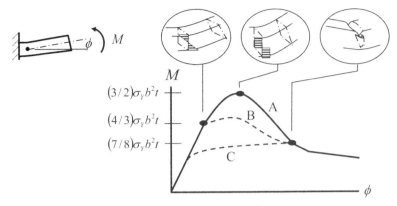

Figure 6.33 Plastic hinge behavior.

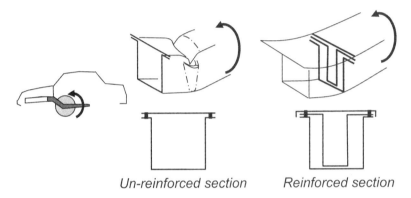

Un-reinforced section   *Reinforced section*

Figure 6.34 Reinforcement to increase plastic moment capacity.

A final assumption for our limit analysis is that as the plastic hinge rotates, the stress level remains constant at $\pm\sigma_Y$ independent of strain. Cold-formed steel approaches this assumed stress-strain behavior. As sheet steel is cold formed by the stamping process, forming stresses exceeds yield, as shown on the stress-strain curve of Figure 6.35a. After forming, the part exhibits a formed shape residual strain. Upon subsequent loading by service loads, the stress-strain curve now approaches the shape shown in Figure 6.35b. Note also that the loads applied to the reaction structure during the crash event induce a high strain rate. For many materials, this increased strain rate will raise the design stress over the quasi-static values [13, 14]. Figure 6.36 shows the dynamic behavior for several metals. For limit analysis, the design stress used should recognize the increases to yield stress due to both cold forming and high strain rate.

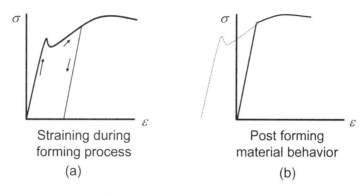

Straining during forming process

(a)

Post forming material behavior

(b)

Figure 6.35 Work hardening.

change from static strain rate (*.0005/sec*)
to dynamic (*9.8/sec*) indicated by arrow

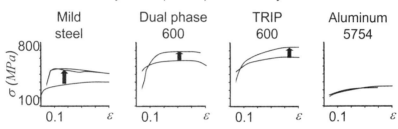

Figure 6.36 Dynamic stress-strain behavior.

Now that we have a means to estimate the plastic moment for a specific section, we can apply limit analysis to the cabin structure. Our objective is to identify the limit load capacity of the cabin structure and ensure it is sufficient to react the crush loads being generated by the midrail structure during crush. A typical design for cabin structure is shown in Figure 6.37. This longitudinal structure connects to the crushable midrail and extends down along the toe pan and under the passenger compartment. It is supported under the passenger compartment by cross members extending out to the rocker beams and is also supported by the floor pan in shear.

We can isolate the longitudinal structure, restrain it at the floor pan connection, and load it with a horizontal force representing the crush load being applied, Figure 6.37a. Applying limit analysis concepts, we view this structure as two rigid beams connected by two plastic hinges. The applied load, F_{ULT}, which induces this state, is the ultimate load for this structure. Attempts to exceed this load will cause the structure to collapse by rotations about the hinges as shown in Figure 6.37b.

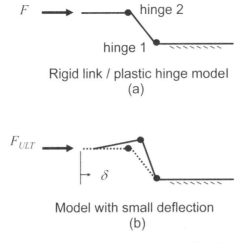

Rigid link / plastic hinge model
(a)

Model with small deflection
(b)

Figure 6.37 Mid rail limit load analysis.

We can determine the load, F_{ULT}, which causes this collapse by considering a small horizontal movement at the point of load application, δ, Figure 6.38. The work done by the force F_{ULT} is $(F_{ULT}\delta)$. This work must be equal to the work dissipated in the plastic hinges, Figure 6.38. The work dissipated in each plastic hinge is the plastic moment times the angular rotation caused by the small deflection δ. For *Hinge 1* this angular rotation is $d\phi_1$, for *Hinge 2* it is $(d\phi_1 + d\phi_2)$:

$$F_{ULT}\delta = M_{P1}d\phi_1 + M_{P2}(d\phi_1 + d\phi_2)$$

where:

$d\phi_1$ and $d\phi_2$ = Angular rotations

M_{P1} and M_{P2} = Plastic moments for each hinge

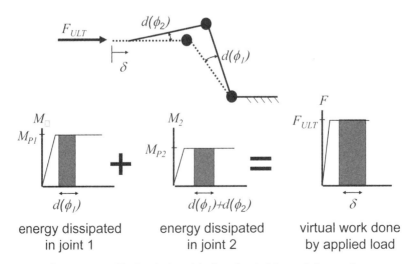

Figure 6.38 Work relationship for plastic hinge deformation.

The mechanism formed by this collapsing structure is kinematically deterministic. That is, knowing the deflection δ, we can calculate values for the angles of rotation, $d\phi_1$ and $d\phi_2$, using geometric relationships, Figure 6.39,

$$d\phi_1 = \frac{\delta}{L_1 \sin(\phi_1)}, \qquad d\phi_2 = \frac{\cos(\phi_1)\delta}{L_2 \sin(\phi_1)}$$

Substituting these angles into the above work balance yields,

$$F_{ULT}\delta = M_{P1}d\phi_1 + M_{P2}(d\phi_1 + d\phi_2)$$

$$F_{ULT}\delta = M_{P1}\left(\frac{\delta}{L_1 \sin(\phi_1)}\right) + M_{P2}\left(\frac{\delta}{L_1 \sin(\phi_1)} + \frac{\cos(\phi_1)\delta}{L_2 \sin(\phi_1)}\right)$$

$$F_{ULT} = M_{P1}\left(\frac{1}{L_1 \sin(\phi_1)}\right) + M_{P2}\left(\frac{1}{L_1 \sin(\phi_1)} + \frac{\cos(\phi_1)}{L_2 \sin(\phi_1)}\right) \tag{6.6}$$

Thus knowing the geometry of the longitudinal structure (L_1, L_2, and ϕ_1) and the plastic moment capacity at the hinge joints (M_{P1} and M_{P2}), we can determine the load capacity, F_{ULT}, for the structure and ensure that it is large enough to react the crush loads.

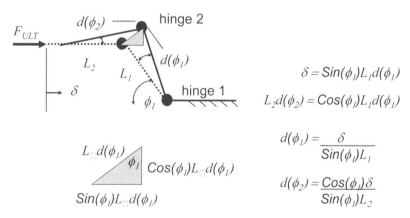

Figure 6.39 Geometric relationships.

6.2.7 Design for reducing vehicle pitch during impact

It has been observed that upon impact with a fixed barrier, some vehicles rotate in the side view with the rear raising upward, Figure 6.40. This rotation or pitch can increase the likelihood of neck injuries, Figure 6.41.

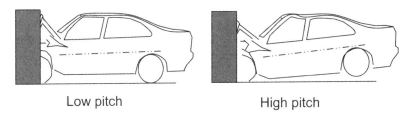

Low pitch   High pitch

Figure 6.40 Vehicle pitch upon impact with rigid barrier

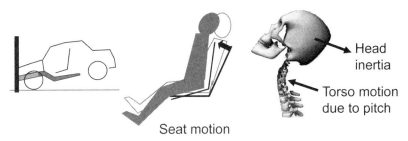

Seat motion

Figure 6.41 Neck injury related to vehicle pitch

To understand this phenomenon, consider a free body of the pitching vehicle at the time of impact, Figure 6.42a. The inertial force of the impact acts through the vehicle center of gravity while the counter-acting barrier face force is applied primarily at the crushing midrail. Summing moments about the CG:

$$F_B(h - h_L) = I\alpha, \quad F_B = Ma$$
$$\frac{Ma}{I}(h - h_L) = \alpha$$

where:

M = Vehicle mass

I = Pitch mass moment of inertia

h = Height of the CG above ground

h_L = Height of the effective load path above ground

α = Pitch acceleration

a = Acceleration of the vehicle during impact

In practice, the CG height is above the bumper height where the midrail is located. The difference in the heights of these equal and opposite forces creates a couple which causes the rotational (pitching) acceleration.

(a) Vehicle pitch without upper load path

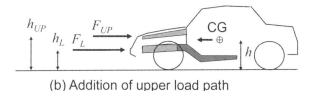

(b) Addition of upper load path

Figure 6.42 Reducing pitch upon impact.

To reduce pitching, we can add another crushable load path above the CG such that the moments about the vehicle CG for the two crushable load paths sum to zero, Figure 6.42b.

$$F_{UP}(h_{UP} - h) - F_L(h - h_L) = 0$$
$$\frac{F_{UP}}{F_L} = \frac{h - h_L}{h_{UP} - h} \qquad (6.7)$$

where:

F_{UP} = Average crush load generated by the upper load path

F_L = Average crush load generated by the lower load path

h = Height of the CG above ground

h_L = Height of the lower load path above ground

h_{UP} = Height of the upper load path above ground

In practice, the upper load path is just under the hood, and to the sides of the motor compartment, Figure 6.43.

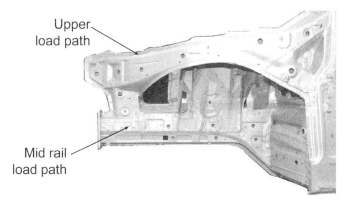

Figure 6.43 Example of upper load path. (Photo courtesy of A2Mac1.com, Automotive Benchmarking)

6.2.8 Summary: Structure for front barrier impact

We can now summarize a procedure for preliminary structure sizing for front impact:

1. Select the maximum allowable cabin acceleration based on occupant injury, a_{MAX}

2. Determine a consistent structural efficiency and crush space, (η, Δ, using Equation 6.3

3. Compute the average and maximum allowable crush forces which the vehicle must generate during impact, F_{AVG}, F_{MAX}, using the maximum acceleration, a_{MAX}, from step 1 and the efficiency, η, from step 2; $F_{MAX} = Ma_{MAX}$, $F_{AVG} = \eta F_{MAX}$

4. Allocate these total forces from step 3 to the structural elements within the vehicle front end, Figure 6.15

5. Size the crushable midrail using the average required crush force requirement of step 4 using Equation 6.4 and Figures 6.19–21

6. If the peak crush load, P_{MAX}, calculated in step 5 exceeds the maximum load requirement from step 4, then consider crush-initiator designs, Figure 6.23

7. The cabin reaction structure capacity must exceed the maximum midrail crush load identified in step 6. Use limit analysis to determine the required plastic moments for the hinges, Equation 6.6

8. Size the reaction structure sections to generate the hinge moments from step 7

Note that the structural requirements flow from vehicle requirements, and any change at the vehicle level must be examined for changes to the structural requirements.

6.3 Side Impact

Side impact plays an important role in sizing vehicle structure. As the strategy for side impact is quite different from that of front impact, this section is devoted to design for side impact. We will first look at the standardized vehicle tests and then flow down the vehicle level requirements to specific body structural elements.

Current FMVSS and NCAP side-impact tests consist of a stationary target vehicle and a moving barrier impacting from the side, Figure 6.44. The barrier moves at a 27° angle to the vehicle lateral axis, simulating the relative velocity of a side impact at an intersection. The face of the moving barrier, Figure 6.45, is deformable and crushes at a uniform *45 psi (0.31 N/mm²)*. The impact speed differs between FMVSS, *33.5 mph (53.6 km/h)*, and NCAP, *38.5 mph (61.6 km/h)* (Figure 6.46) and, as with front impact, FMVSS requires a minimum injury performance while NCAP reports the probability of serious injury using the star scale. The injury criterion is the Thoracic Trauma Index, TTI, with larger values of TTI indicating a more severe injury (a TTI<57 is desirable).

Impact by deformable moving barrier, Occupant injury criteria

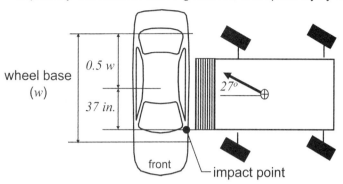

Figure 6.44 Side-impact test.

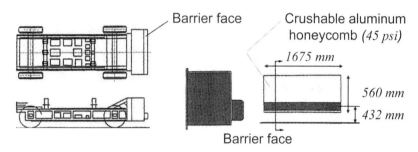

Figure 6.45 Rear moving barrier.

The SINCAP or LINCAP is based on the thoracic trauma index (TTI) using the US-SID dummy. TTI is defined as the average of the peak accelerations in the rib and lower spine.

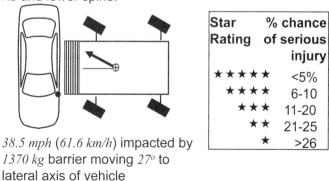

38.5 mph (61.6 km/h) impacted by *1370 kg* barrier moving *27°* to lateral axis of vehicle

Star Rating	% chance of serious injury
★★★★★	<5%
★★★★	6-10
★★★	11-20
★★	21-25
★	>26

Figure 6.46 Side impact: New car assessment program.

Figure 6.47 shows a typical velocity-time history during a side impact (the door velocity and dummy velocity shown are lateral to the vehicle and relative to ground). The door velocity is measured at the inside surface of the front door structure and dummy velocity at the torso. As a first-order model, the velocity-time histories may be idealized as linear, Figure 6.48. In this figure, the barrier velocity and vehicle velocity have been added. Initially, the impacted vehicle is at rest with zero velocity. Its velocity increases as the barrier begins to distort the door and accelerate the vehicle. At some time during the impact (approximately *20 msec* in Figure 6.48) the barrier has fully deformed the door, and both the door panel and barrier move at the same velocity for the remainder of the impact. For several milliseconds, the dummy velocity remains at zero. This is because there is initially some clearance between the dummy shoulder and door trim panel. The vehicle must slide laterally through this distance before the dummy is struck (this occurs at approximately *25 msec*). The dummy is then accelerated by the door trim panel until both the door and dummy reach the same speed (*~50 msec*). The dummy impact event is over at this time, T_{FINAL}.

For body design, we wish to identify which characteristics of this velocity-time history most influence injury. Correlations between observed TTI with measured parameters from the velocity-time history have been examined [15]. A parameter with high correlation to TTI is the total change in velocity the dummy undergoes during the impact, V_{TF}, Figure 6.48. Lower velocity change corresponds to lower TTI. This correlation provides a single performance criterion for preliminary design for side impact—by minimizing V_{TF} we will also minimize TTI. We now analyze impact kinematics to determine what structural parameters affect V_{TF}.

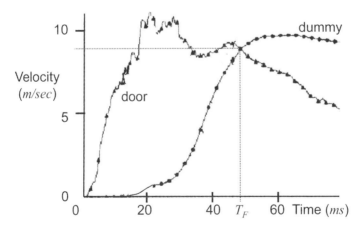

Figure 6.47 Side-impact velocity-time history.

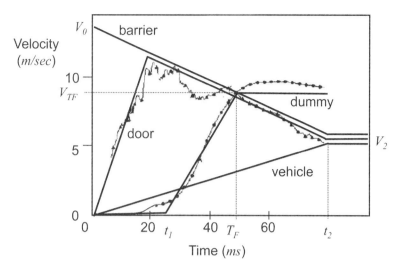

Figure 6.48 Idealized side-impact velocity-time history.

6.3.1 Kinematic and load path analysis of side impact

Consider the impact of the vehicle by a moving barrier [16]. We can model each as a point mass with the impact being perfectly plastic, Figure 6.49. In this linear model, we are looking at motions lateral to the vehicle and will consider the lateral component of the barrier velocity as the initial impact velocity, $V_0 = S_{TEST} \cos \alpha$, where S_{TEST} is the standard test impact speed and α is the angle of impact. As no external forces are applied to the impacting masses, the momentum is unchanged before and after the impact. Equating the momentum yields a relationship for final velocity, V_2.

$$M_1(V_0) = (M_1 + M_2)V_2$$

$$V_2 = \frac{M_1}{M_1 + M_2} V_0 \tag{6.8}$$

where:

M_1 = Barrier mass

M_2 = Vehicle mass

V_0 = Lateral impact speed

V_2 = Final speed of vehicle and barrier

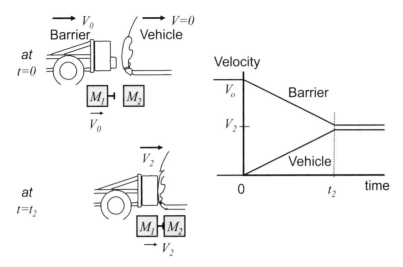

Figure 6.49 Terminal velocity in side impact.

We now idealize the crush characteristics of the side of the vehicle by assuming a square wave for crush load vs. lateral deformation, Figure 6.50. By looking at a free body of each mass during impact, we can find the acceleration during impact for each mass:

$$-F_2 = M_1(a_1), \qquad F_2 = M_2(a_2)$$

$$a_1 = \frac{-F_2}{M_1}, \qquad a_2 = \frac{F_2}{M_2} \qquad\qquad (6.9)$$

where:

a_1 = Barrier acceleration

a_2 = Vehicle lateral acceleration

F_2 = Crush load for the vehicle side (note: $F_2 < 290,000$ N (65,200 lb), the moving barrier face crush capacity)

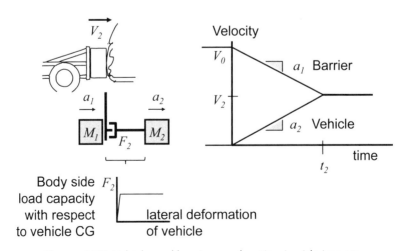

Figure 6.50 Vehicle and barrier acceleration in side impact.

The time for the impact event, t_2, can be found by equating the time it takes to accelerate the vehicle to V_2

$$a_2 t_2 = V_2$$

$$t_2 = \frac{V_2}{a_2} \tag{6.10}$$

where t_2 = Time at the end of the impact event

We can now sketch the velocity-time histories for the barrier and vehicle center of gravity in this idealized model, Figure 6.50.

Remember that the area under the velocity-time history is the distance traveled relative to ground, Figure 6.51.

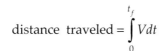

$$\text{distance traveled} = \int_0^{t_f} V dt$$

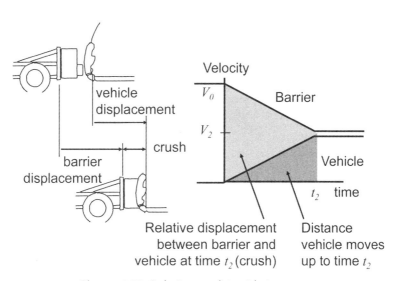

Figure 6.51 Relative crush in side impact.

Thus the area under the *vehicle v-t* graph up to time t_f is the distance the vehicle slides laterally; and the area under the *barrier v-t* graph is the distance it slides. By looking at the area between the two graphs up to time t_2 we find the difference of these two distances or the relative crush of the vehicle and barrier during impact, Figure 6.51. From the geometry of the curve, this crush is

$$crush = \frac{1}{2} V_0 t_2$$

Now we can turn our attention to the interior of the vehicle and the impact between occupant and door. We imagine the occupant sitting on the vehicle mass, but not restrained in the lateral direction, Figure 6.52. We now add a mass-less rigid door side to the outer end of the vehicle crush element, F_2. To this door outer side, we add a crush element at the occupant shoulder level which represents the crush characteristics of the door and trim panel, Figure 6.52. There is a space, Δ_0, between the inside of the door crush element and occupant shoulder, and a door width of Δ.

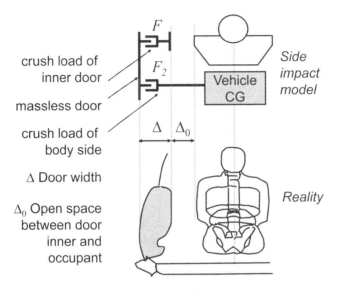

Figure 6.52 Side-impact first-order model.

Upon impact of vehicle and barrier, the vehicle will begin to move to the side, but as no lateral forces are being applied to the occupant he will not move relative to ground. Thus the vehicle will move through the distance Δ_0 before the inner door will strike the occupant, Figure 6.53. The time at which the door impacts the occupant, t_1, can be found by equating the area between the vehicle and occupant v-t histories (shaded area in Figure 6.53) with the distance Δ_0,

$$\frac{1}{2}\left[V_0 + (V_0 + a_1 t_1)\right]t_1 = \Delta_0$$

$$t_1^2 + \frac{2V_0}{a_1}t_1 - \frac{2\Delta_0}{a_1} = 0$$

using the quadratic formula:

$$t_1 = \frac{-\dfrac{2V_0}{a_1} \pm \sqrt{\left(\dfrac{2V_0}{a_1}\right)^2 + 4\left(\dfrac{2\Delta_0}{a_1}\right)}}{2} \tag{6.11}$$

where t_1 = Time at which the occupant begins to strike door.

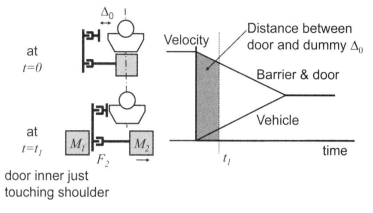

Figure 6.53 Time at occupant impact.

At this time, the door side begins to load the occupant with force and to accelerate him to the side.

The door side now accelerates the occupant laterally, Figure 6.54. Once the velocity of the door and occupant are equal, there will be no more relative deformation between the two, and the impact between the door and occupant is over. This occurs when the door inner has crushed through a distance of Δ. We equate the area under the occupant velocity history between the time t_1 and time T_F (the shaded area in Figure 6.55) with the door crush, Δ.

$$\frac{1}{2}[T_F - t_1)]V_1 = \Delta$$

where V_1 = Velocity at time t_1:

$$V_1 = V_0 + a_1 t_1 \tag{6.12a}$$

$$T_F = \frac{2\Delta}{V_1} + t_1 \tag{6.12b}$$

where T_F = Time at which the occupant and door are moving at same speed

Our objective for this kinematic analysis was to estimate the change in velocity of the occupant, V_{TF} as an indicator of injury. We can now estimate this change in velocity,

$$V_{TF} = V_0 + a_1 T_F \tag{6.13}$$

where V_{TF} = Velocity change of occupant from initial contact with the interior to release

We can also estimate the occupant acceleration as

$$a_{OCC} = \frac{V_{TF}}{T_F - t_1} \tag{6.14}$$

where a_{OCC} = Average acceleration of occupant during contact with the interior

This acceleration represents a minimum value and assumes the deformation characteristic of the door inner has a square wave load-deformation crush curve.

249

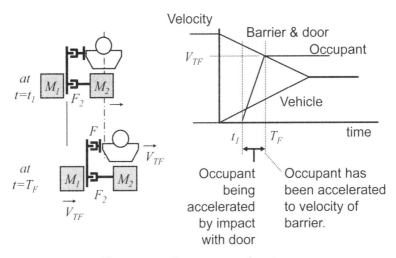

Figure 6.54 Occupant acceleration.

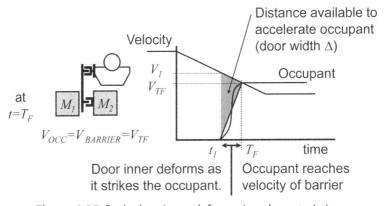

Figure 6.55 Body door inner deformation characteristics.

We can now completely define the velocity-time histories using the above equations, Figure 6.56, when given the impact speed, V_0; the barrier and vehicle masses, M_1 and M_2; the force characteristics for the body side, F_2; and the dimensions for the door crush thickness and space between occupant and door inner, Δ and Δ_0.

Remember that occupant injury in a side impact is correlated with the total velocity change for the occupant, V_{TF}, and reducing V_{TF} reduces occupant injury.

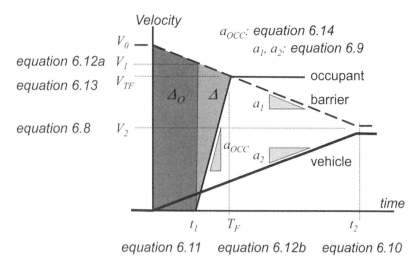

Figure 6.56 Summary of kinematic relationships.

6.3.2 Flow down of requirements for side impact

Using the kinematic relationships derived, we can now look at the influences on V_{TF} for three structural parameters.

Crush load for the vehicle side
The body side crush load, F_2, should be high. Increased side crush force decelerates the barrier quickly before impact with the occupant, Figure 6.57a. Thus the occupant is struck by the door at a lower velocity. Physically this high side crush force can be achieved with rigid side-to-side structural members aligned at the barrier face height. A cross member at the B pillar location or at the front of the rear seat pan is effective. Also, the barrier face must be engaged early by aligning structural members such as the rocker, lower B, and C pillars at the height of the barrier face.

Clearance between the occupant shoulder and door panel
It is desirable to use the space between the occupant and door outer surface, Δ, to minimize the impact acceleration of the occupant. Increasing this space decreases the average slope of the occupant velocity curve, Figure 6.57b, and results in a lower door impact force being applied to the occupant.

Door inner crush characteristic
When the door crush characteristic, F, is a square wave, the peak acceleration of the occupant is minimized, Figure 6.55. A means to accomplish this is by using crushable foam placed in the door trim area.

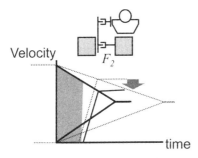

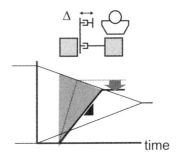

Side load path should have a high load capacity; Barrier decelerates quicker, door impacts occupant at a lower speed.

As much space, Δ, as possible should be used to decelerate the occupant. Door crush force should be square wave

(a) Body side load path characteristic

(b) Occupant to door space allocation

Figure 6.57 Desirable design parameters for side impact.

6.4 Note on Rear Impact

In the standard rear impact test, the stationary target vehicle is impacted by a moving barrier, Figure 6.58. The criterion for this test is to minimize fuel system leakage, so we are interested in absorbing the energy of the barrier by deforming structure rearward of the fuel system. If we could replace this impact with one between a moving vehicle and fixed barrier, we could apply the structure-sizing procedure developed for the front barrier case. To do this, we must identify the equivalent impact velocity which would result in the same work of deformation to be done by the structure.

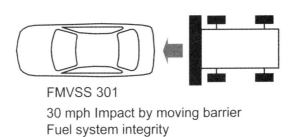

FMVSS 301

30 mph Impact by moving barrier
Fuel system integrity

Figure 6.58 Rear barrier.

We first apply conservation of momentum principles to find the final moving barrier impact speed, Figure 6.59a & b.

$$M_1(0) + M_2V_0 = (M_1 + M_2)V_F$$

$$V_F = \frac{M_2}{M_1 + M_2}V_0$$

where:

M_1 = Struck vehicle mass

M_2 = Moving barrier mass

V_0 = Initial moving barrier speed

V_F = Common final speed of vehicle and barrier

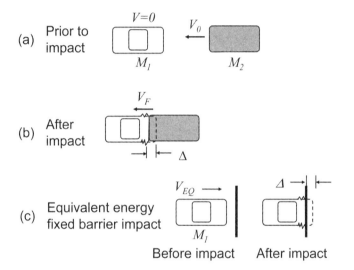

(a) Prior to impact

(b) After impact

(c) Equivalent energy fixed barrier impact

Before impact   After impact

Figure 6.59 Rear impact equivalent fixed barrier test.

Now we can find the work of deformation, W, during the moving barrier impact, Figure 6.59b, by equating work to change of kinetic energy before and after the impact.

$$W = \frac{1}{2}M_2V_0^2 - \frac{1}{2}(M_1 + M_2)V_F^2$$

$$W = \frac{1}{2}\left(\frac{M_1M_2}{M_1 + M_2}\right)V_0^2$$

Finally we consider an impact with a fixed barrier which will cause the same work of deformation, Figure 6.59c and determine the equivalent impact velocity, V_{EQ}.

$$\frac{1}{2}M_1V_{EQ}^2 = W$$

$$\frac{1}{2}M_1V_{EQ}^2 = \frac{1}{2}\left(\frac{M_1M_2}{M_1 + M_2}\right)V_0^2$$

$$V_{EQ} = V_0 \sqrt{\frac{M_2}{M_1 + M_2}} \qquad (6.15)$$

Now using this fixed barrier equivalent velocity, we can identify the needed average rear crush force using

$$\frac{1}{2} M_1 V_{EQ}^2 = F_{AVG} \Delta \qquad (6.16)$$

$$F_{AVG} = \frac{M_1 V_{EQ}^2}{2\Delta}$$

where:

Δ = Available crush space between fuel tank and bumper

M_1 = Vehicle mass

V_{EQ} = Equivalent impact speed (Equation 6.15)

The process to size the rear-energy-absorbing structure and reaction structure is the same as with the front barrier. Because this is a test for fuel tank integrity, we do not use cabin acceleration as the main requirement as in the front barrier. The main requirement here is to maintain the integrity of the fuel tank, so all crush must occur behind the tank. In preliminary design, we assume the fuel tank is between the rear wheels and extends to the back of the rear wheel. The available crush space, Δ, is then the distance from the back of the rear wheel to the end of the car. After obtaining F_{AVG} from the above equation, Equation 6.4 may be used to size the rear rails for energy absorption. The reaction structure for rear impact can be sized using limit analysis as we did in the front impact case.

6.5 Note on Roof Crush

A standard test for roof integrity is the static roof crush test. In this test, a rigid platen is pushed into the front corner of the roof, Figure 6.60. The criterion for this test for FMVSS 216 is to develop a minimum level of crush force, 1½ times the vehicle weight, without deforming beyond a set distance, *5 in. (125 mm)*.

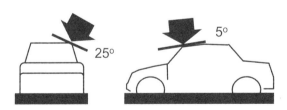

FMVSS 216
Load 1½ times vehicle weight
Criterion: Less than 5 inches of deformation

Figure 6.60 Static roof crush.

For preliminary structure sizing for this condition, we may again use limit analysis. A first-order model of side-view behavior is shown in Figure 6.61. Here the A pillar beam and roof side rail beam are connected by three plastic hinges: one at the belt line, one at the top of windshield, and one at the roof rail-to-B-pillar intersection. We ensure that the limit load exceeds the crush force requirement by applying the analysis procedures discussed earlier and shown in Figure 6.38. Note that this first-order model of roof crush neglects the often significant role the windshield plays in reacting the lateral component of the roof crush load.

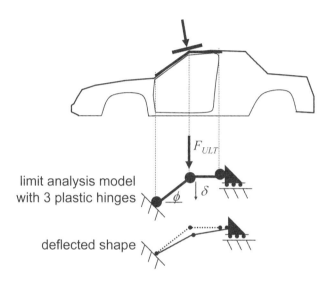

Figure 6.61 Limit analysis of roof crush.

References

1. *United States Federal Motor Vehicle Safety Standards and Regulations*, www.nhtsa. dot.gov/cars/rules/standards/safstan2.htm.

2. *Ultra Light Steel Auto Body-Final Report*, American Iron and Steel Institute, August 16, 1995.

3. Kikuchi, N. & Malen, D., Course notes for ME513 Fundamentals of Body Engineering, University of Michigan, Ann Arbor, MI, 2007.

4. Mahmood, H. F. and Paluszny, A., "Design of Thin Walled Columns for Crash Energy Management," SAE Paper No. 811302, SAE International, Warrendale, PA, 1981.

5. *Cold-Formed Steel Design Manual*, American Iron and Steel Institute, Washington, DC, 1986, Section 3.6.4.

6. Sidhu, R., et al, J., "Improved Vehicle Crashworthiness Via Shape Optimization," IMECE2003–43439, Society of Automotive Engineers, PA, 2003.

7. Sheh, M.Y. and Khalil, T.B. "The Impact Response of a Vehicle Structural Rail by Experiments and Finite Element Analysis," *Symposium on Crashworthiness and*

Occupant Protection in Transportation Systems, ASME Publication, AMD-Vol. 126/BED-Vol. 19, 1991, pp. 195–207.

8. Chang, D., "A Design-Analysis Method for the Frontal-Crush Strength of Body Structures," SAE Paper No. 770593, SAE International, Warrendale, PA, 1977.

9. Hamza, K. and Saitou, K., "Design Optimization of Vehicle Structures for Crashworthiness Using Equivalent Mechanism Approximations," *Journal of Mechanical Design*, AMSE, May 2005, pp. 485–491.

10. Prasad, P. and Belwafa, J. Editors, *Vehicle Crashworthiness and Occupant Protection*, American Iron and Steel Institute, 2004.

11. Van Den Broek, J. A., *Theory of Limit Design*, John Wiley, NY, 1948.

12. Cook, R. and Young, W., *Advanced Mechanics of Materials*, Macmillan Co., NY, 1985, p. 447.

13. Chen, X. M., et al, "Crash Performances of Advanced High Strength Steels of DP780, TRIP780 and DP," SAE Paper No. 2005–01–0354, SAE International, Warrendale, PA, 2005.

14. Shi, M. F. and Meuleman, D. J., "Strain Rate Sensitivity of Automotive Steels," SAE Paper No. 920245, SAE International, Warrendale, PA, 1992.

15. Deb, A. and O'Connor, C. "Impact Using a Regression-Based Approach," *SAE Side Collision Research* (SP-1518), 2000–01–0636, SAE International, Warrendale, PA, 2000.

16. Yamaguchi, S., "Concept of Side Impact Occupant Protection and Evaluation Method for Automobile Development; comments on presentation by J. Kanianthra," *SAE Stapp Car Crash Conference Proceedings*, SAE International, Warrendale, PA, 1993–1994.

Chapter 7
Design for Vibration

The body structure is a resonant system with an infinite number of natural frequencies. If any of these resonances occur at the wrong frequency, objectionable or destructive levels of vibration will result. In this chapter, we will investigate ways to avoid this and identify desirable vibration behavior for the body structure. We will do this from a design perspective—where shall we place these body structure resonances to result in satisfactory vibration performance?

7.1 First-Order Vibration Modeling

Generally during the early design stages we are not interested in estimating the *precise* vibration amplitudes. We wish only to assure that the amplitude will not be objectionably large to the *receiver* of the vibration. We can do this by assuming, for a well-designed vehicle, that the frequency of the vibration *source* does not coincide with a resonance in the vibration *path* (i.e., the vibration is uncoupled). This assumption will allow us to simplify our analysis models by considering the well-designed automobile to be a set of *independent* single-degree-of-freedom oscillators in which the natural frequencies of these oscillators do not coincide. Following is an example of this concept.

Example: First-order vibration model of steering column

In this example, we look at a vibration system consisting of the powertrain, steering column, and driver air cushion restraint, Figure 7.1. The steering column mounting structure, Figure 7.2a, is designed to meet static stiffness requirements—in this case, a vertical stiffness of *200 N/mm* as measured at the steering wheel. This level of stiffness provides a very small static downward deflection (*1 mm*) under a downward load (*200 N*), which the driver may impart to the steering wheel. This deflection is perceived by the driver as a "solid" feel. The steering wheel, mass *m=5 kg*, is mounted on this column structure, and the combination behaves as a single-degree-of-freedom oscillator, Figure 7.2b. The natural frequency of this oscillator is given by (this relationship will be developed in Section 7.3) [1]:

$$\omega_n = \sqrt{\frac{k}{m}}$$

where:

ω_n = Natural frequency in radians /sec

k = Stiffness

m = Mass

For this example, *k=200 N/mm, m=5 kg* and

$$\omega_n = \sqrt{\frac{k}{m}} = \sqrt{\frac{(200 N / mm)(1000 mm / meter)}{5 kg}}$$

$$\omega_n = \sqrt{40000 N / meter\, kg} = 200\, rad / \sec$$

Note units: *1 N=1 kg m/$_s^2$*

It is often convenient to express the frequency in *Hz* (cycles/sec):

$$f_n (Hz) = \omega(rad / s)(1 cycle / 2\pi\, rad)$$
$$f_n = 31.8 Hz$$

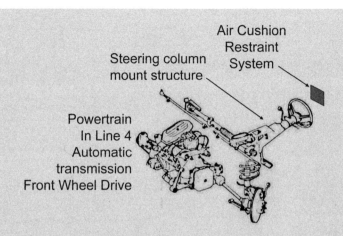

Figure 7.1 Vibration interaction example.

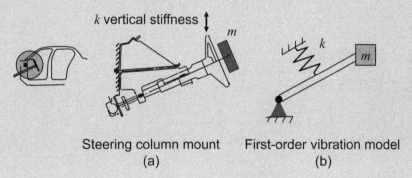

Figure 7.2 Steering column resonant system.

The "goodness" of this resonant frequency depends on whether it coincides with a forcing frequency (source), and also depends on how the driver (receiver) perceives a vibration at this frequency at his hands.

Example: First-order model of vibration source

Consider a source of vibration which may excite the steering column—a four-cylinder engine in a transverse front-wheel-drive configuration with an automatic transmission. Each time a cylinder fires, a torque pulse occurs at the crankshaft, Figure 7.3. For a four-stroke engine, this pulse occurs once every two revolutions for each cylinder. With four cylinders and an engine speed of N revolutions per minute, this torque pulse excitation occurs at a frequency of

$$\Omega = (4\,cylinders)(1\,pulse\,/\,2\,rev)(N\,rev\,/\,min)(1\,min\,/\,60\,sec)$$
$$\Omega = (N\,/\,30)\,Hz$$

where:

Ω = Torque pulse excitation frequency (Hz)

N = Engine rotational speed (rev/min)

When such a powertrain is at idle speed *N=700 rev/min* with the transmission in Drive, these torque pulses are reacted by the engine block. The torque pulse frequency for this example is *N=700 rev/min* and *Ω=700/30=23.3 Hz*.

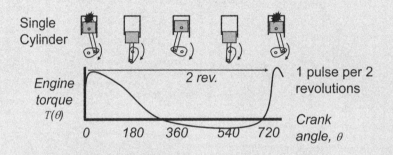

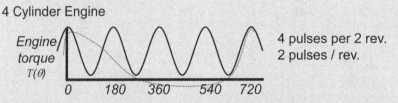

Figure 7.3 Engine torque variation.

This idle condition imposes an oscillating input from the block through the engine mounts, into the body, and ultimately to the steering column mount. However, since the forcing frequency of the torque pulses, *Ω=23.3 Hz*, does not coincide with the resonant frequency of the steering column, *fn=31.8 Hz*, we would not expect a troublesome vibration amplitude.

Example: Coupled resonance

Now consider that a driver side *Air Cushion Restraint System*, ACRS, is mounted on the steering wheel, Figure 7.1. The ACRS adds an additional mass, *4.5 kg*, to the resonant model of the previous example. The steering column resonant frequency now changes:

$$\omega_n = \sqrt{\frac{k}{m}} = \sqrt{\frac{(200N/mm)(1000mm/meter)}{(5+4.5)kg}}$$

$$\omega_n = \sqrt{21053N/meter\,kg} = 145rad/sec$$

$$f(Hz) = 145(rad/s)(1cycle/2\pi\,rad)$$
$$f = 23.1Hz$$

Now the engine torque pulse excitation frequency approximately coincides with the steering column resonance, Figure 7.4. This undesirable condition will result in large steering column vibration levels which will be perceived by the customer, Figure 7.5. It is this condition—coinciding source and path frequencies—which we attempt to avoid during design.

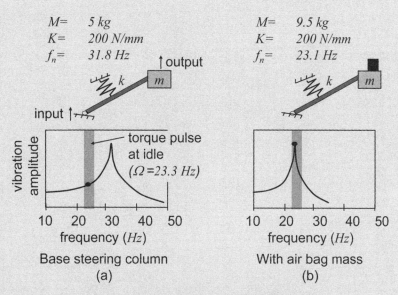

$M=$ 5 kg
$K=$ 200 N/mm
$f_n=$ 31.8 Hz

torque pulse at idle ($\Omega=23.3$ Hz)

Base steering column
(a)

$M=$ 9.5 kg
$K=$ 200 N/mm
$f_n=$ 23.1 Hz

With air bag mass
(b)

Figure 7.4 Source-path interaction.

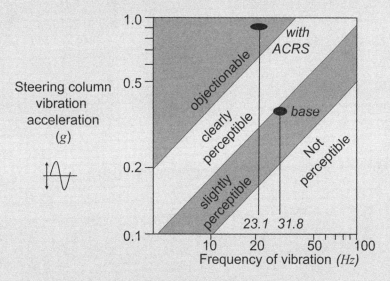

Steering column vibration acceleration (g)

Figure 7.5 Driver tactile perception.

(Note that the addition of the ACRS did not cause any difficulties with the *static* behavior of the steering column/ACRS system; there would be no excessive static deflection with the ACRS installed, nor would the strength of the column mount be exceeded by the addition of the ACRS. It is only when we look at the vehicle system—including the source of vibration energy—that we find this problem.)

7.2 Source-Path-Receiver Model of Vibration Systems

The example of Section 7.1 suggests a way to view vibratory systems for design purposes. The system, Figure 7.6, is viewed as consisting of:

1. A *Source* of vibration energy (engine torque pulses in Section 7.1 example)

2. A *Path* for the vibration consisting of a series of subsystems (the steering column with ACRS)

3. A *Receiver* which determines the acceptability of the vibration level (the driver's hands).

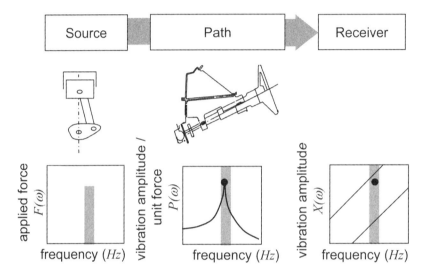

Figure 7.6 General vibration model.

The vibration characteristics for each of these entities may be viewed as a function of vibration frequency (frequency domain), Figure 7.6 bottom. For the source, we have a generalized force amplitude, $F(\omega)$, being applied to the system; for the path, we have a transfer function, $P(\omega)$. This transfer function is the amplitude of the output (deflection) resulting from a unit of input (force amplitude); and for the receiver, a relation for the perception of the vibration at amplitude $X(\omega)$:

$$F(\omega)\left[\frac{X(\omega)}{F(\omega)}\right] = X(\omega)$$

$$F(\omega)[P(\omega)] = X(\omega) \qquad (7.1a)$$

$$\uparrow \qquad \uparrow \quad \uparrow$$

Source Path Sensed by Receiver

where:

    $F(\omega)$ = Source amplitude

    $X(\omega)$ = Response amplitude

    $[P(\omega)]$ = Path transfer function [response, $X(\omega)$, per unit of input, $F(\omega)$]

In the above relationship, the path is viewed as a single subsystem which responds directly to the source. This model we used in the steering column example. In many cases in automobile design, the path is a series of subsystems: for example, a vibration force, $F(\omega)$, applied at the spindle of the front suspension. This applied force is attenuated by the suspension characteristics, $[T(\omega)]$, and a reduced force, $F_T(\omega)$, is transmitted to the body structure through the suspension attachments. This behavior can be described by the relationship,

$$F(\omega)\left[\left(\frac{F_T(\omega)}{F(\omega)}\right)\left(\frac{X(\omega)}{F_T(\omega)}\right)\right] = X(\omega)$$

$$F(\omega)[T(\omega)P(\omega)] = X(\omega) \qquad (7.1b)$$

$$\uparrow \qquad \uparrow \qquad \uparrow$$

Source   Path   Sensed by Receiver

where:

$F(\omega)$ = Source amplitude

$X(\omega)$ = Response amplitude

$T(\omega)$ = Isolation characteristic of a subsystem in the path (force transmitted, $F_T(\omega)$, per unit force applied, $F(\omega)$)

$F_T(\omega)$ = Force transmitted through a subsystem of the path

$P(\omega)$ = Body structure transfer function

Equation 7.1b will be used to develop design requirements for the body structure vibration performance. Later in this chapter we will look closely at the four vibration systems shown in the table below.

Table 7.1 Four automobile vibration systems.

Vibration System	Source $F(\omega)$	Isolator $T(\omega)$	Force into Body $F_T(\omega)$	Body Transfer Function $P(\omega)$	Body Deflection $X(\omega)$
1	Powertrain unbalance force	Mounted powertrain	Force through engine mounts	Body structure	Deflection at seat, steering column
2	Force at suspension spindle	Suspension	Force through shock absorber and ride spring	Body structure	Deflection at seat, steering column
3	Road deflection at tire patch	Suspension	Force through shock absorber and ride spring	Body structure	Deflection at seat, steering column
4	High frequency chassis deflections	Chassis links with end bushings	Body panel vibrations	Passenger compartment acoustic resonances	Interior sound pressure

7.2.1 Automobile vibration spectrum

As each of the vibration characteristics, $F(\omega)$, $T(\omega)$, $F_t(\omega)$, $P(\omega)$, and $X(\omega)$, depend on frequency, ω, it is useful to consider the vibration spectrum for an automobile, Figure 7.7. Below *10 Hz* the body structure acts as a rigid body. Above *100 Hz*, body vibration behavior is more localized and influenced by structural details. For these design details we have the ability to take corrective measures for poor performance later in the design sequence. We are most concerned with the frequency range of *10–100 Hz* in body structure design. Behavior of the structure within this range is that of the primary bending and torsion resonances. These are set by the overall body architecture which, once established, cannot easily be changed in the later design stages. Therefore it is important to design-in proper behavior for this range during the early stages of design.

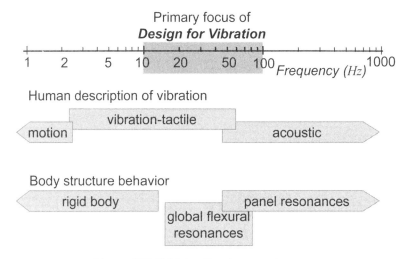

Figure 7.7 Vehicle vibration spectrum.

7.2.2 Human response to vibration

The objective of designing for vibration is to create an acceptable vibration environment for the automobile passengers. There are considerable data which relate human preference to vertical seat vibration level. In a typical test, a sinusoidal vibration level is imposed at a specific frequency to the seat and the subject is asked to subjectively describe the vibration—imperceptible, just perceptible, annoying. The data from these evaluations are plotted on a vibration amplitude vs. frequency plot as lines of constant comfort. A classical criterion is the Janeway curve, Figure 7.8, which describes the boundary of acceptable vibration amplitude [2]. Figure 7.9 is a summary of data from several contemporary sources compared with the Janeway curve [3, 4, 5]. The general character of all these iso-comfort curves is U shaped, with the area least tolerated occurring in the *6–20 Hz* range. These data give us some idea for acceptable vibration levels as we use the source-path-receiver model to design the structure for vibration behavior.

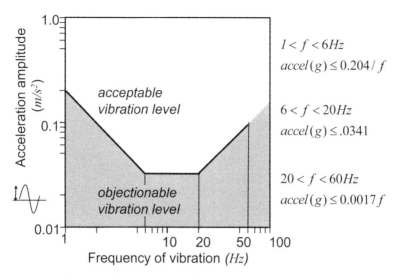

$1 < f < 6Hz$
$accel(g) \leq 0.204 / f$

$6 < f < 20Hz$
$accel(g) \leq .0341$

$20 < f < 60Hz$
$accel(g) \leq 0.0017f$

Figure 7.8 Janeway vertical seat vibration criteria.

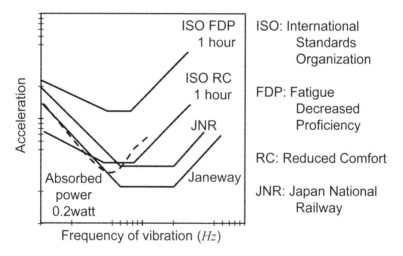

ISO: International
   Standards
   Organization

FDP: Fatigue
   Decreased
   Proficiency

RC: Reduced Comfort

JNR: Japan National
   Railway

Figure 7.9 Comparison of recommended vibration limits. (Courtesy of SAE International)

In the following sections we will look at some important vibration systems of the automobile using this source-path-receiver model of Equation 7.1b, Figure 7.10. We shall model the isolation behavior, $T(\omega)$, and also the body structure transfer function, $P(\omega)$, as single-degree-of-freedom (SDOF) oscillators, Figure 7.11. Therefore in the next section, we will begin by developing the equations of motion for a general SDOF system.

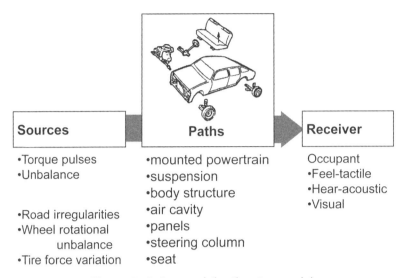

Figure 7.10 Automobile vibration model.

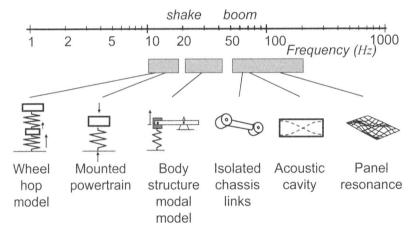

Figure 7.11 Major vibratory systems.

7.3 Frequency Response of a Single-Degree-of-Freedom System

As an example of a SDOF system, consider the vertical motion of a powertrain mounted on engine mounts, Figure 7.12a. (In a SDOF system, only one coordinate is needed to describe the motion. This coordinate may be either a linear or rotational deflection.) We assume the powertrain is mounted to ground and that a vertical sinusoidal force is applied with amplitude F and frequency ω. With a sinusoidal forcing function, the vertical displacement will also be sinusoidal with amplitude X and frequency ω:

$$f(t) = F\sin(\omega t)$$
$$x(t) = X\sin(\omega t)$$

(7.2)

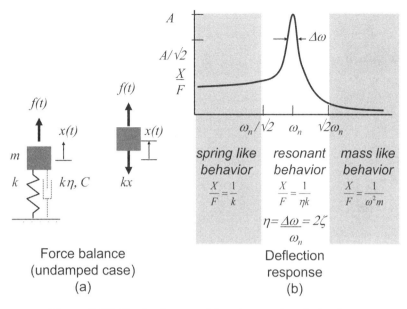

Figure 7.12 Single-degree-of-freedom system behavior.

7.3.1 Equation of motion for SDOF system

Consider a free body of the mass, Figure 7.12a, with a positive deflection of x [6]:

$$\Sigma(\text{forces acting on m}) = m\frac{d^2x(t)}{dt^2}$$

(7.3)

$$f(t) - kx(t) = m\frac{d^2x(t)}{dt^2}$$

$$\frac{d^2x(t)}{dt^2} = -X\omega^2\sin(\omega t)$$

$$F\sin(\omega t) = kX\sin(\omega t) - mX\omega^2\sin(\omega t)$$

(7.4)

$$F = kX - m\omega^2 X$$

$$\frac{X}{F} = \frac{1}{k - m\omega^2}$$

When $\omega^2 = k/m$, the denominator of this expression goes to zero and we have very large deflections. This frequency is the natural or resonant frequency for this system:

$$\omega_n^{\,2} = \frac{k}{m}$$

(7.5)

With this definition the response, $P(\omega)$, is:

$$\frac{X}{F} = \frac{1/k}{1 - \left(\dfrac{\omega}{\omega_n}\right)^2} = P(\omega)$$

(7.6)

where:

X = Amplitude of sinusoidal displacement

F = Amplitude of applied sinusoidal force

k = Spring stiffness

ω_n = Natural frequency

This is the transfer function for this path of vibration. The units are deflection amplitude per unit of applied force amplitude (m/N).

Example: Mounted powertrain vertical bounce

A four-cylinder, automatic transmission powertrain has a mass of *100 kg*. We assume the mount system constrains motion to the vertical. The combined engine mount vertical stiffness is *600 N/mm*. For an evaluation, the mounted powertrain is placed on a bed plate (ground) and a sinusoidal vertical force is applied at the center of mass.

a) Determine the natural frequency for this bounce mode

$$\omega_n{}^2 = \frac{k}{m}$$

$$\omega_n{}^2 = \frac{(600 N/mm)(1000 mm/m)}{100 kg} = 6000 N/kg \left[(kg\, m/s^2)/kg \right]$$

$$\omega_n = 77.46 rad/\sec$$

$$f_n = 38.73 rad/\sec(cycle/2\pi rad) = 12.33 Hz$$

b) At a certain engine speed, the reciprocating masses in the engine apply a vertical sinusoidal force of amplitude of *500 N* at a frequency of *15 Hz*; what is the deflection amplitude of the powertrain?

$$\frac{X}{F} = \frac{1/k}{1 - \left(\dfrac{\omega}{\omega_n}\right)^2} = \frac{1/(6x10^5 N/m)}{1 - \left(\dfrac{\omega}{77.46 rad/s}\right)^2} = \frac{1.67x10^{-6}}{1 - \left(\dfrac{\omega}{77.46 rad/s}\right)^2} m/N$$

$$\frac{X}{F} = \frac{1.67x10^{-6}}{1 - \left(\dfrac{(15 cycle/s)(2\pi rad/cycle)}{77.46 rad/s}\right)^2} m/N = -3.48x10^{-6} m/N$$

$$X = (-3.48x10^{-6} m/N)(500N) = -1.738x10^{-3} m$$

Note, the negative sign for *X* means that the deflection and force sinusoids are out of phase by *180°*. Note also that this dynamic deflection is greater than if the force had been applied statically: *500 N/(600 N/mm)=0.833x10⁻³ m*.

7.3.2 Relation of vibration amplitudes

To gain insight into SDOF systems under sinusoidal motion, consider the relationship between the amplitudes for displacement, $x(t)$, velocity, $v(t)$, and acceleration, $a(t)$. Given a displacement amplitude, X:

$$displacement = x(t) = X \sin(\omega t)$$

$$velocity = \frac{dx(t)}{dt} = X\omega \cos(\omega t) \tag{7.7}$$

$$acceleration = \frac{d^2 x(t)}{dt^2} = -X\omega^2 \sin(\omega t)$$

So for any frequency, ω, these amplitudes are related as

$$displacement\ amplitude = X$$
$$velocity\ amplitude = X\omega \tag{7.8}$$
$$acceleration\ amplitude = X\omega^2$$

Often in vibration testing, an accelerometer is used to measure vibration amplitude. The relations above can be used to convert between amplitudes.

Example: Powertrain vibration test

An accelerometer is applied to a powertrain under test. During a vibration test at a sinusoidal forcing frequency of *40 Hz*, the amplitude recorded by the accelerometer is *10 g*. What is the deflection amplitude?

$$acceleration\ amplitude = X\omega^2$$

$$X = acceleration\ amplitude\ /\ \omega^2$$

$$X = (10g)(9.8m\,/\,s^2)\,/\,\left[40\,Hz(2\pi rad\,/\,cycle)\right]^2$$

$$X = 1.55 x 10^{-3}\,m$$

7.3.3 Regions of vibration behavior

Plotting Equation 7.6 for $P(\omega)$ against frequency, we can identify three important regions, Figure 7.12b; at frequencies much below the resonance frequency, $(\omega<<\omega_n)$ we have spring-like behavior:

$$F = kX$$

$$\left|\frac{X}{F}\right| = \frac{1}{k} \tag{7.9}$$

For frequencies much higher than the resonance frequency, $(\omega>>\omega_n)$, we have mass-like behavior:

$$F = m(acceleration)$$

$$F = m(-X\omega^2)$$

$$\left|\frac{X}{F}\right| = \frac{1}{\omega^2 m} \tag{7.10}$$

At resonance, $(\omega=\omega_n)$, the vibration amplitude grows very large—infinite in this undamped model. For real systems, the vibration amplitude at resonance is large, but is limited by damping.

7.3.4 Amplitude at resonance

For behavior close to the resonant frequency, the deflection of the system is limited only by damping forces, which dissipate vibration energy. A common damping model is *viscous damping*, where the damping force is proportional to, and in phase with, velocity.

$$F_D = C(velocity) \tag{7.11}$$

where C = Viscous damping coefficient

Often, the damping coefficient is expressed in terms of the viscous damping factor, ζ:

$$\zeta = \frac{C}{2\sqrt{km}} \tag{7.12}$$

For this case, the amplitude at resonance is given by:

$$F_D = C(velocity)$$

$$\left|F_D\right| = C(X\omega) = 2\zeta\sqrt{km}(X\omega) = 2\zeta k\sqrt{m/k}(X\omega)$$

$$\left|\frac{X}{F_D}\right| = \frac{1}{2\zeta k(\omega/\omega_n)} \tag{7.13}$$

at resonance $\omega = \omega_n$

$$\left|\frac{X}{F_D}\right|_{\omega=\omega_n} = \frac{1}{2\zeta k}$$

Viscous damping is generally assumed for suspension shock absorbers where a typical value for damping factor, C, is approximately *C=2 Nsec/mm (11 lb sec/in)*.

A second type of damping is *structural damping*, in which the damping force is proportional to deflection, X, but in phase with velocity. The structural damping factor, η, describes the damping force as a fraction of force through the stiffness element:

$$\left|F_D\right| = (\eta k X) \tag{7.14}$$

As with viscous damping, the damping force determines the amplitude at resonance:

$$\left|\frac{X}{F_D}\right|_{\omega=\omega_n} = \frac{1}{\eta k} \tag{7.15}$$

Structural damping is typically used to describe the behavior of metal structures. While structural damping for the base metal is typically quite low *(0.0001<η<0.001)*, significant damping occurs in built-up metal structures at the welded joints. For a spot-welded automobile body, the damping factor is on the order of *(0.03<η<0.1)* [7].

(The same mathematical form is also used to characterize damping behavior of elastomeric mounts [8]. In that case the factor η is called the loss factor, having a typical range from $0.05<\eta<0.2$.)

Comparing Equations 7.13 and 7.15, note the relation of viscous damping and structural damping at the resonant frequency:

$$\eta = 2\zeta \tag{7.16}$$

For either viscous or structural damping, the damping coefficient may be estimated by the bandwidth at the half-power amplitude, as shown in Figure 7.12b:

$$\eta = \frac{\Delta\omega}{\omega_n} \tag{7.17}$$

where $\Delta\omega$ is the bandwidth measured at the half-power amplitude [$\sqrt{2}$(amplitude at resonance)], Figure 7.12b.

Example: Powertrain vibration amplitude at resonance

A powertrain has a mass of *100 kg*. The engine mount system constrains motion to the vertical. The combined engine mount vertical stiffness is *600 N/mm* with a damping ratio of $\eta=0.1$. An oscillating force of *500 N* is applied at the resonance frequency of the system. What is the deflection amplitude?

$$\left.\left|\frac{X}{F_D}\right|\right|_{\omega=\omega_n} = \frac{1}{\eta k}$$

$$\left.\left|\frac{X}{F_D}\right|\right|_{\omega=\omega_n} = \frac{1}{(.1)(600 N/mm)(1000 mm/m)} = 0.0167 \times 10^{-3} m/N$$

$$X = 500 N(0.0167 \times 10^{-3} m/N) = 8.33 \times 10^{-3} m$$

7.3.5 Transfer function as log-log plot

Because the source-path-receiver model, Equation 7.1b, involves multiplying frequency characteristics, it is helpful to display $F(\omega)$, $T(\omega)$, $P(\omega)$, and $X(\omega)$ as log-log graphs. When displayed in this way, the multiplication in Equation 7.1b is visualized by adding the y values of the response plots. For example, the response of a SDOF system, Equation 7.6, is plotted on logarithmic axes in Figure 7.13.

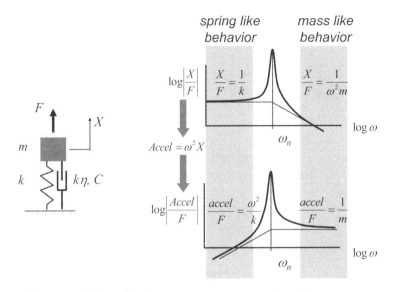

Figure 7.13 Log displacement output vs. acceleration output.

7.4 SDOF Models of Vehicle Vibration Systems

With the background of Section 7.3, we can now look at some specific vehicle systems using the model of Equation 7.1b.

$$F(\omega)\left[\left(\frac{F_T(\omega)}{F(\omega)}\right)\left(\frac{X(\omega)}{F_T(\omega)}\right)\right] = X(\omega)$$

$$F(\omega)[T(\omega)P(\omega)] = X(\omega)$$

$$\uparrow \qquad \uparrow \qquad \uparrow$$

Source   Path   Sensed by Receiver

(7.1b) repeated

In each case, we are interested in how the forces being applied to the body structure, F_T, vary with frequency, Figure 7.14. Understanding this, we can then decide where the structural resonances of the body should be placed.

7.4.1 Powertrain path: Reciprocating unbalance

The first system we will look at is the Powertrain path, Table 7.2, Figure 7.15.

Table 7.2 Powertrain path vibration system.

Source F(ω)	Isolator T(ω)	Force into Body F_T(ω)	Body Structure Transfer Function P(ω)	Body Deflection X(ω)
Powertrain unbalance force	Mounted powertrain	Force through engine mounts	Body structure	Deflection at seat, steering column

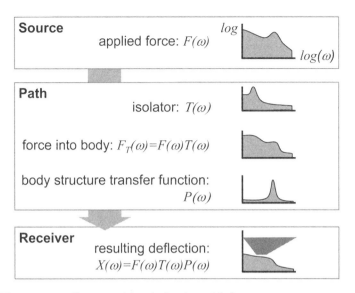

Figure 7.14 Characterizing behavior with frequency response.

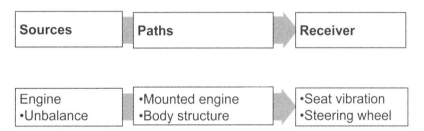

Figure 7.15 Powertrain-driven vibration.

The reciprocating elements of the engine—piston and connecting rod—impose forces on the crankshaft, Figure 7.16. For each cylinder, the oscillating forces are given by [9]:

$$f(t) = \left[mr\Omega^2 \frac{r}{l} \right] \sin(2\Omega t)$$

$$f(t) = \quad F \quad \sin(\omega t)$$

(7.18)

where:

$f(t)$ = Oscillating force applied to the crankshaft

m = Reciprocating mass

r = Crank shaft offset shown in Figure 7.16

l = Connecting rod length

Ω = Engine speed in rad/s [$\Omega = N$ *(rpm) (2π rad/rev)(1 min/60 sec)*]

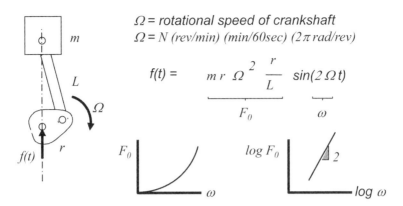

Figure 7.16 Unbalance forcing function for single cylinder engine.

Note that this oscillating force occurs at two times the engine speed due to the kinematics of the piston motion, and that the amplitude increases with the square of the engine speed. For engines with multiple cylinder configurations—L4, V6, V8—the unbalance forces of the individual cylinders combine to the resultant force or moment shown in Figure 7.17 [9] for the most common configurations.

	In line 4	V6	V8
	$F_{VERTICAL}$ (planar crankshaft)	60° $M_{ROTATING}$ (even 120° crankshaft)	90° (even 90° crankshaft)
Excitation amplitude (2 x engine speed)	$F_{VERTICAL} = 4mr\dfrac{r}{l}\Omega^2$	$M_{ROT} = \dfrac{3}{2}mr\dfrac{r}{l}\Omega^2 a$ cylinder spacing a	None
Balance strategy	$F_{VERTICAL}$ may be eliminated with dual counter rotating balance shafts at 2 x engine speed	crankshaft counter weights balance the primary rotating couple leaving the above moment	crankshaft counter weights balance the primary rotating couple

Figure 7.17 Unbalance forcing function for multiple cylinder engines.

With the source-path-receiver model shown in Figure 7.15, we have engine unbalance forces applied to the powertrain. The path into the body is through the mounted powertrain acting as an isolator. For body design, we are interested in the force transmitted through the engine mounts into the body structure. The mounted powertrain is a SDOF system, Figure 7.18, with frequency response given by Equation 7.4. The transmitted force though the engine mounts is then:

$$F_T = kX$$

$$\frac{X}{F} = \frac{1/k}{1 - \left(\dfrac{\omega}{\omega_n}\right)^2}$$

$$\frac{Xk}{F} = \frac{1}{1 - \left(\dfrac{\omega}{\omega_n}\right)^2} \qquad (7.19)$$

$$\frac{F_T}{F} = \frac{1}{1 - \left(\dfrac{\omega}{\omega_n}\right)^2} = T(\omega)$$

where:

F_T = Force transmitted to body through engine mounts

F = Force applied to powertrain

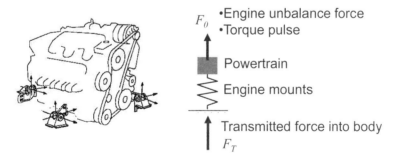

Figure 7.18 First order model for powertrain on rigid base.

In this expression, F is the magnitude of the unbalance force, and F_T is the force applied to the body structure through the engine mounts. When the magnitude of (F_T/F) is greater than one, the engine mount system is amplifying the unbalance forces; when it is less than one, it is reducing—isolating—the unbalance forces. This isolation begins to occur when:

$$\frac{F_T}{F} = abs\left[\frac{1}{1-\left(\dfrac{\omega}{\omega_n}\right)^2}\right] < 1$$

$$\pm 1 < 1 - \left(\frac{\omega}{\omega_n}\right)^2$$

$$2 < \left(\frac{\omega}{\omega_n}\right)^2 \qquad\qquad (7.20)$$

$$\sqrt{2} < \frac{\omega}{\omega_n}$$

$$\omega > \omega_n \sqrt{2}$$

Example: Isolation frequency of a mounted powertrain

A four-cylinder, automatic-transmission powertrain has a mass of *100 kg*. The engine mount system constrains motion to the vertical. The combined engine mount vertical stiffness is *600 N/mm*. For an evaluation, the mounted powertrain is placed on a bed plate (ground) and a sinusoidal vertical force is applied to the center of mass.

a) Determine the bounce natural frequency; in the example of Section 7.3 we found that

$$\omega_n = 77.46 \, rad/\sec$$
$$f_n = 38.73 \, rad/\sec(cycle/2\pi rad) = 12.33 Hz$$

b) At what frequency does isolation of unbalance forces begin?

From Equation 7.20:

$$\omega > \omega_n \sqrt{2}$$
$$\omega > 77.46\sqrt{2}$$
$$\omega > 109.54 \, rad/s$$
$$f_n = 17.43 Hz$$

c) What is the engine speed at which isolation begins?

The unbalance forces occur at two times the engine speed, Equation 7.18:

$$\omega = 2\Omega$$
$$\Omega = (109.54 \, rad/s)/2 = 54.77 \, rad/s$$
$$or$$
$$N = 54.77 \, rad/s(1rev/2\pi rad)(60s/\min) = 523 RPM$$

The equation above is for an undamped system. If we consider engine mounts with loss factor η, we replace the stiffness k with $k^*=k+i\eta k$ in Equation 7.4. (See the note at end of this chapter for the use of complex numbers in solving forced vibration problems.) The force transmitted into the body through a mount with damping is:

$$\left|\frac{F_T}{F}\right| = \frac{\sqrt{1+\eta^2}}{\sqrt{\left(1-\left(\frac{\omega}{\omega_n}\right)^2\right)^2 + \eta^2}} = T(\omega) \tag{7.21}$$

where:

F_T = Force transmitted to body through engine mounts

F = Force applied to powertrain

η = Mount structural damping coefficient

As with the undamped powertrain model, isolation begins when $\omega = \omega_n\sqrt{2}$. The peak force transmitted at resonance $\omega = \omega_n$ is reduced as mount damping increases. However, for the isolation region $\omega > \omega_n\sqrt{2}$, as damping increases the force transmitted increases, Figure 7.19. This *control vs. isolation* trade-off requires adjusting damping to balance controlling the amplitude at resonance against the level of vibration isolation achieved.

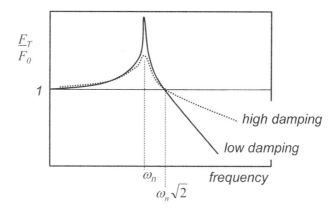

Figure 7.19 Force transmitted into body: Powertrain on rigid base.

The behavior of the powertrain vibratory system is summarized in Figure 7.20. The unbalance force, $F(\omega)$, applied to the powertrain increases as the square of the frequency (slope of +2 on log-log plot). The mounted powertrain acts as an isolator, $T(\omega)$, above ($\omega > \omega_n\sqrt{2}$). Multiplying these two functions yields the force transmitted to the body through the engine mounts, F_T. During design of the body structure, an objective is to position the primary structure resonances—both bending and torsion—in the isolation region.

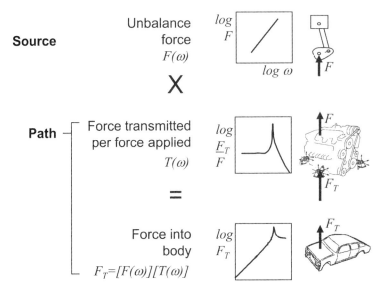

Figure 7.20 Transfer function model of powertrain using log-log plots.

7.4.2 Suspension path: Load at spindle

An important vibration path into the body is through the suspension elements, primarily through the shock absorber and ride spring, Figure 7.21. In general, the suspension system acts as an isolator for vibration energy at frequencies above the wheel hop resonance. The source-path-receiver model for this vibration system is shown in Table 7.3.

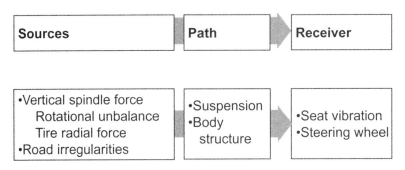

Figure 7.21 Suspension-driven vibration.

Table 7.3 Suspension path vibration system: Load at spindle

Source $F(\omega)$	Isolator $T(\omega)$	Force into Body $F_T(\omega)$	Body Structure Transfer Function $P(\omega)$	Body Deflection $X(\omega)$
Force at suspension spindle	Suspension	Force through shock absorber and ride spring	Body structure	Deflection at seat, steering column

For the purpose of investigating the loads into the body structure, we can model the suspension as a SDOF system, Figure 7.22 [10]. Here the spring is the parallel combination of the ride spring of the suspension and the radial rate of the tire, as measured at the spindle. The mass consists of all elements of the suspension which participate in the vertical motion of the spindle. This mass is primarily the wheel, knuckle, brakes, and the outer portion of the control arms.

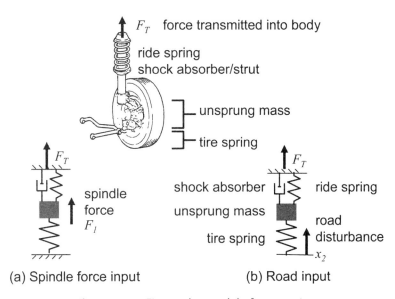

Figure 7.22 First-order model of suspension.

As shown in Figure 7.22a this SDOF may be excited by a load at the spindle. When a wheel unbalance is present, a centrifugal force is applied to the suspension spindle. Generally we are interested in the vertical component of this rotating force, as this is the direction the suspension is most compliant. Figure 7.23a shows this case, with the resulting applied force given by:

$$f(t) = mr\omega^2 \sin \omega t \tag{7.22}$$

where:

$f(t)$ = Vertical force at the spindle

m = Equivalent unbalance mass at wheel rim

r = Wheel rim radius

ω = Rotation speed of wheel [$V = \omega R$, V: vehicle speed, R: tire rolling radius]

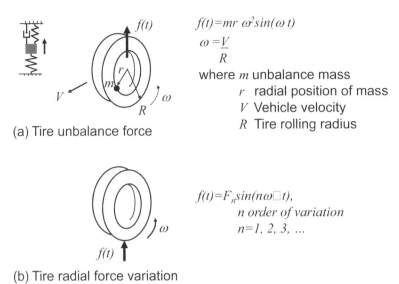

$$f(t)=mr\ \omega^2 sin(\omega t)$$

$$\omega =\frac{V}{R}$$

where m unbalance mass
r radial position of mass
V Vehicle velocity
R Tire rolling radius

(a) Tire unbalance force

$$f(t)=F_n sin(n\omega t),$$

n order of variation
$n=1, 2, 3, ...$

(b) Tire radial force variation

Figure 7.23 Suspension vibration sources.

Example: Suspension unbalance excitation

A suspension with rolling radius $R=300\,mm$ has an unbalance of *1 oz* (*28.35 g*) at the tire rim of radius $r=170\,mm$. The vehicle is traveling at *70 mph*. What is the unbalance force frequency and magnitude?

The frequency is given by:

$$\omega =\frac{V}{R}=\frac{70\,mi/hr(1.609\,km/mi)(1000\,m/km)(1hr/3600\sec)}{300\,mm(1m/1000mm)}=104.29\,rad/\sec$$

$$f=\frac{\omega}{2\pi}=16.6\,Hz$$

At $\omega=104.29$ rad/s, the force amplitude is given by:

$$F= mr\omega^2 = (0.02835kg)(0.170m)(104.29\,rad/s)^2 = 52.42N$$

A second source for a force applied to the spindle is radial force variation of the tire. Imagine a tire on a test fixture where the spindle is at a fixed distance from an ideally smooth surface, Figure 7.23b. As the tire rolls, the measured vertical force is not constant as expected. Rather we would see some oscillation about the mean force due to imperfections in the tire material and construction. This variation generally has constant amplitude independent of speed, with frequency expressed as multiples (orders) of the rotational speed:

$$f(t) = F_n \sin n\omega t \qquad (7.23)$$

where:
n = Vibration order, n=1, 2, 3, . . .

all other variables are the same as Equation 7.22

Under either an unbalance or a radial tire force at the spindle, we may find the resulting spindle deflection for a unit force at the spindle using a free body of the unsprung mass, Figure 7.24 (the use of complex number notation is covered in the note at the end of this chapter):

$$\sum forces\ on\ unsprung\ mass = m(acceleration)$$

$$-X_1 k_1 - X_1 k_2 - iX_1 C\omega + F = m(-X_1\omega^2)$$

$$\frac{X_1}{F} = \frac{1}{k_1 + k_2 - m\omega^2 + iC\omega}$$

$$\frac{X_1}{F} = \frac{\left(\dfrac{1}{k_1 + k_2}\right)}{\left(1 - \left(\dfrac{\omega}{\omega_n}\right)^2\right) + i\left(\dfrac{C\omega}{k_1 + k_2}\right)}$$

$$\left|\frac{X_1}{F}\right| = \frac{\left(\dfrac{1}{k_1 + k_2}\right)}{\sqrt{\left(1 - \left(\dfrac{\omega}{\omega_n}\right)^2\right)^2 + \left(\dfrac{C\omega}{k_1 + k_2}\right)^2}} \qquad (7.24)$$

where:

F = Force applied to spindle

k_2 = Ride spring rate

k_1 = Tire spring radial rate

m = Unsprung mass

C = Shock absorber viscous damping factor (typical value is 1000–2000 Ns/m)

ω_n = Wheel hop frequency, $\omega_n = \sqrt{\dfrac{k_1 + k_2}{m}}$

$i = \sqrt{-1}$

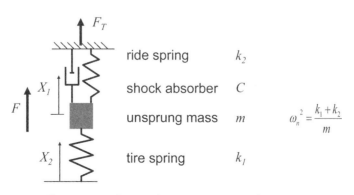

Figure 7.24 First-order suspension analysis.

Example: Wheel hop frequency

For a McPherson strut front suspension as shown in Figure 7.22, a typical tire radial rate is k_1=175 N/mm, ride rate k_2=17.5 N/mm, and unsprung mass m_1=40 kg. The wheel hop frequency for this case is

$$\omega_n = \sqrt{\frac{k_1+k_2}{m_1}} = \sqrt{\frac{(175+17.5)N/mm(1000mm/m)}{40kg}} = 69.37 \, rad/sec$$

$$f_n = \frac{\omega_n}{2\pi} = \frac{69.37 \, rad/sec}{2\pi} = 11.04 Hz$$

This resonance would be excited by a wheel unbalance at a vehicle speed of (rolling radius of R=300 mm):

$$V=\omega R=(69.37 rad/s)(0.3m)(1km/1000m)(3600s/hr)=74.9kph \ (46.8mph)$$

The suspension acts as an isolator for force transmitted into the body. The force transmitted through the ride spring and shock absorber into the body is $F_T=X_1(k_2+i\omega C)$. Substituting Equation 7.24 into this expression gives:

$$\left|\frac{F_T}{F}\right| = \frac{\left(\dfrac{k_2}{k_1+k_2}\right)\sqrt{1+\left(\dfrac{C\omega}{k_2}\right)^2}}{\sqrt{\left(1-\left(\dfrac{\omega}{\omega_n}\right)^2\right)^2+\left(\dfrac{C\omega}{k_1+k_2}\right)^2}} = |T(\omega)| \tag{7.25}$$

where:

F_T = Force transmitted to body through shock absorber and ride spring

other variables the same as Equation 7.24

Example: Wheel hop frequency (continued)

For the suspension in the previous example moving at the vehicle speed corresponding to wheel hop ($\omega=\omega_n$), the force transmitted to the body per unit force at the spindle is given by Equation 7.25. For this case the parameters are:

$$k_1 = 175000N/m, \quad k_2 = 17500N/m, \quad m_1 = 40kg, \quad C = 1500Ns/m, \quad \omega_n = 69.37 r/s$$

Substituting these values into Equation 7.25 gives:

$$\left|\frac{F_T}{F}\right| = \frac{\left(\dfrac{17500N/m}{175000N/m+17500N/m}\right)\sqrt{1+\left(\dfrac{(1500Ns/m)(69.37r/s)}{17500N/m}\right)^2}}{\sqrt{\left(1-(1)^2\right)^2+\left(\dfrac{(1500Ns/m)(69.37r/s)}{175000N/m+17500N/m}\right)^2}} = |T(\omega)|$$

$$\left|\frac{F_T}{F}\right| \cong 1$$

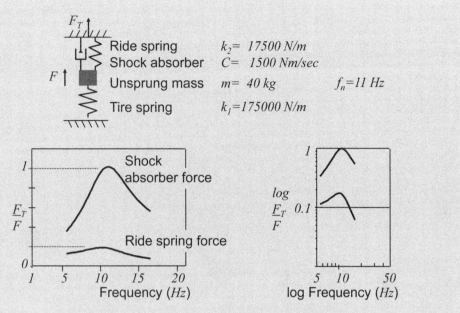

Figure 7.25 Force into body due to unit force at spindle.

This is a peak for the suspension transfer function, and shows that a force applied to the spindle will be passed unattenuated into the body at the wheel hop frequency. Above this frequency, the suspension will attenuate spindle forces (See Figure 7.25 for frequency response in the range *5 Hz<f<15 Hz*. Note that the primary path is though the shock absorber.)

The behavior of this vibratory system is summarized in Figure 7.26. The tire variation spindle force is constant with frequency, while unbalance force increases as the square of the frequency (slope of +2 on the log-log plot). Multiplying these two functions yields the force transmitted to the body through the ride spring/ shock absorber, F_T. Above the wheel hop resonance, the force into the body is less than the force applied to the spindle, and isolation occurs. The primary path for this transmitted force is the shock absorber. During design of the body structure, an objective is to place the primary structure resonances—both bending and torsion— above the wheel hop frequency in the isolation region.

It should be noted that the SDOF suspension model of Figure 7.24 is attached to ground at the ride spring and shock absorber rather than attached to the vehicle rigid-body mass (often called a *quarter car model*). Therefore the suspension model used here does not exhibit the vehicle ride motions occurring at approximately *1 Hz* and should be applied only for frequencies greater than *5 Hz*. Our model does predict the forces transmitted to the body in the region of wheel hop frequency and above. As these forces are much larger than those occurring at the ride frequencies, our SDOF model is sufficient to understand the isolation behavior of the suspension with respect to the body.

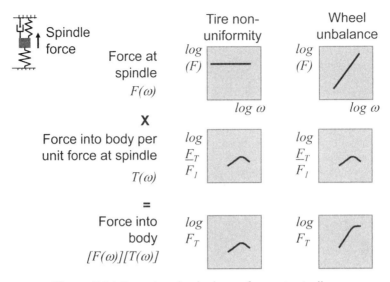

Figure 7.26 Force into body due to force at spindle.

7.4.3 Suspension path: Deflection at tire patch

A second source of vibration flowing through the suspension into the body structure is due to road deflections at the tire patch, Table 7.4.

Table 7.4 Suspension path vibration system: Road input

Source $F(\omega)$	Isolator $T(\omega)$	Force into Body $F_T(\omega)$	Body Structure Transfer Function $P(\omega)$	Body Deflection $X(\omega)$
Road deflection at tire patch	Suspension	Force through shock absorber and ride spring	Body structure	Deflection at seat, steering column

Dynamic characteristics for typical roads have been measured and characterized by the Power Spectral Density (PSD) of the displacement, Equation 7.26 [10]:

$$G = G_0 \frac{\left[1 + \left(\dfrac{v_0}{v}\right)^2\right]}{(2\pi v)^2} \tag{7.26}$$

where:

G = Power Spectral Density [$(m^2)/(cycle/m)$]

v = Wave number ($cycle/m$)

G_0 = 1.35×10^{-4} Rough Roads, 1.35×10^{-5} Smooth Roads (m^2)

v_0 = 0.015 Bituminous Roads, 0.0061 Concrete Roads ($cycle/m$)

$$f\,(Hz) = \upsilon\,(cycle/m)\,\,V\,(m/sec) \tag{7.27}$$

where:

υ = Spatial frequency (*cycle/m*)

V = Vehicle speed (*m/sec*)

f = Temporal frequency (*Hz*)

The PSD is a means to characterize a random signal, and may be visualized as the mean-square value of the signal as filtered through a *1 Hz* bandwidth filter at a center frequency f. The center frequency is varied over the range of interest to arrive at the full spectrum. For maintained paved roads, Equation 7.26 predicts that the deflection amplitude rapidly diminishes with increasing frequency, Figure 7.27.

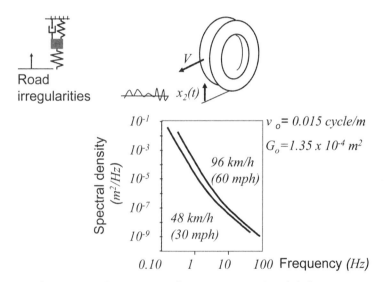

Figure 7.27 Suspension vibration source: Road deflections.

Again, a SDOF model of the suspension may be used to characterize the force transmitted into the body, Figure 7.24. In this model, the input is the displacement at the tire patch, X_2, which may be characterized by the PSD, G, of Equation 7.26. The amplitude of the spindle displacement, X_1, for a unit displacement at the tire patch, X_2, is determined by a free body of the unsprung mass:

$$\sum forces\ on\ unsprung = m(acceleration)$$

$$-(X_1 - X_2)k_1 - X_1k_2 - iX_1C\omega = m(-X_1\omega^2)$$

$$X_2k_1 = (k_1 + k_2 + iC\omega - m\omega^2)X_1$$

$$\frac{X_1}{X_2} = \frac{k_1}{(k_1 + k_2 + iC\omega - m\omega^2)}$$

$$\left|\frac{X_1}{X_2}\right| = \frac{\left(\dfrac{k_1}{k_1 + k_2}\right)}{\sqrt{\left(1 - \left(\dfrac{\omega}{\omega_n}\right)^2\right)^2 + \left(\dfrac{C\omega}{k_1 + k_2}\right)^2}} \tag{7.28}$$

where:

X_1 = Deflection of the unsprung mass

X_2 = Imposed deflection at tire patch

k_2 = Ride spring rate

k_1 = Tire spring radial rate

m = Unsprung mass

C = Shock absorber viscous damping factor

The force transmitted through ride spring and shock absorber into body is $F_T = X_1(k_2 + i\omega C)$ and substituting into Equation 7.28 yields:

$$\left|\frac{F_T}{X_2}\right| = \frac{\left(\dfrac{k_1 k_2}{k_1 + k_2}\right)\sqrt{1 + \left(\dfrac{C\omega}{k_2}\right)^2}}{\sqrt{\left(1 - \left(\dfrac{\omega}{\omega_n}\right)^2\right)^2 + \left(\dfrac{C\omega}{k_1 + k_2}\right)^2}} = |T(\omega)| \tag{7.29}$$

where variables are those of Equation 7.28

The behavior of this vibratory system is summarized in Figure 7.28. Tire patch deflections, $F(\omega)$, rapidly diminish with increasing frequency (slope of –4 on log-log plot). Again the suspension acts as an isolator, $T(\omega)$, and the product of these two functions, $F(\omega)T(\omega)$, yields the force transmitted to the body through the ride spring/shock absorber, F_T. During body structure design, we take advantage of the isolation properties of the suspension by positioning the primary structure resonances—both bending and torsion—above the wheel hop frequency.

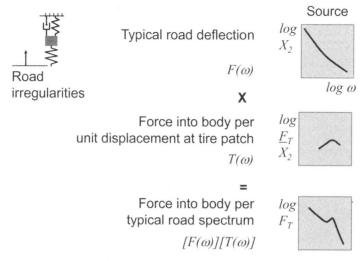

Figure 7.28 Force into body due to road disturbance.

7.5 Strategies for Design for Vibration

We have now characterized some of the major vibration systems for the vehicle. Our design objective is to minimize the source vibration energy flowing to the receiver with undesirable results. Three of the most important strategies to meet this objective are 1) Reduce amplitude of the source; 2) Block the flow of energy using isolators in the path; and 3) Detune resonances in the system [11]. Table 7.5 provides some examples for each of these strategies, and Figure 7.29 illustrates the strategies as log-log frequency response plots.

Table 7.5 Design for vibration strategies.

Strategy	Examples
Reduce amplitude of the vibration source	*Powertrain*
	• Minimize reciprocating mass in engine
	• Add balance shafts to in-line 4 cylinder engine
	Suspension
	• Balance tires
	• High quality tires with low radial force variation
	• Minimize shock absorber forces using a linkage ratio~1
Block the flow of vibration energy using isolators in path	• Mounted powertrain at isolator
	• Suspension as isolator
	• Rubber bushings in chassis links at acoustic frequencies (treated in Section 7.8)
Detune resonances in the system	Positioning body primary bending and torsion resonances to avoid peaks in vibration sources and to take advantage of isolation of mounted powertrain and suspension

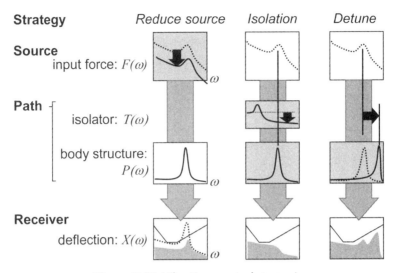

Figure 7.29 Vibration control strategies.

7.5.1 Mode map of vehicle vibratory systems

One of the most important strategies in setting body structure requirements is that of detuning resonances of the body from sources and responders. If this can be achieved during the early design phases, many vibration problems can be avoided, and those remaining may be treated by minor tuning. Using the first-order vibration models of Section 7.4, we can summarize the source, isolator, and responder characteristics for a typical vehicle on a *mode map*, Figure 7.30. This map clearly shows the desirable structural resonance band of *22–25 Hz*. This band falls in the isolation region of the suspension and mounted powertrain, and yet below resonances of downstream paths.

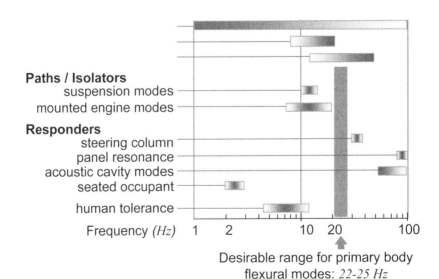

Figure 7.30 Noise and vibration mode map.

7.6 Body Structure Vibration Testing

Often the vibration behavior of an automobile body is evaluated by test. The result of a vibration test is the transfer function, $P(\omega)$, and the deflected shape (mode shape) for each of the resonances. Figure 7.31 shows a typical test set-up [12]. The body shell is suspended on very soft springs such that the rigid body modes occur at low frequencies (<3 Hz). This ensures that the rigid modes do not interfere with the flexural resonances to be measured. Often, these support springs are either inflated inner tubes or elastic cords. An electromagnetic or hydraulic shaker is attached to the structure. The forcing location is determined by two criteria: first, the location will excite major modes of vibration (i.e., is not near a nodal point of any important mode). Second, the location is locally stiff (the vibration energy will be directed to the overall mode, not to locally flexing the structure). A typical forcing location is at a front bumper attachment. The shaker is connected to the body through a force transducer and a quill—a very-small-diameter shaft which transmits axial force but does not constrain the body with side loads. An accelerometer is attached to the body at the shaker attachment location to measure the driving point frequency response.

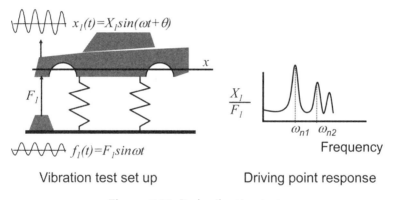

Figure 7.31 Body vibration test.

In practice, a randomly varying force is applied, and the Fourier Transform is taken of the input and output signals resulting in the frequency response. However for visualization purposes, view the input as a sine wave force at a specific frequency. At each frequency, the accelerometer amplitude is plotted while holding the force amplitude fixed. After the frequency has been incremented over a range, the result is a plot of *driving point* response. The peaks on this plot are the resulting body structure resonances.

Now holding the forcing frequency fixed at one of the resonance frequencies, Figure 7.32, the accelerometer location may be moved about the structure and the amplitude recorded for each location. Plotting the amplitude at each location results in a plot of the deflected shape for that resonance—the mode shape, ϕ. The mode shape may be considered a snap-shot taken at the moment when there is maximum deflection throughout the structure. This shape will contain a number of locations with no deflection—nodal points, and a number of locations with greatest deflection—anti-nodes. Note that when a lightly damped structure like the automobile body is excited at a resonance frequency, the motion of all points on the structure are either moving in-phase or $180°$ out-of-phase with the driving point.

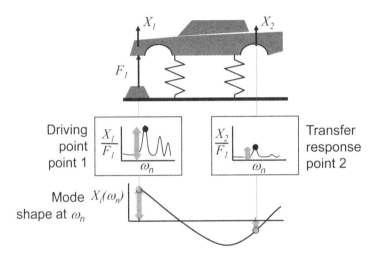

Figure 7.32 Mode shape at resonance.

7.7 Modeling the Body Structure Resonant Behavior

The body structure is a continuous structure with an infinite number of resonant modes, Figure 7.33. The structure's modal density—number of modes occurring in a fixed bandwidth—increases with increasing frequency. For lower frequencies from *10–150 Hz*, individual modes may be characterized using a modal model which we will discuss discussed below. At high frequencies above *1000 Hz*, the modal density is high, with the specific resonant frequencies determined by very small structural details. As these details are subject to manufacturing variability, the precise frequency for the resonance is random. In this high-frequency region, a statistical approach describing the vibration amplitude envelope rather than individual resonances is useful. (The reader is referred to *Machinery Noise and Diagnostics* by R. H. Lyons [13] for more on this approach.) Here we will describe the modal model method for characterizing the lower-frequency flexural modes of the body.

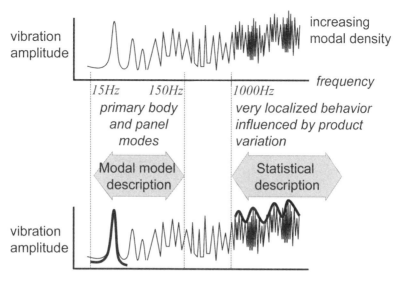

Figure 7.33 Appropriate body model depends on frequency range.

7.7.1 Modal model

The vibration behavior of the primary modes of vibration may be modeled using a *modal model*. In this technique, the behavior around a structural resonance, ω_n, is viewed as a SDOF system.

Consider a body structure of mass, m, with a measured resonant frequency, ω_n, and mode shape ϕ, Figure 7.34a. We can model the behavior of this system with the SDOF system shown in Figure 7.34b. To simulate a load $F_{PHYSICAL}$ applied to the real body structure, we apply a force, F_{MODAL}, to the modal model such that

$$F_{MODAL} = F_{PHYSICAL}\phi_{INPUT} \tag{7.30}$$

where:

$F_{PHYSICAL}$ = Force applied to the physical body structure at the input location

F_{MODAL} = Force applied to the modal model

ϕ_{INPUT} = Influence coefficient at the output (determined from mode shape at resonance at ω_n)

For example, if the physical force is applied near a nodal point *(very small ϕ_{INPUT})*, the modal force will be very small. If the force is applied near an anti-node *(large ϕ_{INPUT})*, then the modal force is large.

In response to this applied modal force, the modal mass oscillates at a deflection amplitude X_{MODAL}. To arrive at the corresponding deflection for the real structure,

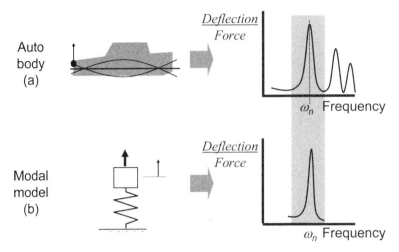

Figure 7.34 First-order body model using a SDOF modal model.

we multiply this modal deflection by the influence coefficient at the output location:

$$X_{PHYSICAL} = X_{MODAL}\phi_{OUTPUT} \qquad (7.31)$$

where:

$X_{PHYSICAL}$ = Deflection of the physical body structure at the output location

X_{MODAL} = Deflection of the modal model

ϕ_{OUPUT} = Influence coefficient at the output (determined from mode shape at resonance at ω_n)

Using the relationship for a SDOF transfer function, Equation 7.6, but with the addition of influence coefficients for the input location and output location, the following model results, Figure 7.35:

$$\frac{X_{OUTPUT}}{F_{INPUT}} = \frac{\phi_{OUTPUT}\phi_{INPUT} / k_{MODAL}}{\sqrt{\left(1-\left(\frac{\omega}{\omega_n}\right)^2\right)^2 + (\eta)^2}} \qquad (7.32)$$

where:

k_{MODAL} = Modal stiffness: $k-\omega_n^2$

ω_n = Model frequency

m_{MODAL} = Modal mass (often set equal to the mass of the body)

η = Modal damping

ϕ_{INPUT} = Input influence coefficient for mode at ω_n

ϕ_{OUTPUT} = Output influence coefficient for mode at ω_n

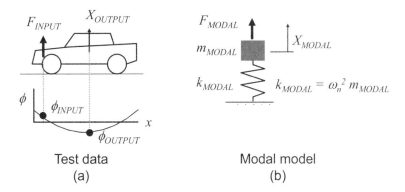

Test data
(a)

Modal model
(b)

Figure 7.35 Modal model of body.

This modal model represents the behavior in the region of the resonance. We may add the rigid body behavior to extend the model below the resonance region, Figure 7.36. The parameters for this model: ω_n, m_{MODAL}, k_{MODAL}, η, ϕ_{INPUT}, ϕ_{OUTPUT}, may be determined from experimental data.

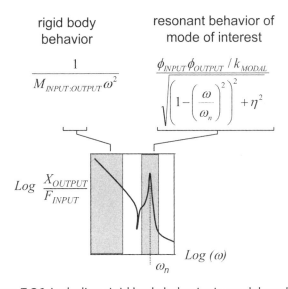

Figure 7.36 Including rigid body behavior in modal model.

A physical interpretation of Equation 7.32 is shown in Figure 7.37 [14]. The modal oscillator is connected to a lever of unit length. The physical force is applied at the lever position ϕ_{INPUT}, and the physical deflection is measured at position ϕ_{OUTPUT}.

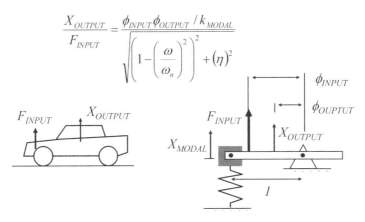

$$\frac{X_{OUTPUT}}{F_{INPUT}} = \frac{\phi_{INPUT}\phi_{OUTPUT}/k_{MODAL}}{\sqrt{\left[1-\left(\dfrac{\omega}{\omega_n}\right)^2\right]^2 + \left(\eta\right)^2}}$$

Figure 7.37 Lever analogy for modal model.

Example: Effect of mass placement on structural resonance

A typical design requirement is to place the primary body resonances in the *22–25 Hz* range. Often this involves design changes to increase the resonant frequency from a lower value. We can do this through increased body stiffness, but also by careful placement of subsystem masses. Consider the selection of battery location. Feasible locations include front corner, at the dash, and in the trunk. A modal model of the primary bending resonance may be used to assess these options.

Tests for the body shell without battery provide the following data for this modal model:

Primary bending resonance: f_n=47 Hz (ω_n=295.3 rad/sec)

Body shell mass: *M=250 kg (we set m_{MODAL} = M)*

Influence coefficients for bending resonance determined from mode shape at 47 Hz:

at front corner: $\phi_i = 0.90$

at dash: $\phi_i = -0.20$

at trunk floor: $\phi_i = 0.15$

From these data, we can calculate the modal stiffness:

$$k_{MODAL} = \omega^2 m_{MODAL} = (295.3 rad/sec)^2 250 kg = 21.8 \times 10^6 N/m$$

When we add the battery, the mass of the SDOF modal model is increased, and the new resonant frequency will be

$$f_n = \frac{1}{2\pi}\sqrt{\frac{k_{MODAL}}{m_{MODAL} + m_{EFFECTIVE}}} \qquad\qquad\text{(A)}$$

where $m_{EFFECTIVE}$ is the effective mass of the battery for this mode of vibration, Figure 7.38. We can determine the effective battery mass by considering the force which a battery of mass, *m*, applies to the car body when at location *i*:

$$F_i = m(acceleration\ at\ location\ i)$$

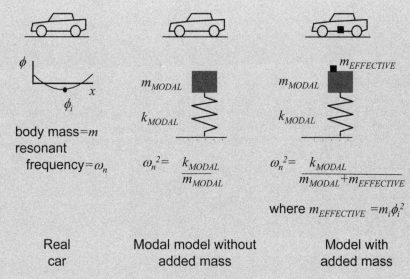

body mass=m
resonant
frequency=ω_n

$$\omega_n^2 = \frac{k_{MODAL}}{m_{MODAL}}$$

$$\omega_n^2 = \frac{k_{MODAL}}{m_{MODAL} + m_{EFFECTIVE}}$$

where $m_{EFFECTIVE} = m_i\phi_i^2$

| Real car | Modal model without added mass | Model with added mass |

Figure 7.38 Effect on resonant frequency of an added mass.

At the resonant frequency, the acceleration at the battery location is $(X_i\,\omega_n^2)$, where X_i is the deflection amplitude at location i, so

$$F_i - (X_i\,\omega_n^2)$$

Now consider the modal force and acceleration by substituting Equation 7.30 and 7.31 into the above:

$$\frac{F_{MODAL}}{\phi_i} = m(\phi_i X_{MODAL}\omega_n^2)$$

$$F_{MODAL} = m\phi_i^2(X_{MODAL}\omega_n^2)$$

$$F_{MODAL} = m\phi_i^2 \text{(modal acceleration)}$$

Inspection of this equation indicates that the effective battery mass for the modal model is

$$m_{EFFECTIVE} = m\phi_i^2$$

Substituting this expression for effective battery mass into Equation A we get an equation for natural frequency of the body structure with the battery mounted:

$$f_n = \frac{1}{2\pi}\sqrt{\frac{k_{MODAL}}{m_{MODAL} + \phi_i^2 m}}$$ (B)

Substituting the data given, and with a battery mass, $m=16.8\,kg$, we get the following estimates for the primary bending frequency for the battery at each location:

at front corner:

$$f_n = \frac{1}{2\pi}\sqrt{\frac{21.8 \times 10^6\,N/m}{250kg + (0.95)^2 16.8kg}} = 45.63\text{Hz}$$

at dash:
$$f_n = \frac{1}{2\pi}\sqrt{\frac{21.8\,x10^6\,N/m}{250kg+(-0.2)^2 16.8kg}} = 46.93\text{Hz}$$

at trunk floor:
$$f_n = \frac{1}{2\pi}\sqrt{\frac{21.8\,x10^6\,N/m}{250kg+(0.15)^2 16.8kg}} = 46.96\text{Hz}$$

This example demonstrates the large influence of component location on primary frequency, Figure 7.39.

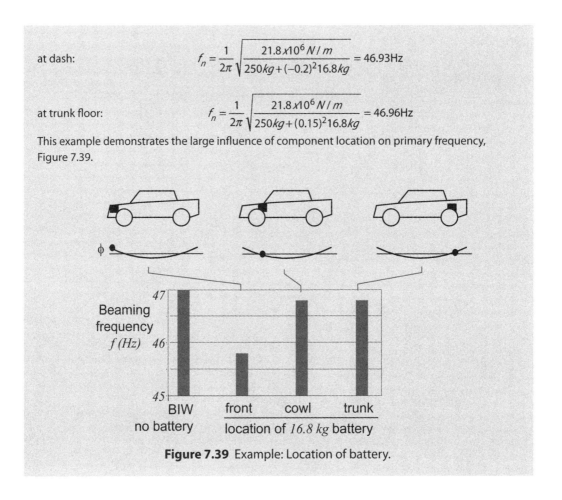

Figure 7.39 Example: Location of battery.

7.8 Vibration at Frequencies Above the Primary Structure Modes

We have been looking at the interaction of vibration sources with the primary body structure resonances in the region *18 Hz to 50 Hz*. At these frequencies, the vibration at the receiver is tactile. At higher frequencies—*50 Hz to 400 Hz*—the response of the body structure is more localized and the vibration is sensed by the receiver acoustically. In this section, we will look at a vibration system in this high-frequency region, Table 7.6, where structure-borne panel vibration excites the air cavity within the vehicle, Figure 7.40.

Table 7.6 Panel vibration system.

Source F(ω)	Body Structure Transfer Function P(ω)	Acoustic Deflection X(ω)
Body panel vibrations	Passenger compartment acoustic resonances	Interior sound pressure

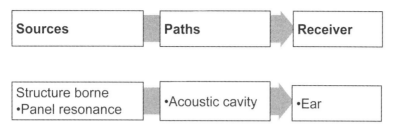

Figure 7.40 Structure- borne vibration.

7.8.1 Body panel vibration

We can model the vibration behavior of a body panel as a flat plate, Figure 7.41. The plate equation below relates the normal deflection of the plate, $w(x, y)$, the plate stiffness, D, and the applied loads, q.

$$\frac{\partial^4 w}{\partial x^4} + 2\frac{\partial^4 w}{\partial x^2 \partial y^2} + \frac{\partial^4 w}{\partial y^4} + \frac{q}{D} = 0 \tag{7.33}$$

where:

$w(x,y)$ = Normal deflection of the plate

$D = \dfrac{Et^3}{12(1-\mu^2)}$ Plate stiffness

$q(x,y)$ = Normal load per unit of area at location (x, y)

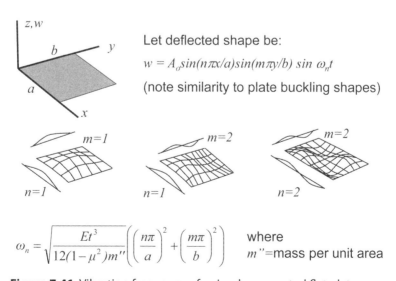

Let deflected shape be:

$w = A_o \sin(n\pi x/a)\sin(m\pi y/b) \sin \omega_n t$

(note similarity to plate buckling shapes)

$$\omega_n = \sqrt{\frac{Et^3}{12(1-\mu^2)m''}\left(\left(\frac{n\pi}{a}\right)^2 + \left(\frac{m\pi}{b}\right)^2\right)}$$

where
m'' = mass per unit area

Figure 7.41 Vibration frequency of a simply supported flat plate.

During vibration at a resonance, $\omega_{m,n}$, we assume the plate deformation is described by half sine waves in both the x and y direction:

$$w = A_{m,n}\left(\sin\frac{n\pi}{a}x\right)\left(\sin\frac{m\pi}{b}y\right)\sin\omega t \qquad (7.34)$$

where:

$A_{m,n}$ = Vibration amplitude

a = Plate length in the x direction

b = Plate width in the y direction

n = Number of half sine waves in the x direction, $n=1, 2,...$

m = Number of half sine waves in the y direction, $m=1, 2, ...$

ω = Vibration frequency (rad/s)

This assumed deformation is appropriate for simply supported boundary conditions. The normal load, q, is that of the inertia force of a small patch of plate:

$$q = m''\frac{\partial^2 w}{\partial t^2} \qquad (7.35)$$

where m'' = Plate mass per unit area

Substituting the deflected shape, Equation 7.34, and 7.35 into Equation 7.33 yields:

$$\left[\left(\frac{n\pi}{a}\right)^4 + 2\left(\frac{n\pi}{a}\right)^2\left(\frac{m\pi}{b}\right)^2 + \left(\frac{m\pi}{b}\right)^4 - \frac{\omega^2 m''}{D}\right]\left[A_{m,n}\left(\sin\frac{n\pi}{a}x\right)\left(\sin\frac{m\pi}{a}y\right)\sin\omega\right] = 0 \qquad (7.36)$$

To satisfy this equality, the first bracketed term must be zero at the resonant frequencies. Solving for the frequency yields

$$\omega_{m,n} = \pi^2\sqrt{\frac{D}{m''}}\left[\left(\frac{n}{a}\right)^2 + \left(\frac{m}{b}\right)^2\right] \qquad (7.37a)$$

where:

$\omega_{m,n}$ = Resonant frequency

other variables defined in Equations 7.33, 7.34, & 7.35

Rarely are automotive panels flat. Generally they are crown or ribbed to enhance structural performance. Figure 7.42 provides guidance in predicting the resonant frequency for these formed conditions. Addition of these formations to a panel will increase resonance frequency, as given by Equation 7.37b

$$\frac{f_{FORMED}}{f_{FLAT}} = C\sqrt{\frac{H_C}{t}} \qquad (7.37b)$$

where:

f_{FORMED} = Formed panel primary frequency

f_{FLAT} = Flat panel primary frequency given by Equation 7.37a with $m=n=1$

H_C = Crown height

t = Panel thickness

C = Constant ($C=1.25$ for crown, $C=1$ for ribbed panel)

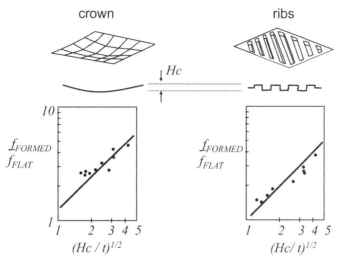

Figure 7.42 Experimental plate stiffening study.

Example: Vibration frequencies for floor pan

The floor pan at the rear foot well may be considered a flat panel of dimensions $a=500$ mm and $b=300$ mm. For a steel panel with thickness $t=1$ mm, the plate stiffness is:

$$D = \frac{Et^3}{12(1-\mu^2)} = \frac{(207000N/mm^2)(1mm)^3}{12(1-0.3^2)}$$

$$= 18956 Nmm(1m/1000mm) = 18.956 Nm$$

and the mass area density is:

$$m'' = \rho t = (7.83 \times 10^{-6} kg/mm^3)(1mm)$$

$$m'' = (7.83 \times 10^{-6} kg/mm^2)(1000mm/m)^2 = 7.83 kg/m^2$$

Using Equation 7.37 yields:

$$\omega_{m,n} = \pi^2 \sqrt{\frac{D}{m''}} \left[\left(\frac{n}{a}\right)^2 + \left(\frac{m}{b}\right)^2 \right] = \pi^2 \sqrt{\frac{18.956 Nm}{7.83 kg/m^2}} \left[\left(\frac{n}{0.5m}\right)^2 + \left(\frac{m}{0.3m}\right)^2 \right]$$

$$\omega_{m,n} = 15.356 \left[\left(\frac{n}{0.5}\right)^2 + \left(\frac{m}{0.3}\right)^2 \right] rad/\sec$$

for $m=1$, $n=1$:

$$\omega_{1,1} = 15.356 \left[\left(\frac{1}{0.5}\right)^2 + \left(\frac{1}{0.3}\right)^2 \right] rad/\sec = 232 rad/\sec (36.9 Hz)$$

for $m=2, n=1$:

$$\omega_{1,2} = 15.356\left[\left(\frac{1}{0.5}\right)^2 + \left(\frac{2}{0.3}\right)^2\right] rad/\sec = 744 rad/\sec(118.4 Hz)$$

If a crown panel of crown height $Hc=20\,mm$ is substituted for this flat panel, the frequency of the first mode, using Equation 7.37b, is increased to

$$\frac{f_{FORMED}}{f_{FLAT}} = 1.25\left(\frac{H_c}{t}\right)^{1/2} = 1.25\left(\frac{20mm}{1mm}\right)^{1/2} = 5.59$$

$$f_{FORMED} = 5.59(36.9\,Hz) = 206\,Hz$$

Automotive panels in practice are both crown and ribbed, Figure 7.43, largely due to this ability to increase panel resonant frequencies [15].

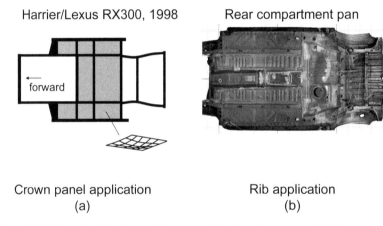

Harrier/Lexus RX300, 1998   Rear compartment pan

forward

Crown panel application   Rib application
(a)   (b)

Figure 7.43 Panel stiffening to increase vibration frequency. (Photo courtesy of A2Mac1.com, Automotive Benchmarking)

7.8.2 Acoustic cavity resonance

The closed air cavity of the passenger compartment can resonate with a standing acoustic wave similar to an organ pipe, Figure 7.44. In this case, the boundary conditions at either end are closed. For this condition the acoustic resonant frequency is given by:

$$f_n \lambda = c \tag{7.38}$$

where:

f_n = Resonant frequency (Hz)

λ = Wavelength

c = Speed of sound in air $(\sim 330\ m/sec)$

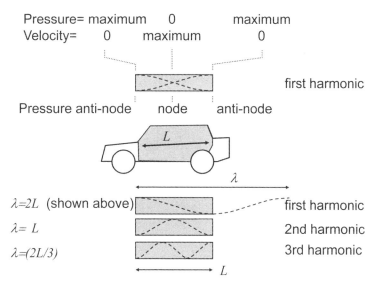

Figure 7.44 Cavity resonance.

The standing wavelength is

$$\lambda = \frac{2L}{n} \tag{7.39}$$

where:

   L = Cabin length

   n = Number of half cosine waves along cabin length, $n=1, 2, 3 \ldots$

Substituting Equation 7.39 into 7.38 yields:

$$f_n = c\left(\frac{n}{2L}\right) \tag{7.40}$$

At each resonance the *pressure* mode shape is given by the cosine function, $cos(\pi n x/L)$, where x is the length coordinate with origin at the front of the cavity. For example, at the first acoustic resonance, $n=1$, the pressure amplitude is maximum (anti-node) at either end of the passenger compartment and zero (node) at the midpoint. The pressure mode shape gives some notion of the sound level. For this case, with a midpoint node the sound pressure level at the driver's ear is low.

The air *velocity* mode shape is given by $sin(\pi n x/L)$. The velocity mode shape provides an indication of the sensitivity of the cavity mode to excitation by a panel. A panel located near the velocity anti-node will excite the cavity resonance to a high degree.

Example: Vibration modes of sedan interior cavity

A sedan has an interior cavity length of $L=3\,m$. Using Equation 7.40, the frequency for first acoustic resonance is:

$$f_n = c\left(\frac{n}{2L}\right) = (330\,m/s)\left(\frac{1}{2\cdot 3m}\right) = 55\,Hz$$

with pressure mode shape $cos(\pi x/L)$, and velocity mode shape $sin(\pi x/L)$.

The second acoustic resonance is at

$$f_n = 330\,m/s\left(\frac{2}{2\cdot 3m}\right) = 110\,Hz$$

with pressure mode shape $cos(\pi 2x/L)$ and velocity mode shape $sin(\pi 2x/L)$. Note that for this second mode, there are two pressure nodal points at $L/4$ and $3L/4$, and anti-nodes at the ends as well as at the midpoint.

The two examples above illustrate the source-path-receiver system in which panel vibration becomes a vibration source, exciting cavity resonances, causing sound pressure at the driver's ear, Figure 7.45. For these examples, the flat panel resonance at *118.4 Hz* will couple with the air cavity resonance at *110 Hz* resulting in high sound pressure levels at the driver's ear. By crowning the panel, the panel resonant frequency will increase and would become detuned from the acoustic resonance resulting in a reduced sound level.

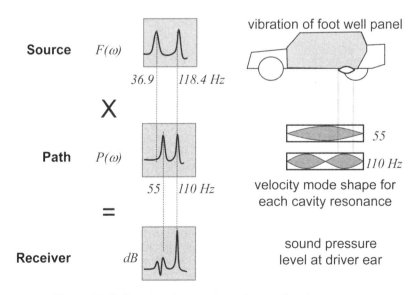

Figure 7.45 Panel and acoustic cavity as vibration system.

Table 7.7 Suspension element noise path vibration system.

Source $F(\omega)$	$T(\omega)$	$F_T(\omega)$	$P(\omega)$	$X(\omega)$
High-frequency chassis deflections	Chassis links with end bushings	Body panel vibrations	Passenger compartment acoustic resonances	Interior sound pressure

7.8.3 Vibration isolation through elastomeric elements

We can now expand the vibration system of the previous section by including a typical isolator in the path, Table 7.7 and Figure 7.46. High frequency deflections may occur in suspension elements due to road impacts. To isolate these higher-frequency vibrations, the suspension elements have elastomeric bushings at the body connections, Figure 7.47.

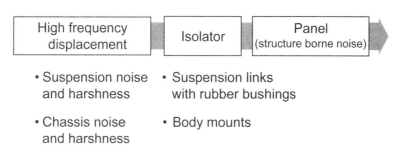

Figure 7.46 Role of isolators in higher-frequency vibration.

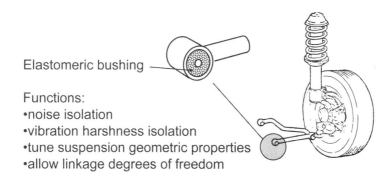

Figure 7.47 Lower control arm bushing.

To understand how these bushings help reduce vibration into the body, consider the model shown in Figure 7.48 [16]. A suspension lower control arm is shown. At either end of the arm, the connection is through an elastomeric bushing. The bushings may be viewed as a stiffness element and damping element in parallel. This may be modeled as a complex stiffness, k^*, with the imaginary part indicating the bushing damping. *(See note at end of this chapter on using rotating phasors to model vibration systems.)* The bushing behavior can then be written as:

$$F = kX + i\eta kX = k^* X$$

$$\frac{F}{X} = (k + i\eta k) = k^*$$
(7.41)

$$\frac{F}{X} = (Stiffness\ Portion) + i(Damping\ Portion)$$

where:

F = Force through the bushing

X = Deflection across the bushing

η = Loss factor for the elastomeric material.

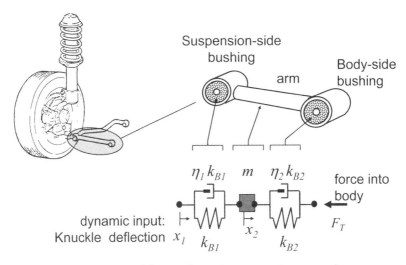

Figure 7.48 Modeling isolators in suspension control arm.

We consider the vibration source to be the higher-frequency displacements of the suspension knuckle, and we are interested in the force being transmitted through the body-side bushing into the body structure. This transfer function, $T(\omega)=F_T/X$, is given by:

$$\frac{F_T}{X} = \frac{\left(\dfrac{k^*_{B1}k^*_{B2}}{k^*_{B1} + k^*_{B2}}\right)}{1 - \omega^2 \dfrac{m}{\left(k^*_{B1} + k^*_{B2}\right)}}$$
(7.42)

where:

k^*_{B1} = Suspension-side bushing stiffness $k^*_{b1}=(k_{B1}+i\eta_1 k_{B1})$

k^*_{B2} = Body-side bushing stiffness $k^*_{B2}=(k_{B2}+i\eta_2 k_{B2})$

m = Mass of the chassis link

F_T = Vibration force amplitude transmitted into body structure

X = Vibration displacement amplitude imposed by suspension knuckle

At very low frequencies ($\omega \approx 0$), Equation 7.42 tells us that the combined link stiffness is the series combination of the two bushing stiffnesses:

$$\left.\frac{F_T}{X}\right|_{\omega \approx 0} = \left(\frac{k_{B1}k_{B2}}{k_{B1}+k_{B2}}\right) \tag{7.43}$$

At a frequency of

$$\omega_n = \sqrt{\frac{\left(k_{B1}+k_{B2}\right)}{m}} \tag{7.44}$$

the chassis link resonates against the two bushings. We can expand Equation 7.42 by substituting in the complex stiffness expressions of Equation 7.41 and arrive at

$$\left|\frac{F_T}{X_1}\right| = \left(\frac{k_{B1}k_{B2}}{k_{B1}+k_{B2}}\right)\left(\frac{\sqrt{1+\eta^4}}{\sqrt{\left(1-\left(\frac{\omega}{\omega_n}\right)^2\right)^2 + \eta^2}}\right) = |T(\omega)| \tag{7.45}$$

where:

η = Loss factor at each bushing

other variables are those of Equation 7.42

The first bracketed expression in Equation 7.45 is the static stiffness. The second expression is the gain at frequency ω. For the damping levels usually seen in bushings ($\eta < 0.2$), the η terms in Equation 7.45 vanish. We then see that for frequencies above $\sqrt{\omega_n}$ the force transmitted through the link begins to be attenuated over that transmitted statically (*gain<1*). The chassis link acts as an isolator for the higher-frequency vibrations, Figure 7.49. Notice also in this figure that there is a trade-off due to bushing damping; high damping controls the amplitude at resonance, but increases the force transmitted at higher frequencies.

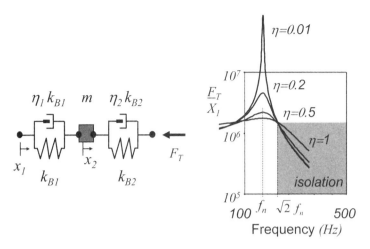

Figure 7.49 Response of isolators in suspension control arm.

Example: Suspension lower control arm

A source of high-frequency vibration is due to gear meshing in the transmission. This vibration is transmitted through the front wheel drive shaft to the suspension knuckle. The knuckle vibration is then transmitted though the suspension control arm (chassis link) into the body structure.

For this example consider a mesh frequency of *400 Hz*. To see how the lower control arm with bushings can attenuate this vibration, consider the following values: Identical bushings at either end of the control arm with stiffness $k_{B1}=k_{B2}=175000$ N/m and loss factor $\eta=0.2$, control arm mass of *m=5 kg*, and excitation frequency *f=400 Hz*.

For very low frequencies (*f≈0 Hz*), the force transmitted to the body per unit of knuckle deflection is given by:

$$\frac{F_T}{X} = \left(\frac{k_{B1}k_{B2}}{k_{B1}+k_{B2}} \right)$$

$$\frac{F_T}{X} = \left(\frac{(1750000N/m)(1750000N/m)}{1750000N/m+1750000N/m} \right) = 875000N/m$$

The natural frequency of this bushing/arm SDOF is

$$\omega_n = \sqrt{\frac{\left(k_{B1}+k_{B2}\right)}{m}}$$

$$\omega_n = \sqrt{\frac{(1750000N/m+1750000N/m)}{5kg}} = 836.7r/s \quad (133Hz)$$

Now consider the force transmitted per unit of knuckle deflection at $f=400\,Hz$ given by:

$$\left|\frac{F_T}{X_1}\right| = \frac{\left(\dfrac{k_{B1}k_{B2}}{k_{B1}+k_{B2}}\right)\sqrt{1+\eta^4}}{\sqrt{\left(1-\left(\dfrac{\omega}{\omega_n}\right)^2\right)^2 + \eta^2}}$$

$$\left|\frac{F_T}{X_1}\right| = \frac{\left(\dfrac{k_{B1}k_{B2}}{k_{B1}+k_{B2}}\right)\sqrt{1+0.2^4}}{\sqrt{\left(1-\left(\dfrac{400\,Hz}{133\,Hz}\right)^2\right)^2 + 0.2^2}} = 0.125\left(\frac{k_{b1}k_{b2}}{k_{b1}+k_{b2}}\right)$$

The arm attenuates the dynamic force into the body by a factor of *0.125* at *400 Hz*.

Example: High-frequency powertrain vibration through engine mount

On some vehicles, the powertrain is mounted directly to the body structure through engine mounts (rather than to a softly mounted cradle). In this direct-mounted case, high-frequency vibration of the engine block may be transmitted through the engine mount to the body, resulting in structure-borne noise. In this example, it was desired to increase the engine spark timing for improved fuel economy. This change in timing increased the dynamic block deflections in the *400–2000 Hz* range, Figure 7.50. To isolate these acoustic vibrations from the body, an engine mount consisting of a free mass was used, Figure 7.51b [17] (a conventional mount is shown for comparison in Figure 7.51a). The model for this mount, also shown in Figure 7.51b, is identical to that of the suspension link discussed above. The desired static stiffness was *200 N/mm*. From Equation 7.43 the stiffness for the two identical bushings can be found:

$$\frac{F_T}{X} = \left(\frac{k_{B1}k_{B2}}{k_{B1}+k_{B2}}\right) = 200 N/mm$$

$$k_{B1} = k_{B2} = 400 N/mm$$

It was desired to begin isolation at *270 Hz*. Therefore

$$270\,Hz = \sqrt{2}f_n$$
$$f_n = 190\,Hz$$

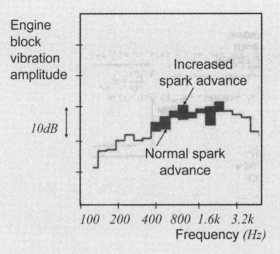

Figure 7.50 Acoustic deflections of powertrain.

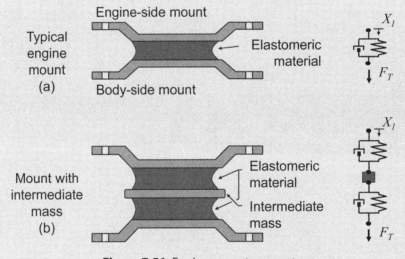

Figure 7.51 Engine mount concepts.

Equation 7.44 can be rearranged to calculate the needed intermediate mass:

$$\omega_n = \sqrt{\frac{\left(k_{B1} + k_{B2}\right)}{m}}$$

$$m = \frac{2k_B}{\omega_n^2} = \frac{2(400000 N/m)}{(2\pi 190 rad/\sec)^2} = 0.56 kg$$

This mount will support the powertrain at the required static rate and additionally act as a filter for the higher-frequency vibration, Figure 7.52.

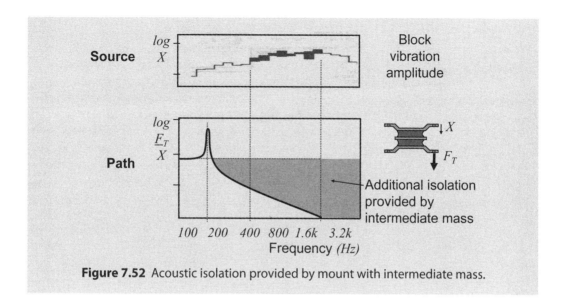

Figure 7.52 Acoustic isolation provided by mount with intermediate mass.

Many additional connections within the vehicle may be viewed as elastomeric elements with an intermediate mass. For example, Figure 7.53 shows a body-frame interface connected through an elastomeric body mount.

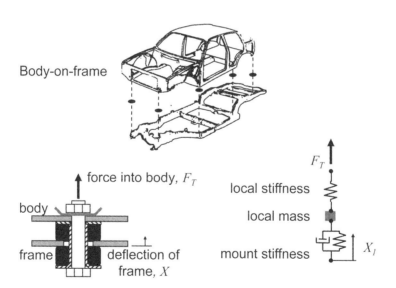

Figure 7.53 Body mount isolation behavior.

7.8.4 Local stiffness effect on vibration isolators

In designing chassis links with bushings, we must select the bushing material (k_b and η) which gives the desired high-frequency-isolation behavior. We expect the bushing stiffness and damping properties to be fully effective. However, when the bushing is attached to the body structure rather than to ground, we have seen that there is localized flexing of the structure. The flexing can be described as a local stiffness, K_L. This localized flexing will cause both the bushing stiffness and damping to be less than fully effective. Consider a bushing attached to a structure with local stiffness, K_L, Figure 7.54. Now the total deflection of this system is the sum of the localized structure deflection and the deflection across the bushing:

$$X = X_{LOCAL} + X_{BUSHING}$$

$$X = \frac{F}{K_L} + \frac{F}{k_B + i(\eta k_B)}$$

$$\frac{F}{X} = \frac{kK_L + i(K_L k\eta)}{(k_B + i\eta k_B + K_L)}$$

Which can be rearranged to

$$\frac{F}{X} = \frac{k_B\left[(k_B / K_L) + 1 + \eta^2 (k_B / K_L)\right] + i(\eta k_B)}{\left[(k_B / K_L) + 1\right]^2 + \left[\eta(k_B / K_L)\right]^2} \tag{7.46}$$

The loss factor, η, is typically $0<\eta<0.2$ and we can make the approximation $\eta^2 \sim 0$ giving:

$$\frac{F}{X} = \frac{k_B}{\left[(k_B / K_L) + 1\right]} + i\frac{\eta k_B}{\left[(k_B / K_L) + 1\right]^2} \tag{7.47}$$

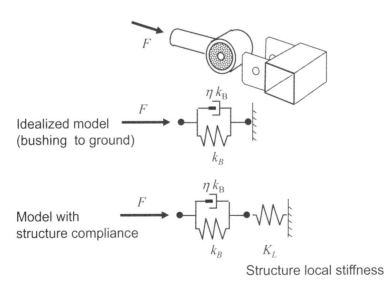

Figure 7.54 Isolator attachment to flexible structure.

Compare this final expression to the expression for a bushing attached to ground:

$$\frac{F}{X} = k_B + i\eta k_B$$

We can see that the stiffness portion of Equation 7.47—the real component—is reduced by a factor of:

$$\frac{1}{\left[(k_B / K_L) + 1\right]}$$

However the damping portion of Equation 7.47—the imaginary component—is reduced by a factor of:

$$\frac{1}{\left[(k_B / K_L) + 1\right]^2}$$

Thus, when the body structure has local flexibility at a bushing attachment, the damping qualities of the bushing are reduced more than that of the stiffness, Figure 7.55. This result has implications for structure design at attachments where we expect high-frequency inputs; the local stiffness of the structure, K_L, should be at least five times the bushing stiffness, k_b, to maintain 70% of the bushing damping.

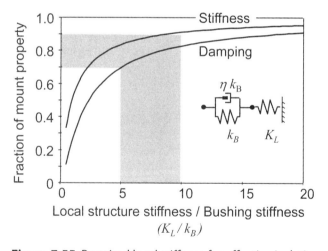

Figure 7.55 Required local stiffness for effective isolation.

7.8.5 Summary: Design for vibration

In this chapter, we have looked at a few of the important vibratory systems in the automobile. Our focus has been on determining the required qualities of the body structure which will minimize objectionable vibration levels. During early design layout, it is important to detune structural resonances from source vibrations, and also to take advantage of the isolation characteristics of other systems, suspension and mounted powertrain, for example. For preliminary design, this detuning can be achieved using the first-order models presented in this chapter.

7.9 Note on Use of Rotating Phasors to Solve Damped Vibration Problems

Earlier in this chapter we used complex variables to describe damping behavior. In this section we will further develop this useful concept. Consider a vector of length X rotating about a fixed point at a rotation speed of ω rad/sec, Figure 7.56a [18]. The projection of this vector onto the vertical axis can be described by $x(t)=X\sin\omega t$, Figure 7.56b. Now consider the horizontal axis to be the *Real axis* and the vertical axis to be the *Imaginary axis*. The vector may be described as the rotating phasor $x(t)=Xe^{i\omega t}$. If we also let the amplitude of the phasor X be complex $X^*=(X_{Re})+i\,(X_{Im})$, then the phase angle, θ is given by $\tan^{-1}(X_{Im}/X_{Re})$, Figure 7.57. This visualization offers some simplifications in solving damped vibration problems.

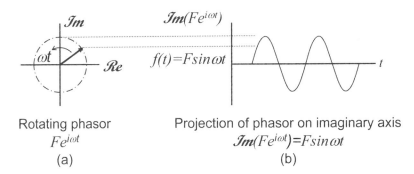

Rotating phasor
$Fe^{i\omega t}$
(a)

Projection of phasor on imaginary axis
$\mathcal{Im}(Fe^{i\omega t})=F\sin\omega t$
(b)

Figure 7.56 Representing a sinusoidal function with a rotating phasor.

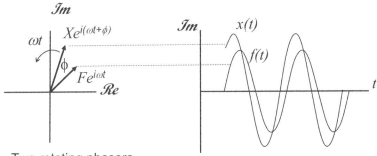

Two rotating phasors
($Fe^{i\omega t}$ reference phasor)

$$Xe^{i(\omega t+\phi)}= (Xe^{i\phi})e^{i\omega t}= (X_{Re}+iX_{Im})e^{i\omega t} = X^* e^{i\omega t}$$

Complex amplitude X^* gives information about magnitude of X and phase relative to the force phasor

Figure 7.57 Complex amplitude X^*.

Example: SDOF model with viscous damping

Consider the SDOF model of Figure 7.12 with viscous damping. As before, we begin with a force balance on the mass:

$$\Sigma(\text{forces acting on m}) = m\frac{d^2 x(t)}{dt^2}$$

$$f(t) + kx(t) + C\frac{dx(t)}{dt} = m\frac{d^2 x(t)}{dt^2}$$

But now we let the force and displacement be represented by phasors:

$$let \ f(t) = Xe^{j\omega t}, \ x(t) = Xe^{j\omega t}, \ then \ \frac{d^2 x(t)}{dt^2} = -X\omega^2 e^{j\omega t}$$

$$Fe^{j\omega t} = kXe^{j\omega t} + CXi\omega e^{j\omega t} - mX\omega^2 e^{j\omega t}$$

$$F = kX + CXi\omega - m\omega^2 X$$

$$\frac{X}{F} = \frac{1}{k + Ci\omega - m\omega^2} = \frac{k - m\omega^2 - Ci\omega}{\left(k - m\omega^2\right)^2 + \left(C\omega\right)^2}$$

$$\frac{X}{F} = \frac{\dfrac{1}{k}\left[1 - \left(\dfrac{\omega}{\omega_n}\right)^2 - \dfrac{Ci}{k}\omega\right]}{\left(1 - \left(\dfrac{\omega}{\omega_n}\right)^2\right)^2 + \left(\dfrac{C}{k}\omega\right)^2}$$

$$\frac{X}{F} = \frac{\dfrac{1}{k}\left[1 - \left(\dfrac{\omega}{\omega_n}\right)^2 - \dfrac{Ci}{\sqrt{km}}\omega\sqrt{\dfrac{m}{k}}\right]}{\left(1 - \left(\dfrac{\omega}{\omega_n}\right)^2\right)^2 + \left(\dfrac{C}{\sqrt{km}}\omega\sqrt{\dfrac{m}{k}}\right)^2}$$

$$\frac{X}{F} = \frac{\dfrac{1}{k}\left[1 - \left(\dfrac{\omega}{\omega_n}\right)^2 - 2\zeta i\dfrac{\omega}{\omega_n}\right]}{\left(1 - \left(\dfrac{\omega}{\omega_n}\right)^2\right)^2 + \left(2\zeta\dfrac{\omega}{\omega_n}\right)^2} \qquad (7.48)$$

Note that we can immediately go from the differential equation of motion to an algebraic expression in the frequency domain by making the following substitutions:

$$x(t) \rightarrow X, \ \frac{dx(t)}{dt} \rightarrow i\omega X, \ \frac{d^2 x(t)}{dt^2} \rightarrow \left(i\omega\right)^2 X = -\omega^2 X, \ f(t) \rightarrow F$$

The substitution results in the following for the viscously damped SDOF system above:

$$f(t) + kx(t) + C\frac{dx(t)}{dt} = m\frac{d^2 x(t)}{dt^2} \rightarrow F + kX + CXi\omega = -m\omega^2 X$$

Solving this algebraic equation results in a complex expression for the transfer function, (X/F), Equation 7.48, in which the phasor angle represents the phase angle of displacement relative to the force. Often, we are only interested in the magnitude of this complex number.

Example: SDOF model with structural damping

In Equation 7.4, we substitute for k the complex value $k^*=k+i\eta k$. This expression includes the imaginary component, $i\eta k$, representing the structural damping force which is in phase with velocity but proportional to displacement.

$$F = (k + \eta ki)X - m\omega^2 X$$

$$\frac{X}{F} = \frac{1}{k + \eta ki - m\omega^2} = \frac{k - m\omega^2 - k\eta i}{\left(k - m\omega^2\right)^2 + \left(k\eta\right)^2}$$

$$\frac{X}{F} = \frac{\frac{1}{k}\left[1 - \left(\frac{\omega}{\omega_n}\right)^2 - \eta i\right]}{\left(1 - \left(\frac{\omega}{\omega_n}\right)^2\right)^2 + \left(\eta\right)^2} \tag{7.49}$$

$$\left|\frac{X}{F}\right| = \frac{1/k}{\sqrt{\left(1 - \left(\frac{\omega}{\omega_n}\right)^2\right)^2 + \left(\eta\right)^2}}$$

7.10 Note on Mechanical Impedance Technique

The use of complex variables to describe mechanical vibration offers a very useful tool in developing models of more complex vibratory systems. The mechanical impedance technique is based on an analogy with alternating current electrical circuits [19]. The impedance of an electrical element is defined as the voltage amplitude, E, measured across the element divided by the current amplitude, I, flowing through the element:

$$Z = \frac{E}{I} \tag{7.50}$$

For example, the impedance for a resistor is $Z=R=E/I$. To analyze electrical networks, we apply *loop equations* (the voltage around any closed loop is zero), and *node equations* (the sum of current into any node is zero). Applying these equations allow the prediction of the current through and voltage drop across any element.

Note that the impedance in an electrical element is the voltage difference measured *across* the element, divided by the current flowing *through* the element. For an analogous mechanical element, the *across variable* is the deflection across the element (we could equivalently use velocity or acceleration) divided by the force flowing *through* the element. Note that this analogy maintains the validity of both node equations—forces sum to zero at a point, and loop equations—relative deflections around a closed path sum to zero. This electrical-mechanical analogy is summarized in Figure 7.58.

Impedance in an electrical network	Mechanical Impedance analogy
E (voltage) is measured <u>across</u> an element	*X* displacement (we could alternatively take velocity or acceleration as the across variable)
I (current) flows <u>through</u> an element	*F* Force
Impedance definition $Z\equiv$ (across variable)/(through variable) $Z=E/I$ *e.g. resistor* $E=IR$ $R=E/I$ *(impedance of a resistor, Z=R)*	**Impedance definition** $Z\equiv$ (across variable)/(through variable) $Z=X/F$ $Z_K=1/k$ *(spring)* $Z_M=1/(-w^2m)$ *(mass)* $Z_D=1/(i\eta k)$ *(structural damping)*
Note: •at a node, currents must sum to zero •around a closed loop, the net voltage drop is zero	**Note:** •at a point, forces must sum to zero •around a closed loop the relative displacements sum to zero

Figure 7.58 Mechanical impedance analogy.

The impedance of a general mechanical system element is:

$$Z = \frac{X}{F} \tag{7.51}$$

Table 7.8 provides the impedance for common mechanical elements.

Table 7.8 Impedance of mechanical elements.

Linear elastic stiffness element	Mass element	Viscous damping element	Stiffness element with structural damping
$Z_K = \dfrac{1}{k}$	$Z_M = \dfrac{1}{-m\omega^2}$	$Z_D = \dfrac{1}{i\omega C}$	$Z_K = \dfrac{1}{k+i\eta k}$

To solve a mechanical network, we apply node and loop equations to determine deflections and forces just as we would with an electrical network. We can also use standard impedance rules for solving electrical circuits to arrive at a result. Two important rules are:

1. For two elements, Z_1, Z_2, in series, the equivalent impedance is

$$Z_{EQ} = \frac{Z_1 Z_2}{Z_1 + Z_2} \tag{7.52}$$

2. For two elements in parallel, the equivalent impedance is

$$Z_{EQ} = Z_1 + Z_2 \tag{7.53}$$

Example: Impedance approach

As an example of this technique, consider the suspension link with identical bushings at each end, Figure 7.48. The impedance diagram for this system is shown in Figure 7.59. Note that one terminal of a mass element is always connected to an inertial reference—ground. In this example, we are interested in finding the driving point impedance: $Z_{11} = X_1/F_1$.

Elements Z_2 and Z_3 are in parallel with equivalent impedance:

$$Z_{EQ} = \frac{Z_2 Z_3}{Z_2 + Z_3}$$

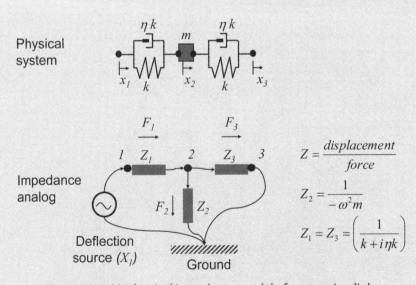

$$Z = \frac{displacement}{force}$$

$$Z_2 = \frac{1}{-\omega^2 m}$$

$$Z_1 = Z_3 = \left(\frac{1}{k + i\eta k} \right)$$

Figure 7.59 Mechanical impedance model of suspension link.

Element Z_1 is in series with this combined impedance:

$$Z_{11} = Z_1 + \left(\frac{Z_2 Z_3}{Z_2 + Z_3} \right)$$

Substituting the variables: $Z_2 = \dfrac{1}{-m\omega^2}$, and $Z_1 = Z_3 = \dfrac{1}{k + i\eta k}$ gives:

$$Z_{11} = \left(\frac{1}{k + i\eta k} \right) + \left(\frac{\left(\dfrac{1}{-m\omega^2} \right)\left(\dfrac{1}{k + i\eta k} \right)}{\left(\dfrac{1}{-m\omega^2} \right) + \left(\dfrac{1}{k + i\eta k} \right)} \right)$$

which can be simplified to:

$$|Z_{11}| = \left(\frac{2\sqrt{\left(1 - \left(\dfrac{\omega}{\omega_2} \right)^2 \right)^2 + (\eta)^2}}{k\sqrt{1 + (\eta)^2}\sqrt{\left(1 - \left(\dfrac{\omega}{\omega_1} \right)^2 \right)^2 + (\eta)^2}} \right)$$

$$\omega_1 = \sqrt{\frac{k}{m}}$$

$$\omega_2 = \sqrt{\frac{2k}{m}}$$

References

1. Thomson, W., *Vibration Theory and Applications*, Prentice-Hall, NJ, 1965.

2. Gillespie, T. D., *Fundamentals of Vehicle Dynamics*, SAE International, Warrendale, PA, 1992, p. 465.

3. Lee, R. and Pradko, F., "Analytical Analysis of Human Vibration," SAE Paper No. 680091, SAE International, Warrendale, PA, 1968.

4. Stikeleather, L., Hall, G. and Radke, A., "A Study of Vehicle Vibration Spectra as Related to Seating Dynamics," SAE Paper No. 720001, SAE International, Warrendale, PA, 1972.

5. Kamal M. and Wolf, J., *Modern Automotive Structural Analysis*, Van Nostrand Reinhold, NY, 1982, p. 55.

6. Meirovitch, L., *Elements of Vibration Analysis*, McGraw Hill, NY, 1986, pp. 45–72.

7. Malen, D. E., *Applied Damping in the Automobile Body*, Masters Thesis, Massachusetts Institute of Technology, MA, 1970.

8. Flugge, W., *Viscoelasticity*, Blaisdell Publishing, MA, 1967, p. 54.

9. Den-Hartog, J., *Mechanical Vibrations*, McGraw Hill, NY, 1956, Chapter 5.

10. Gillespie, T. D., *Fundamentals of Vehicle Dynamics,* SAE International, Warrendale, PA, 1992, pp. 129–147.

11. Kikuchi, N. and Malen, D., *Course notes for ME513 Fundamentals of Body Engineering*, University of Michigan, Ann Arbor, MI, 2007.

12. Elliott, W., "Plastic Models for Dynamic Structural Analysis, SAE Paper No. 710262, SAE International, Warrendale, PA, 1971.

13. Lyon, R., *Machinery Noise and Diagnostics*, Butterworths, MA, 1987.

14. Davis, J., "Modal Modeling Techniques for Vehicle Shake Analysis," SAE Paper No. 720045 SAE International, Warrendale, PA, 1972.

15. Kato, T., Hoshi, K. and Umemura, E., "Application of Soap Film Geometry for Low Noise Floor Panels," *Proceedings of the 1999 Noise and Vibration Conference*, P-342, SAE 1999-01-1799, SAE International, Warrendale, PA, 1999. Also "Harrier/Lexus RX300 Uniform-curvature panels to isolate road noise," *Automotive Engineering International*, SAE International, Warrendale, PA, April, 1998, p24.

16. Snowdon, J. C., *Vibration and Shock in Damped Mechanical Systems*, John Wiley, NY, 1968, Chapter 5.

17. Cogswell, J. and Malen, D., E., "Engine Mount for Integral Body Vehicle," SAE Paper No. 830258, SAE International, Warrendale, PA, 1983.

18. Kennedy, C. and Pancu, C., "Use of Vectors in Vibration Measurement and Analysis," *Journal of the Aeronautical Sciences*, V14, N11, November, 1947.

19. Shearer, J., Murphy, A., and Richardson, H., *Introduction to System Dynamics*, Addison-Wesley, MA, 1967.

Chapter 8
Design for Vehicle and Styling Integration

In previous chapters we have considered design for major conditions individually—bending, torsion, crashworthiness, and vibration. Now we consider the task of creating a new body structure concept which will best meet all these conditions. By *best*, we usually mean the lightest body structure meeting all requirements (although there are many objectives we may be interested in such as manufactured cost). The layout of this best body concept is called the *topology* problem, with the objective of creating the arrangement of structural elements to meet requirements in the most efficient manner. In addition to meeting all structural requirements, the structure must also fit within envelope of all vehicle components: suspensions, powertrain, occupants, etc. The structure must also fit within the desired external surface of the vehicle, which is largely defined by styling or appearance considerations.

8.1 Designing the Best Body Structure

The topology problem may be stated more formally as:

Objective: Determine the best spatial arrangement of structure elements (often "best" means minimizing mass)

Decision Variables: Layout, shape and size of structural elements

Constraints:

1. Meet all structural requirements of strength, stiffness, and energy absorption
2. Meet spatial constraints imposed by vehicle packaging and body exterior surface

Designing to the constraints imposed by vehicle and styling reminds us that the objective of body layout is to achieve the best *vehicle configuration* (not the most ideally efficient structure).

Often the constraints imposed by vehicle packaging and the styling surface are *soft constraints* in that they may be negotiable as the vehicle design progresses. For this reason, as body designers we need to know the origins of these constraints and how best to exploit them.

8.2 Vehicle Layout

In this section, we will provide an overview of the vehicle layout process in which the vehicle package is established [1, 2, 3]. This results in the spatial constraints on body topology. We will then summarize the *pinch points*—the traditional areas in which the constraints most influence the body topology. For this, we consider the layout for a new passenger vehicle using an *Inside-out sequence* [4]:

1. Establish occupant package including seating, entry egress, and vision.
2. Establish the relation of ground with respect to occupants.
3. Locate those subsystems whose location is mandated by government standards, e.g., bumper location.
4. Locate the remaining subsystems for functionality.

5. Enclose the vehicle in an envelope defined by the exterior body surface, defined by styling and aerodynamics.

We can consider this sequence first in the side view of the vehicle.

8.2.1 Side-view vehicle layout

The fundamental purpose of a passenger car is to transport people, and we begin with orienting occupants in a comfortable configuration. Given a population of occupants of varying sizes, the seating package is based on a statistical sample. The occupant dimensions shown in Figure 8.1 provide an envelope containing 95% of the population. (These dimensions vary from study to study, and those shown are for illustrative purposes.) Depending on the vehicle type—sport car, sedan, sport utility vehicle—we have a choice of *chair height*, the distance from the floor to a seating reference point located at the hip (SRP), Figure 8.2. Once a chair height is selected, foot position may be established for comfort. Studies on seating comfort give a relationship between chair height and the horizontal distance from the ball of foot to seating reference (BOF):

$$BOF=913.6+0.6723(FCH)-0.001956(FCH)^2 \qquad (8.1)$$

where

FCH = Front chair height (mm), *170 mm<FCH<455 mm*

BOF = Ball of foot to seating reference point horizontal dimension (mm)

This relationship provides leg room sufficient for 95% of the total population. With chair height and ball of foot position established, the occupant leg position is now fixed. Orienting the torso at an angle of approximately 26° from the vertical based on comfort studies, gives the final orientation of the driver.

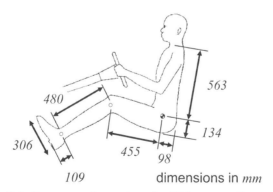

Figure 8.1 Occupant dimensions for 95th percentile envelope.

The rear-seat passenger is now oriented relative to the front-seat occupant, Figure 8.3. Generally, the rear SRP is slightly higher than the front and, again based on the vehicle type, the desired knee room will establish the rear seat package. With both front and rear occupants oriented, the eye position is established and defined by an ellipse containing 95% of the population. In a similar way, the head profile is added, Figure 8.4.

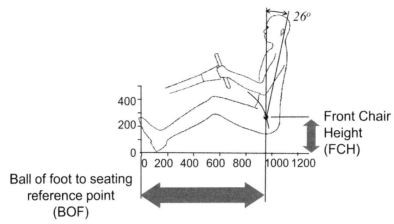

Figure 8.2 Setting up most-compact vehicle length: Front seat position.

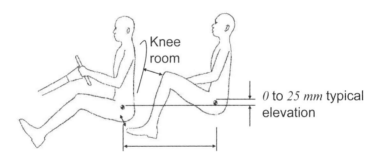

Figure 8.3 Setting up most-compact vehicle length: Rear seat position.

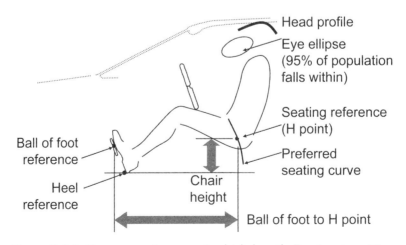

Figure 8.4 Setting up most compact vehicle length: Front seat position.

We are attempting to find the most compact package for the vehicle, that is the vehicle with the smallest footprint, and we can now begin to build the vehicle package outward from this occupant package. Based on a preliminary mass estimate for the vehicle, a tire size may be selected and oriented with respect to the occupants, Figure 8.5. The tire is positioned with a minimum fore-aft clearance from the driver's foot to the tire. This clearance will depend on whether this is to be a front-wheel-drive or rear-wheel-drive configuration, or if tire chains will be needed. Similarly, the rear wheel is positioned as closely as possible to the rear occupant. This rear clearance will depend on the suspension travel and trajectory of the wheel, as determined by the suspension type.

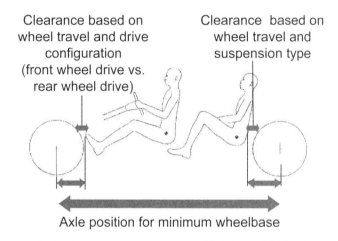

Figure 8.5 Setting up most-compact vehicle length: Wheel position.

With the wheels positioned, the powertrain may be oriented in the fore-aft position. In Figure 8.6, a front-wheel-drive configuration is shown where the transmission drive shaft orients the powertrain with respect to the wheel center. The front of the

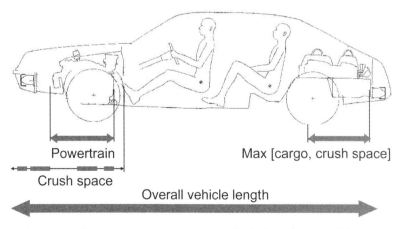

Figure 8.6 Setting up most-compact vehicle length: Overall length.

car is determined by laying in the needed crush space for front crashworthiness. In the rear, space for the design cargo, or space for rear crashworthiness is provided—whichever is the greater. This procedure results in building up the package to the minimum overall vehicle length.

We can now establish the vertical orientation of the seating package with respect to ground. In orienting the package, we imagine the coordinate system to be fixed to the vehicle body. In doing so, the ground line "moves" relative to the fixed body, Figure 8.7, depending on the loading on the vehicle. Figure 8.8 shows a front suspension where four conditions are defined: the Design position in which the vehicle is loaded with two occupants, fluids, and some fraction of the maximum cargo mass; the Curb position in which the vehicle is loaded with fluids only; the Jounce position in which the wheel is at the full upward travel; and the Rebound position in which the wheel is at the full downward travel.

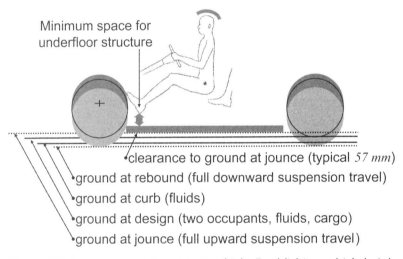

Figure 8.7 Setting up most-compact vehicle: Establishing vehicle height.

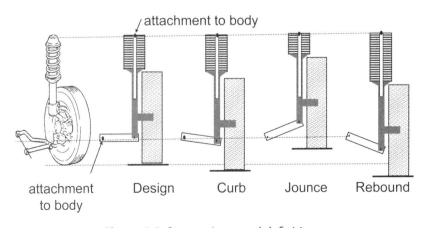

Figure 8.8 Suspension travel definitions.

The primary ground clearance requirement on underbody components is on the order of 57mm for passenger cars. From the front heel, the minimum vertical space for the under-floor body structure is measured downward, Figure 8.9. From this, the ground clearance is added. This results in the ground position at full jounce. Measuring upward from the ground at full jounce by the loaded tire radius results in the wheel center height at jounce. Measuring downward from this wheel center by the suspension jounce travel, approximately *125 mm (5 in.)* for passenger cars, results in the position of the wheel center at Design. The ground at design is found by measuring downward by the loaded tire. Generally, the ground line at Design is taken as a horizontal grid line.

Once the ground lines have been established, other ground clearance conditions may be applied. Figure 8.10 shows several of these, including departure angles,

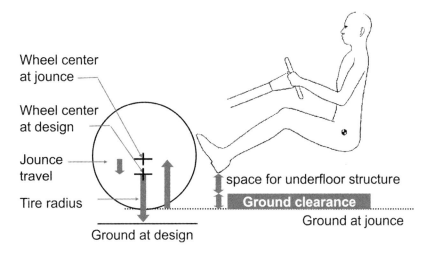

Figure 8.9 Setting up most-compact vehicle: Establishing ground line.

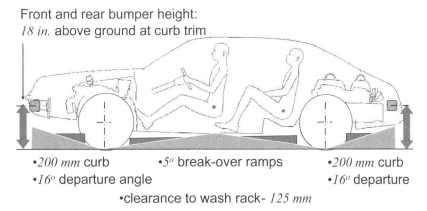

Front and rear bumper height:
18 in. above ground at curb trim

- *200 mm* curb
- *16°* departure angle
- *5°* break-over ramps
- *200 mm* curb
- *16°* departure
- clearance to wash rack- *125 mm*

Figure 8.10 Setting up most-compact vehicle height: Typical ground clearance constraints.

curbs, break-over ramp, wash rack criteria, and bumper vertical location. The side-view vehicle package is shown in Figure 8.11. Now consider the vehicle configuration in the front view.

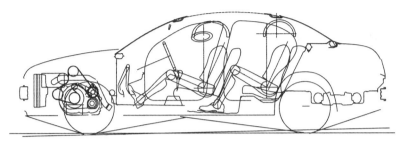

Figure 8.11 Preliminary side-view package. (Courtesy of the American Iron and Steel Institute, UltraLight Steel Auto Body)

8.2.2 Front-view vehicle layout

To establish the most-compact vehicle width, Figure 8.12, we look at three criteria: 1) the width required to seat occupants abreast comfortably, 2) the width required to package the powertrain—particularly for transverse front wheel drive, and 3) the width (track) required for rollover stability. The criterion requiring the largest width will establish the smallest or most compact package.

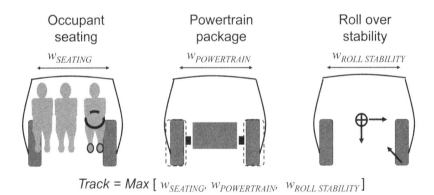

$$Track = Max\,[\,w_{SEATING},\ w_{POWERTRAIN},\ w_{ROLL\ STABILITY}\,]$$

Figure 8.12 Setting up most-compact vehicle width.

1. *Width required to seat occupants abreast comfortably-*To determine the width required to seat occupants abreast comfortably, we imagine selecting n people at random (n=number of passengers in front row) and consider the probability distribution of their combined shoulder room. Figure 8.13 shows torso dimensions for the *99th* percentile person. As an example, consider a two-passenger vehicle, n=2, and the probability of any two adults having a given combined width is shown in Figure 8.14. When space for clothing and controls operation is included, roughly *1200 mm (47.25 in.)* shoulder room is needed in this case. A vehicle width may be determined based on this shoulder room criterion.

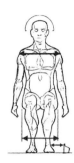

head clearance *99%* contour

shoulder width *99% ~ 493 mm*

hip width *99% ~ 411 mm*

foot clearance *~260 mm*

Figure 8.13 Front view human accommodation dimensions.

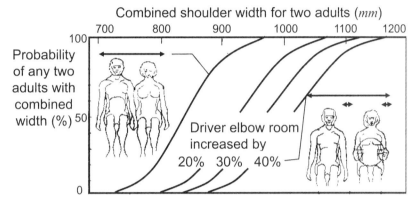

Figure 8.14 Probability of two adults having a given total combined shoulder width.
(University of Michigan Transportation Research Institute)

2. *Width required to package the powertrain*-To determine the width required to package the powertrain, consider Figure 8.15. The cross-car distance between wheels—track—must accommodate the sum of the powertrain width, the width of two motor compartment side rails, the clearance between rail and powertrain (usually established by clearance for assembly), and the tire envelope as it moves from full right to left turn (including tire chains if used). For transverse front-wheel vehicles, this criterion is often the determining case for minimum vehicle width.

In determining width based on the powertrain package, the tire envelope is required. This envelope depends on the turn angles, and these are related directly to the curb-to-curb turn diameter requirement. The turn diameter is an important customer attribute for which smaller is better. A typical value for passenger cars is *10 m (32.8 ft)*. Figure 8.16 shows the geometry to determine the required inside turn angle to achieve a specified turn circle diameter [5]:

$$\tan\theta = \frac{l}{r} = \frac{l}{(D/2)-t}$$

$$\theta = \tan^{-1}\left(\frac{2l}{D-2t}\right)$$

(8.2)

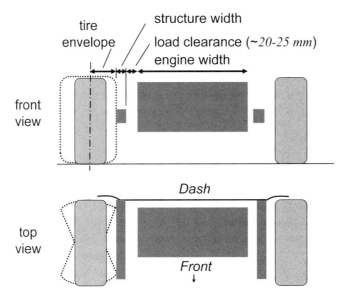

Figure 8.15 Setting up most-compact vehicle width: Track based on powertrain packaging.

where

θ = Required turn angle for inside wheel

t = Track width

l = Wheelbase

D = Required curb-to-curb turn diameter

3. *Width required for rollover stability*-The gross cornering ability—rollover threshold—depends on the track width. Consider a vehicle turning at a steady speed, Figure 8.17. A lateral force due to centrifugal acceleration is applied at the center of gravity and directed outward from the center of the turn. When the line of action of the vector sum of the lateral force and the vehicle weight falls outside the stance of the wheels, the vehicle is at incipient rollover. The track required to allow a steady-state lateral acceleration, *(a/g)*, is given by [6]:

$$t = 2h\left(\frac{a}{g}\right) \qquad (8.3)$$

where

t = Required track for incipient rollover at lateral acceleration, *a*

h = Height of center of gravity above ground

a = Lateral acceleration

g = Acceleration of gravity

328

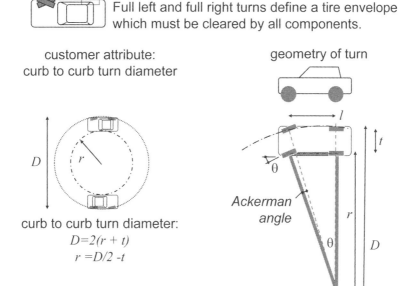

Figure 8.16 Defining the tire envelope.

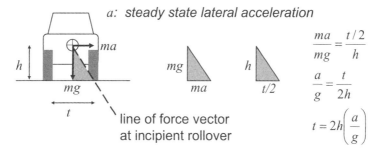

Figure 8.17 Setting up most-compact vehicle width: Track based on rollover threshold.

The design lateral acceleration ratio, *(a/g)*, depends on the type of vehicle and the expected maximum dynamic lateral acceleration. Figure 8.18 shows typical values of *(a/g)* by vehicle type. For passenger cars the range is *1.25<(a/g)<1.50.*

The largest track required of the above three criteria will define the most-compact vehicle width. Once the track is determined, layout of components may proceed. An example of a front-view package is shown in Figure 8.19.

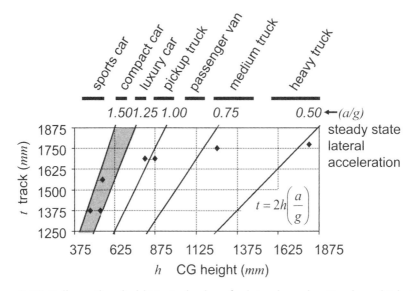

Figure 8.18 Rollover threshold: Typical values for lateral acceleration by vehicle type.

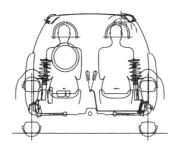

Figure 8.19 Consolidated package: Front view. (Courtesy of the American Iron and Steel Institute, UltraLight Steel Auto Body)

8.2.3 Plan-view vehicle layout

Finally consider vehicle layout in the plan view. Important constraints on body topology are the limits imposed by vision needs on the upper structure. Figure 8.20 illustrates the allowable vision obscuration of the A pillar and B pillar. The example angles given are the maximum obscuration in the field of view with binocular vision. The angles encompass the total pillar including structural section, trim, and door frame. Figure 8.21 illustrates typical constraints for forward and rearward vision which limit the position of roof headers, cowl, and C pillars. An example of a plan view package is shown in Figure 8.22.

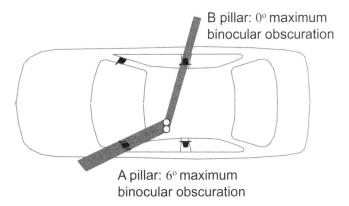

B pillar: 0° maximum binocular obscuration

A pillar: 6° maximum binocular obscuration

Figure 8.20 Vision constraints on pillars.

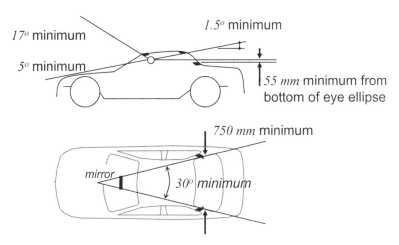

17° minimum

1.5° minimum

5° minimum

55 mm minimum from bottom of eye ellipse

750 mm minimum

mirror

30° minimum

Figure 8.21 Forward and rearward vision constraints.

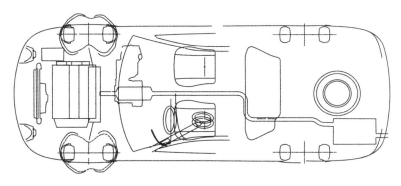

Figure 8.22 Consolidated package: Plan view. (Courtesy of the American Iron and Steel Institute, UltraLight Steel Auto Body)

8.3 Exterior Body Surface

We now have a preliminary layout as we proceeded from the passenger package outward. Now we consider the outer surface of the vehicle. The outer surface is defined by vehicle styling and by aerodynamics.

8.3.1 Basic proportions for styling

An extended discussion of styling of the body surface is beyond the scope of this book. In the following, we will look at the influence of styling on some of the basic proportions of the vehicle: fore-aft location of wheels, tire size, and body-to-wheel-opening relationship. Each of these has implications to the packaging of body structure.

The fore-aft location of wheels influence many vehicle characteristics: front/rear axle mass distribution, turn circle, bending moments on the body structure, but also the visual appearance of the vehicle. Consider Figure 8.23 where we consider a center of visual mass for the vehicle. When the front and rear wheels are equally distant from this center, Figure 8.23a, the vehicle appears to be balanced and static. In figure 8.23b, the wheels are not equally distant from the center with the overall effect of a dynamic appearance—a desirable visual queue in a product intended for motion.

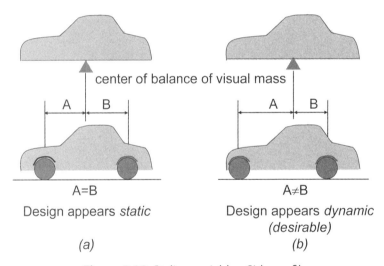

Figure 8.23 Styling variables: Side profile.

The implication of this is that the wheel position may not be that defined by the tightest package as outlined in the previous sections. This may influence the body structure by increased bending moment loading due to an increased wheelbase, or by reduced packaging space for structural sections.

A second area in which styling proportions may deviate from the tightest package defined in Section 8.2 is in the area of track and tire size. Consider Figure 8.24 where we are concerned with the visual firmness of stance of the vehicle in the rear view. This firmness depends upon tire width and track, and their relationship to the visual mass of the body. As seen in Figure 8.24 a and b, the tire size and track determined solely on the engineering approach of Section 8.2, may lead to a design which lacks the desirable visual firmness shown in Figure 8.24c, in which the track and tire size has been increased beyond those determined by engineering requirements. The larger tires and track can make the packaging of structural sections more challenging.

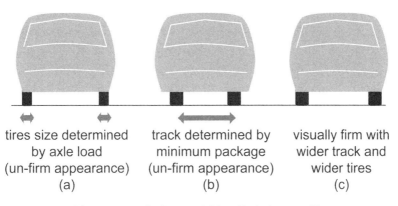

Figure 8.24 Styling variables: End-view profile.

Finally, consider the visual relationship of the front-wheel-to-body surface. Figure 8.25 illustrates the desirability of tight wheel openings relative to the tire outer diameter. Generally a maximum side-view gap of 50mm and a maximum offset in plan view is desirable to achieve this visually tight relationship. Visual appearance is sensitive to the tire relative to other features as shown in Figure 8.26.

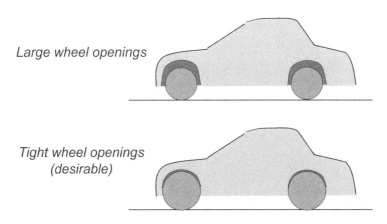

Figure 8.25 Styling variables: Wheel openings.

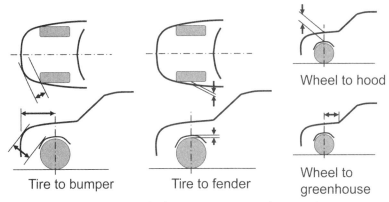

Figure 8.26 Styling constraints on front end.

8.3.2 Aerodynamics

As a vehicle moves relative to the surrounding air, aerodynamic forces are applied to the body. These forces are important for vehicle handling and fuel consumption. In this section, we limit the discussion to the rearward drag force, as this is directly related to fuel economy. As a rule of thumb for midsize passenger cars, a *10%* increase in *CDA* results in an approximate *2%* reduction in fuel economy (EPA fuel economy schedule).

The aerodynamic drag force acting on a vehicle traveling at speed *V* is given by:

$$F_D = AC_D \frac{\rho V^2}{2} \qquad (8.4)$$

where:

F_D = drag force

A = Vehicle area projected on to a frontal plane

C_D = Coefficient of drag

ρ = Density of air

V = Velocity of the vehicle

Our focus is to understand the constraints on body structure imposed by the need for low drag forces. Figure 8.27 offers a convenient means to estimate drag coefficient based on visual characteristics of the body shape [7]. To estimate the drag coefficient, match the visual characteristics of the body with pictographs in each of the nine rows of the table of Figure 8.27. Corresponding ratings are found at the top of the figure. The drag coefficient estimate is then given by:

$$C_D = 0.16 + .0095 \, \Sigma \qquad (8.5)$$

where Σ = Sum of the ratings for all nine rows. Note that $9<\Sigma<39$.

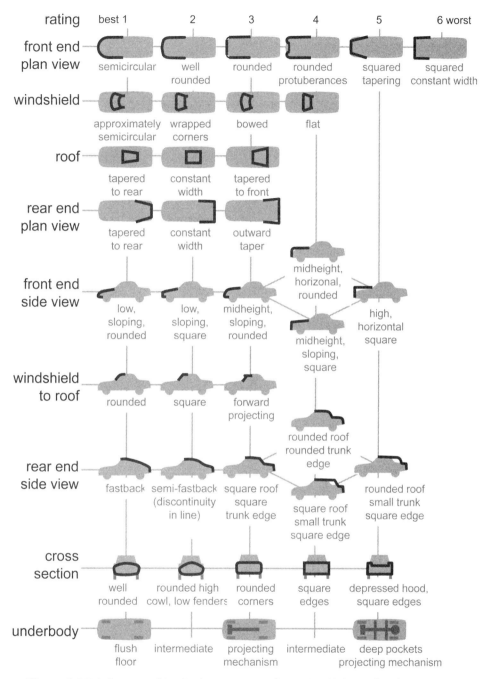

Figure 8.27 Influence of body shape on aerodynamics. (Adapted with permission from MIRA, Ltd)

Example: Aerodynamic influence on upper motor compartment rail

Consider the upper motor compartment rail, the highlighted area in Figure 8.28. One function of this structural member is to generate a crush force to prevent vehicle pitching during a front impact, Chapter 6. The most efficient way to do this is with a straight rail perpendicular to the barrier face, shown at the right side of Figure 8.28. This configuration will load the member axially and generate an axial force by crumpling. Now considering aerodynamics, we see from Figure 8.28 that the square plan view shape offers the worst configuration for drag with a resulting reduction in fuel economy. Conversely to minimize drag, the motor compartment plan view should approach a semicircular shape. However, with the semicircular shape, the upper rail will be loaded with an off-axis force, left side of Figure 8.28. This would generate bending moments resulting in collapse by plastic hinge formation—a very inefficient collapse mode and correspondingly heavy structure. This example of an engineering trade-off—structure mass vs. increased drag force—is typical of the multitude of decisions necessary to achieve a well-balanced vehicle concept. In making these trade-off decisions occurring in the early design stages, simple, quick first-order models are used to provide information. For example, using Equation 8.5 and Figure 8.27, we can estimate the increase in drag for the design concept on the right side:

$$\Delta C_D = 0.0095 \ (5 \text{ rating points}) = 0.0475$$

We can now decide if the improvement in structural efficiency balances this increase in drag coefficient.

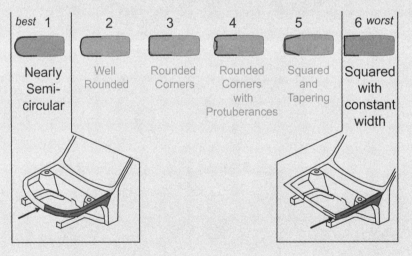

Figure 8.28 Influence of front structure shape on aerodynamics.

8.4 Constraints on Body Structure from Vehicle Layout and Exterior Surface

We can now summarize the primary spatial constraints on body topology imposed by vehicle layout, styling, and aerodynamics. The following are recurring "pinch points" where the desired structural shape and section size are severely limited by vehicle constraints. (These points are based on integral body construction and transverse front-wheel-drive configuration. Points would differ for other powertrain configurations such as longitudinal rear drive.)

Motor compartment midrail in plan view—Constraints 1 to 7 are illustrated in Figure 8.29 and apply to the motor compartment rails. The motor compartment midrail must fit between the powertrain on the inside surface and the tire envelope on the outer surface (1). At the front end, the midrail must also be outboard of the radiator. These constraints can limit both the section width and also the shape of the midrail in the plan view. For efficient energy absorption, a straight midrail is desirable with roughly square section proportions. Further, clearing the tire envelope also forces the rear of the midrail to be inboard of the rocker; hence, the reaction of midrail axial forces cannot be directly to the rocker.

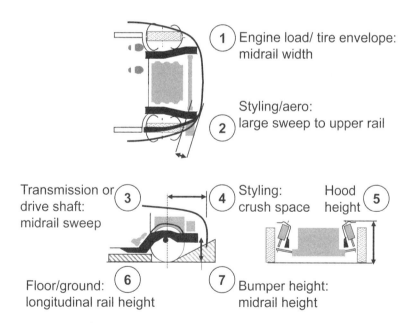

Figure 8.29 Motor compartment constraints.

Motor compartment midrail in side view—At the front, the motor compartment midrail must be at the bumper height (7). Above the axle, the midrail must clear (be above) the drive shaft envelope in the full jounce condition (3). For manual transmissions, the gear case extends below the midrail and the rail must be above. Further, at the rear of the midrail, the shape must be between the floor defined by the occupant and ground clearance plane (6). These constraints impose an arched shape to the midrail rather than the more-efficient straight section, optimizing axial crush. Finally, the crushable space for the midrail is limited by the layout of other components (4).

Motor compartment upper rail—The motor compartment upper rail must fit within the exterior surface. For both styling and aerodynamic reasons, this surface is often highly curved in the plan view approaching a semicircular shape (2). As with the midrail, a straight shape for the rail is desirable for efficient energy absorption. The upper rail section is also constrained between the exterior surface setting the top of the section (5), and the strut setting the lower boundary for the section.

Floor pan cross member shape—Constraints 8 and 9 are illustrated in Figure 8.30 and apply to floor pan cross members and rocker position. Cross members span laterally across the floor pan from rocker to rocker with the most desirable shape being straight. Packaging the exhaust and drive shaft for rear-wheel drive (8) constrains the cross member to sweep up at the tunnel while being under the seat for seat packaging (9).

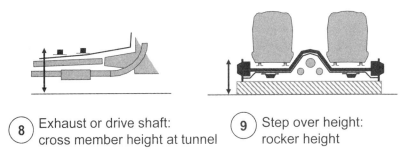

8 Exhaust or drive shaft: cross member height at tunnel

9 Step over height: rocker height

Figure 8.30 Floor pan constraints.

Rocker height—The rocker section is constrained at the bottom by the ground clearance plane, at the top by passenger step-over height (9), on the outboard side by the exterior surface, and on the inboard side by foot room.

Rear compartment rail shape in side view—Constraints 10 to 13 are illustrated in Figure 8.31 and apply to the rear compartment. The rear rail side-view shape is constrained at the rear by the bumper height (10). The fuel fill tube must be accommodated (11). This is usually accomplished by arching the rear rail, with the tube passing below. This accommodation is preferred as it allows assembly of the fuel system from below. Finally, the rail must sweep downward under the rear seat (12). Like the motor compartment midrail, these constraints impose an arched shape rather than the more-efficient straight section.

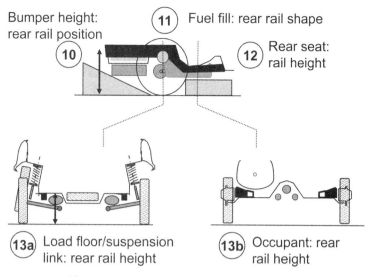

Figure 8.31 Rear compartment constraints.

Rear compartment rail sections—At the rear axle (13a), the rear rail section is constrained at the top by the load floor, which has generally been set as low as possible, and constrained at the bottom by the travel of the rear suspension link. It is also constrained at the outboard surface by the wheel envelope, and on the inboard surface by the exhaust or spare tire. Under the rear seat (13b), the rear rail section is constrained in height by the occupant, above, and the ground clearance plane, below.

Constraints 14 to 19, illustrated in Figure 8.32, apply to the upper structure (greenhouse).

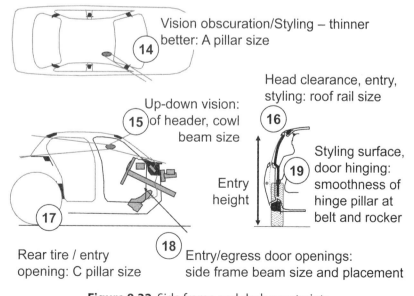

Figure 8.32 Side frame and dash constraints.

Greenhouse pillars—The greenhouse pillars, most notably the A and B pillars, are constrained by vision obscuration and styling (14). The objects causing obscuration include not only the structural pillar, but also door seals, window frame, and interior trim. Therefore, this area requires a careful balance of all these subsystems for optimization.

Roof perimeter members—Upward vision constrains the size and position of the front and rear roof headers (15). Roof side rails are constrained from the interior by adequate clearance to the occupant's head, and from the outside by the body surface and door frame (16).

Side frame beams—These beams, A, B, and C pillars; rocker; and roof rails are constrained by the door opening line established by occupant entry requirements (18). In the case of the lower C pillar, the front surface is constrained by the rear door opening line, and the rear surface is constrained by the rear tire envelope (17).

Hinge pillar outer plane—As discussed in Chapter 4, the front view of the outer A pillar should smoothly transition to hinge pillar outer surface. To achieve this condition, constraints from styling surface and door hinge construction must be carefully coordinated during preliminary design (19).

Now that the vehicle constraints have been examined, we can consider some of the options for layout of the structure members, which adhere to constraints yet efficiently react to structural loads.

8.5 Body Structure Topology

In creating an efficient structure topology, there must be a *Load Path* which supports equilibrium for each of the major load cases, Figure 8.33. That is, each external load must be reacted by a chain of internal forces, each creating equal opposite reactions without excessive deflection. A common visualization for load path is a chain. Each link is necessary, with the effectiveness of the chain determined by the weakest link. For example, consider the model used in Chapter 4 for body bending, Figure 8.34. A load path is provided for the vertical reaction load, R_F, through the motor

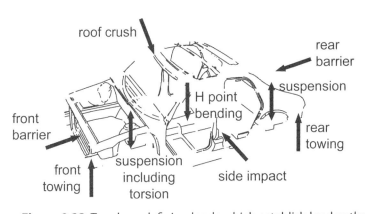

Figure 8.33 Topology defining loads which establish load paths.

compartment shear panel to an upper fore-aft load, A. This load has a reaction path through the cowl member. Now consider an alternative topology, Figure 8.35a. In this topology, the motor compartment shear panel is less tall than that in Figure 8.34. Now the load A is applied to the interior of the dash panel. As a very thin plate, the dash panel lacks the ability to react the normal load A, so a valid load path has not been provided with this topology. If we are constrained to have the lower shear panel, we can create a load path by adding a cross member beam, Figure 8.35b, which now provides a load path for load A to be reacted.

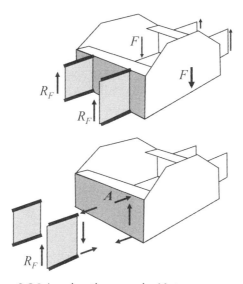

Figure 8.34 Load path example: Motor compartment.

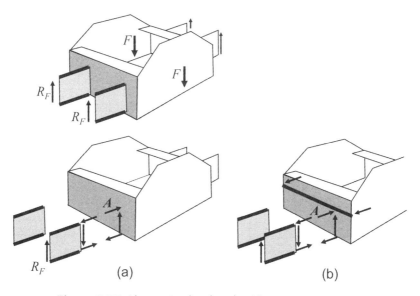

Figure 8.35 Alternative load paths: Motor compartment.

From solely structural considerations, the most efficient topology enables a load path in which structural members are stressed fully. This means a load path in which structural members are axially loaded beams and shear- or membrane-loaded panels. However, spatial constraints imposed by the vehicle prevents this ideal load path, and bending is introduced into the topology. Whenever bending occurs, we are not stressing material fully with the resulting greater structure mass. Keeping these structural objectives in mind, we continue to focus on body-integral construction and will now investigate the topology of the underbody, and the topology surrounding the suspension attachments.

Example: Underbody topology alternatives

Figure 8.36 illustrates three typical options for underbody topology. Each of these provides a load path for fore-aft and vertical loads applied at the bumpers. Each meets the vehicle spatial constraints. Each represents a balance between section sizing and vehicle spatial needs which is unique to a specific vehicle type—sedan, low sport car, and sport utility vehicle.

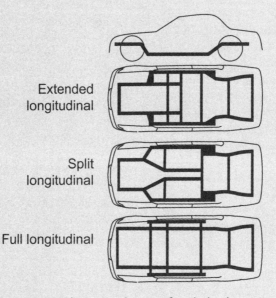

Figure 8.36 Alternative layout of underbody structure.

To gain more insight into this balance, consider the underbody structure, Figure 8.37. In the plan view, layout constraints place the front midrail inboard from the edge of the vehicle, while other constraints place the rocker at the outboard edge of the vehicle. This plan view spatial misalignment requires a load path to be constructed to allow the rocker to react the fore-aft and vertical bumper loads shown. A similar misalignment exists in the side view where constraints place the mid rail and rocker at different heights. The designer has several options for load paths between midrail and rocker in this motor compartment transition area.

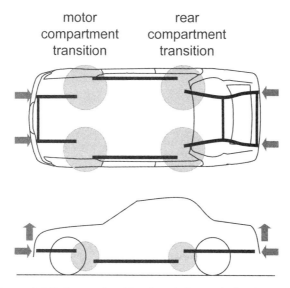

Figure 8.37 Constrained load path for underbody members.

Some of the more common topologies are illustrated in Figure 8.38. One of the most common transitions is by extending the midrail under the floor with a longitudinal rail ending at the number three cross member, Figure 8.38a. A torque box is often used instead of a full number-one cross member. This topology, widely used in sedans, is a good compromise between low floor height and structural efficiency. A similar topology—full longitudinal, Figure 8.38b, extends the longitudinal rail under the floor to the rear rail. The full longitudinal is used on cross-over and sport utility vehicles where the higher floor height allows the package space for the rail section. In the split longitudinal topology, the midrail is extended to the rocker and down the tunnel, Figure 8.38c. This configuration allows a lower floor height and is common on sport sedans and coupes.

Two less frequently seen but viable topologies are shown in Figure 8.38d and e. In the swept rail topology, the midrail is not extended under the floor as in the extended longitudinal configuration. All fore-aft bumper load is transmitted by a moment to the rocker. This makes this topology relatively inefficient, but it does achieve a low floor height. Finally, a vertical member may be added at the dash to react midrail loads, Figure 8.38e. This topology reduces internal bending moments, but at the expense of requiring package space within the motor compartment.

A similar transition occurs at the rear compartment, Figure 8.37. Three typical topologies for this area are shown in Figure 8.39. Again, each topology results from a balance of vehicle and structural needs. Both the shear box, Figure 8.39a, and splayed rail, Figure 8.39b, are seen in sedans. The splayed rail offers the lowest rear seated height and is frequently used for sport sedans and coupes. The continuous longitudinal, Figure 8.39c, is a balance between increased load capability but at an increase in floor height. The continuous longitudinal topology is frequency used in vans and cross-over vehicles.

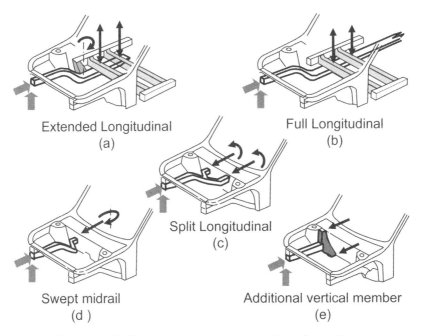

Figure 8.38 Motor compartment transition alternatives.

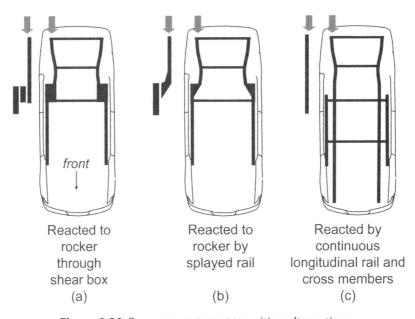

Figure 8.39 Rear compartment transition alternatives.

344

8.6 Load Path Design at Suspension Attachments

Ground vehicles, unlike ships or airplanes, are unique in that support loads are highly concentrated at the suspension attachment points rather than distributed over a large area. Care must be taken to provide a load path which transfers these concentrated point loads to the greater monocoque structure. In this section, we consider the load paths for the tire patch loads, Figure 8.40. First we will look at the motor compartment topology, followed by the rear compartment [8].

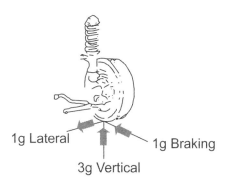

Figure 8.40 Secondary loads: Topology for suspension attachments.

8.6.1 Motor compartment topology

For McPherson strut suspensions, the primary suspension load of interest for topology is the 3-g vertical load. This vertical tire patch load is primarily reacted at the shock absorber attachment point, Figure 8.41, where both a vertical and lateral component exist. In the context of the integral body structure, Figure 8.42, these components are reacted by a set of shear panels and bars.

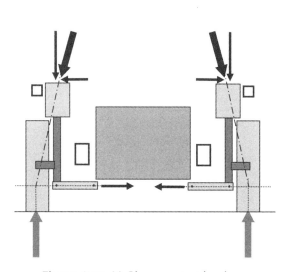

Figure 8.41 McPherson strut loads.

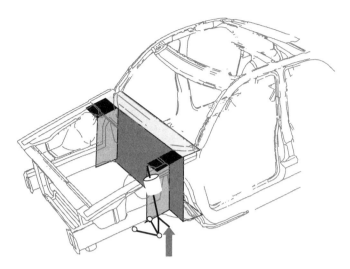

Figure 8.42 Front compartment topology.

An idealization of a typical motor compartment is shown in Figure 8.43. The motor compartment topology includes a horizontal panel, A, a horizontal dash panel, B, a motor compartment side panel, C, a dash, D, and a floor panel, E. The point load at the shock absorber attachment consists of a lateral component, L, and vertical component V. Specific numerical values for these components may be calculated using the approach discussed in Chapter 2. In this section, we are only concerned with providing a path to react the loads, and for this purpose all that is needed is the load direction and point of application.

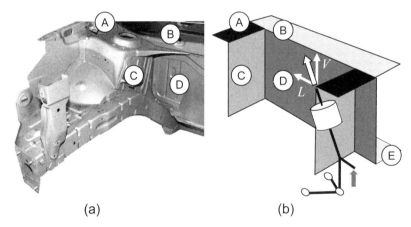

(a)   (b)

Figure 8.43 Load applied at strut attachment point. (Photo courtesy of A2Mac1.com, Automotive Benchmarking)

First consider the vertical component of the attachment load, V, Figure 8.44. As shown in Chapter 4, by placing each element into static equilibrium we can show that a valid load path exists. The load V flows through the motor compartment and

346

is effectively reacted by the cabin monocoque, shown by the highlighted vectors at the hinge pillar. Because the load path consists of shear panels and bars under uniform stress, this is an efficient topology.

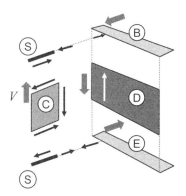

Figure 8.44 Reaction of vertical component.

Now consider the lateral component of the suspension load, L, Figure 8.45, with the same motor compartment topology. Figure 8.46a illustrates the load path with the suspension in a hop mode, both wheels moving symmetrically upward and lateral loads on opposite sides opposing each other. In this mode, shear panel A at the level of the suspension attachment provides a load path through the cowl panel, B. For the suspension hop mode, all forces applied to the cowl panel are in equilibrium and no load is passed along to the cabin monocoque. For the case of the suspension in a tramp mode with wheels moving asymmetrically, up on one side and down on the other, the lateral loads now act in the same direction as shown in Figure 8.46b. For this mode, the lateral loads are passed through the cowl panel and appear as a lateral load on the cabin monocoque. Again the path consists of elements in shear and direct stress and is efficient.

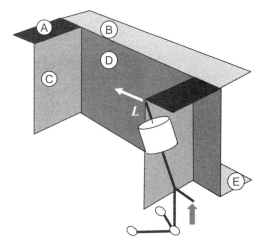

Figure 8.45 Reaction of lateral component.

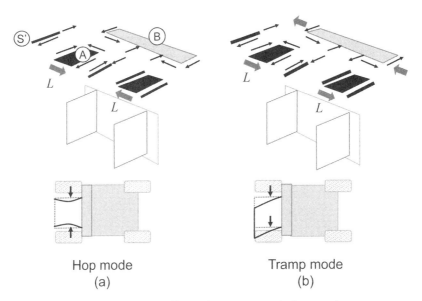

Figure 8.46 Reaction of lateral component in hop and tramp.

The motor compartment topology discussed above and shown in Figure 8.43 requires the ability to package flat shear panels A and C. Often, vehicle layout or manufacturing constraints prevent this condition. Figure 8.47 illustrates an alternative topology in which the motor compartment side panel, C, is now highly shaped to accommodate these constraints. For this topology, the reaction to both the vertical and lateral component results in bending of the upper rail, R1, and the midrail, R2, Figure 8.47.

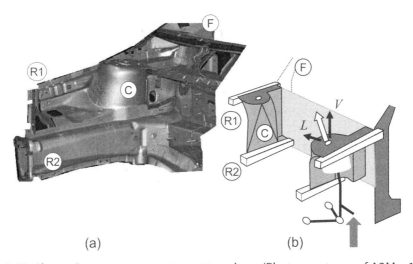

Figure 8.47 Alternative motor compartment topology. (Photo courtesey of A2Mac1.com, Automotive Benchmarking)

Both rails provide a load path for the vertical component, Figure 8.48a. While not as efficient as the shear panel topology of Figure 8.43, this topology is adequate for vertical loads. For the lateral component, Figure 8.48b, the load path is through bending in the upper rail, R1. For some cases, this load path does not provide adequate stiffness given the constraints on upper rail section size. When this occurs, an additional load path can be provided, Figure 8.49, in which a truss structure spans the motor compartment width. This topology increases the lateral stiffness, as measured at the suspension attachment, and can improve vehicle handling.

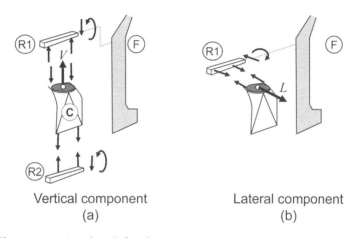

Vertical component
(a)

Lateral component
(b)

Figure 8.48 Load path for alternative motor compartment topology.

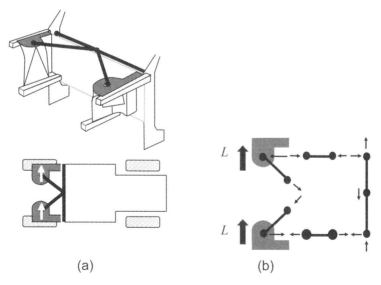

(a)

(b)

Figure 8.49 Reaction of lateral component: Cross-car member at suspension attachments.

We have been considering load paths for the shock absorber load resulting from a vertical load at the tire patch, Figure 8.42. Now consider the load path for lateral and fore-aft tire patch loads. For a McPherson strut, the majority of these loads are reacted by the lower control arm. Figure 8.50 illustrates a common topology for lower control arm attachment at the engine cradle. Typically the engine cradle is isolated from the body through rubber mounts at its four corners, Figure 8.51. The body structure load path is then through the motor compartment midrail at these attachments.

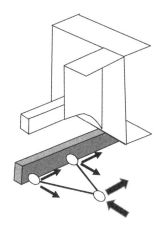

Figure 8.50 Reaction of lower control arm loads by cradle.

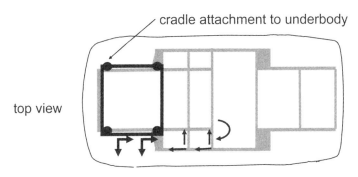

Figure 8.51 Reaction of lower control arm loads: Load path to underbody.

Finally, consider motor compartment load paths for a short and long arm suspension, SLA, rather than a McPherson strut. For the shock absorber loads, the load path is similar to that for the McPherson strut. An important difference for the SLA suspension is the lateral load applied to the body through the upper arm, Figure 8.52. This load is difficult to react using shear panels as the height of the arm is at the mid-height of the motor compartment side, Figure, 8.53a. A typical solution to this constraint is to provide a beam structure, B1, spanning from the upper rail, R1, to the midrail, R2, Figure 8.53b and c. Due to this additional structural element in bending, body structure for SLA suspension is typically greater than that for a McPherson strut. Figure 8.54 illustrates this topology for a typical sedan.

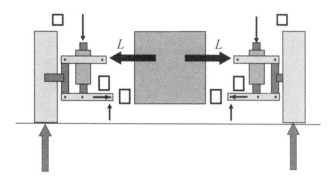

Figure 8.52 Short and long arm suspension: Lateral load at upper arm.

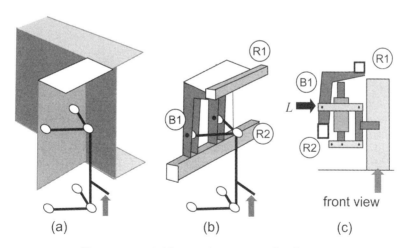

Figure 8.53 Addition of upper arm load path.

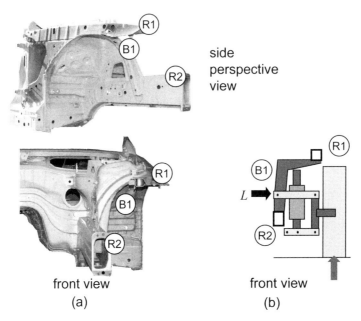

side
perspective
view

front view

(a)

front view

(b)

Figure 8.54 Example: Upper arm load path. (Photo courtesy of A2Mac1.com, Automotive Benchmarking)

8.6.2 Rear compartment topology

In this section, we consider the topology for a rear compartment with a Chapman strut suspension. Again, the focus will be on the shock absorber load resulting from a vertical load at the rear tire patch. Figure 8.55 illustrates a typical rear compartment topology consisting of the inner quarter panel, A, a package shelf, B, the rear seat panel, C, a rear rail, D, a rear cross member, E, and the strut tower, F, where the shock absorber load is applied. In this topology, the top of the strut tower is directly connected to the upper corner of the rear seat panel.

The lateral component, L, during suspension tramp, Figure 8.56, flows through the upper corners of the seat back panel due to the direct connection. The panel is then in a state of shear, with the right and left edge loads reacted by the quarter panel inner. As a variant of this topology, a cross car lateral beam, Figure 8.57, provides a load path to the cabin monocoque. While this variant is structurally less efficient, it allows an open pass-through between passenger compartment and trunk.

The vertical component, V, is applied to the strut tower, F, Figure 8.58. The two insets show the end view of a section through the strut tower, F, the rear rail, D, and the quarter inner panel, A. Two alternative topologies for this area are shown. In topology 1, the strut-tower-to-rear-rail connection is through an off-set flange. This construction allows spot weld access during assembly, but is very flexible for the vertical load shown. For this topology, there is not a direct load path between the applied load, V, and the rail structure, D, which can react the load. The role of the rear cross member, E, is to provide this load path. For topology 1, the load

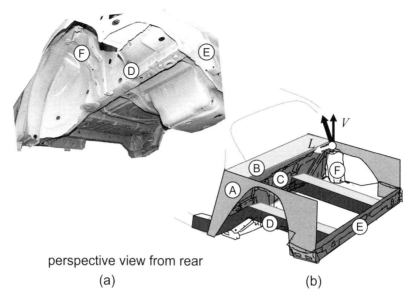

perspective view from rear

(a)   (b)

Figure 8.55 Rear compartment topology: Strut load. (Photo courtesy of A2Mac1.com, Automotive Benchmarking)

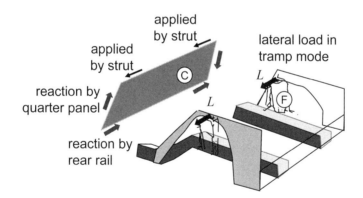

Figure 8.56 Rear compartment lateral strut load: Seat back panel.

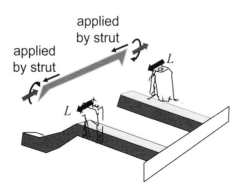

Figure 8.57 Rear compartment lateral strut load: Seat lateral beam.

V is reacted by the inner quarter panel, A, generating load R_A. Load R_A flows from the rear of the inner quarter panel to the rear cross member. This load then flows from the cross member to the end of the rear rail as load R_D. The rear cross member thus effectively connects the outer monocoque to the underbody rails. Alternative topology 2 is shown in the right inset of Figure 8.58. Here the vertical load is transmitted directly from the strut tower to the rear rail through a shear flange.

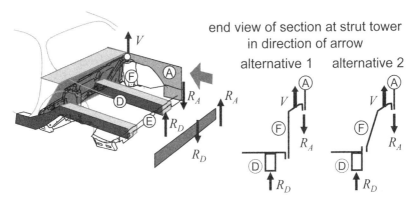

Figure 8.58 Rear compartment vertical load: Load path between quarter inner panel to rear rails.

8.7 Summary

Body structure topology refers to the overall layout of structural elements. This layout defines load paths for the primary structural requirements discussed in previous chapters. An efficient load path will uniformly stress elements, and usually this implies an avoidance of bending within the structure. Once the layout is defined, the structural elements may then be sized based on the principles discussed in Chapter 3.

Because vehicle spatial constraints restrict the load path shape, the ideal state of uniform stress is rarely realized. In this chapter we have discussed some of these constraints due to component layout, styling, and aerodynamics. With an understanding of the origin of the constraints, the structural designer can better reach a balance between vehicle and structural needs.

Some typical topologies for the underbody, the motor compartment, and the rear compartment were discussed. While this discussion did not exhaust all possibilities, an approach for reasoning through the creation of load paths was provided.

References

1. Fenton, J., *Vehicle Body Layout and Analysis*, Mechanical Engineering Publications, Ltd, London, 1980, p. 40–51.

2. Pawlowski, J., *Vehicle Body Engineering*, Business Books, London, 1969, Chapter 6.

3. Torenbeek, E., *Synthesis of Subsonic Airplane Design*, Delft University Press, The Netherlands, 1988, Chapter 2.

4. Kikuchi, N & Malen, D., Course notes for ME513 Fundamentals of Body Engineering, University of Michigan, Ann Arbor, MI, 2007.

5. Milliken, W. & Milliken, D., *Chassis Design, Principles and Analysis*, SAE International, Warrendale, PA, 2002, pp. 74–76.

6. Gillespie, T. D., *Fundamentals of Vehicle Dynamics*, SAE International, Warrendale, PA, 1992, pp. 309–314.

7. "Table for Estimating Aerodynamic Drag Based on Visual Characteristics," Motor Industry Research Association, Society of Automotive Engineers Journal, June, 1969, SAE International, Warrendale, PA. (reprinted in Industrial Design, 1977).

8. Brown, J.C., Robertson, A. J., Serpento, S.T., *Motor Vehicle Structures*, SAE International, Warrendale, PA, 2001, pp. 101–118.

9. Longo, S., Moss, E., and Deutschel, B., "The 1997 Chevrolet Corvette Structure Architecture Synthesis", SAE Paper No. 970089, SAE International, Warrendale, PA, 1989.

10. Morgan, R. & Drees, H., "Lecture 5–1; GM Sunraycer Case History," M-101, SAE International, Warrendale, PA, 1980.

Chapter 9
Material Selection and Mass Estimation in Preliminary Design

In this chapter we discuss two tools which support the automobile body design task. The first treats body materials and their properties and provides a methodology for selection of the best material for a particular function. Second, we cover mass estimation during the preliminary design stage. This topic is important due to the dependence of structural requirements on mass.

9.1 Materials for the Body-In-White

The most common material for the contemporary body structure is steel. Historically, the predominant steel grade was mild steel, which exhibited a favorable balance of strength, formability, and cost. As structural requirements have become more demanding, primarily due to evolving safety standards, steel application has expanded to higher-strength grades, Figure 9.1.

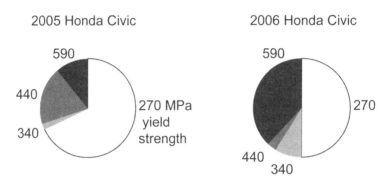

Figure 9.1 Evolving steel-grade application in body-in-white.

9.1.1 Alternative steel grades

As steel grade varies, yield strength spans a wide range of approximately *180 N/mm²* *(26,100 psi)* to *1200 N/mm²* *(174,000 psi)*. However, the increase of strength comes at a cost. Figure 9.2 illustrates the strength and elongation at failure for a range of steel grades [1, 2]. An important trade-off is clear; increased strength comes at a reduced elongation at failure. While the value of higher yield is evident, higher elongation is important both for shaping parts during metal forming, and also for the large deformations required of energy-absorbing structures. The role of advanced high-strength steels such as Dual Phase and TRIP is to provide increased elongation as compared with traditional high-strength steels, effectively moving the elongation-strength trend line upward. Thus, for a particular structural member with required strength and elongation a region of this graph is indicated. For example, a highly shaped semi-structural part such as a quarter panel falls in the upper left of Figure 9.2. A strength-dominant structure, not highly shaped, such as B pillar reinforcements, falls in the lower right. An energy-absorbing structure, such as the motor compartment midrail, falls in the center. Generally as yield stress increases, the cost per unit mass increases, Table 9.1 [3].

The reader is referred to the AISI *Automotive Steel Design Manual* [3] for a more in-depth description of these steel grades.

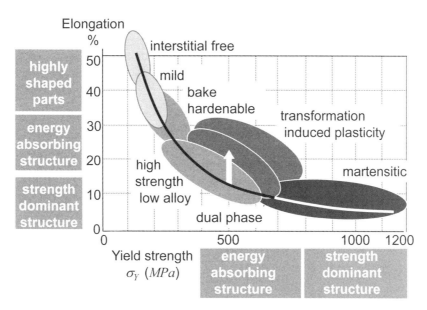

Figure 9.2 Properties for various steel grades. (Courtesy of the American Iron and Steel Institute, Automotive Steel Design Manual)

Table 9.1 Approximate relative steel cost.

Steel Type	Relative Cost
Cold Rolled (Mild)	*1.00*
Bake Hardenable	*1.10*
HSLA	*1.15*
Dual Phase	*1.40*
Martensitic	*1.50*

Example: Steel grade substitution for B-pillar reinforcement

As an example of substituting a higher-strength steel grade for mild steel, consider a B-pillar reinforcement in the form of a hat section, Figure 9.3. In this example, the section width, b, is set by vehicle constraints. Under side impact, this member is loaded by a bending moment, M_{DESIGN}, which places the outer plate in compression. For the original mild steel design, the thickness, t, was chosen such that buckling stress for this plate is at yield, as shown in the illustration. When high-strength steel is substituted, the plate buckling stress must now be increased to the higher yield stress, Figure 9.3. It can be seen that the b/t ratio must be reduced to stabilize the plate at the higher yield stress. We can do this by reducing the plate width by adding ribs even as the thickness is reduced to save mass [4]. In general, as with this example, when we substitute an alternative material in an existing design we must also adjust section shape as well as thickness to take full advantage of the properties of the new material.

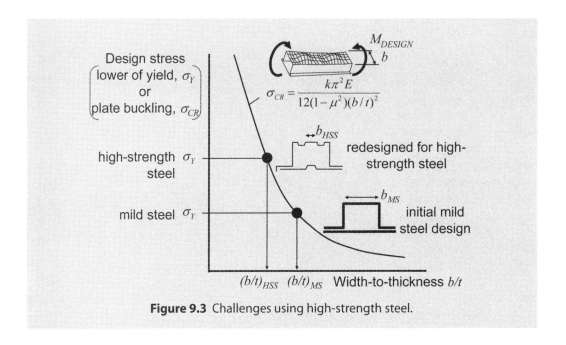

Figure 9.3 Challenges using high-strength steel.

9.1.2 Alternative materials for the body in white

In this section we expand material alternatives to include the most common material alternatives for body structure: mild steel, advanced high-strength steels (AHSS), aluminum alloys, magnesium alloys, glass fiber reinforced plastic (GFRP) in the form of a woven mat in epoxy matrix, carbon fiber reinforced plastic (CFRP) in the form of a woven mat in epoxy matrix. We include two other options with serious flaws for automotive structure for completeness 1) sheet molding compound, a chopped glass fiber reinforced polyester matrix (SMC) which may be molded in a matched die set to enable a cycle time compatible with automotive line rates, but has a low modulus; and 2) titanium, a lightweight metal which has very high strength and a very high cost. Figure 9.4 provides representative values for some of the important properties for these materials [5, 6, 7, 8, 9].

To compare these materials in a typical automotive application, consider the rocker sizing exercise from Chapter 3, Figure 9.5. The rocker is supported at the ends and loaded center span by a load, *P=3335 N (750 lb)*. In Example 1 the rocker dimensions are constrained by the vehicle layout, Figure 9.5 top. As alternative materials are substituted, only the thickness will be adjusted to meet the dominant stiffness requirement (a deflection of *1 mm* or less under the load *P*), and the resulting rocker mass will be compared for each material. In Example 2, we will relax the dimensional constraints on the rocker and optimize both the width and thickness dimensions to minimize mass for each alternative material, Figure 9.5 bottom. The thickness will be adjusted to meet both the stiffness and strength requirements (no failure under the load *P*) and the resulting mass and material cost compared. Six materials will be compared: steel, aluminum, CFRP, GFRP, magnesium, and titanium. The material properties are those given in Figure 9.4.

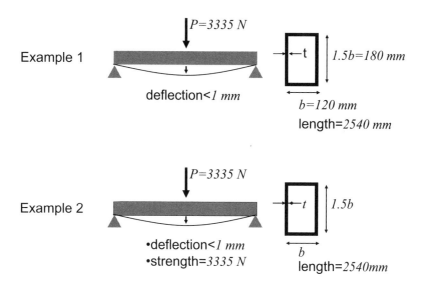

Figure 9.4 Typical properties: Body structure materials.

Materials	Modulus E GPa	Strength Yield σ_Y MPa	Tensile σ_F MPa	Strain ε_F	Density ρ kg/m³	Cost C_P $/kg
Steel						
Mild Steel	200	220	400	0.30	7860	0.80
AHSS	200	1030	1200	0.18	7860	1.00
Nonferrous Metals						
Aluminum	70	260	290	0.05	2710	2.80
Magnesium	45	150	276	0.06	1830	4.80
Titanium	120	920	1000	0.06	4430	100.00
Reinforced Plastics						
SMC	10	-	80	0.01	2000	6.00
GFRP	25	-	130	0.01	1850	20.00
CFRP	160	-	800	0.01	1600	40.00

Figure 9.5 Example: Rocker material selection.

Example 1: Material substitution for rocker-constrained section size

In this example, the vehicle package constrains the rocker width to a maximum of $b=120\,mm$, and constrains the height to $1.5b=180\,mm$.

The deflection for a center-loaded beam supported at the ends is:

$$\delta_0 = \frac{PL^3}{48EI} \tag{9.1}$$

where:

P = applied load (3335 N for this example)

L = beam length (2540 mm for this example)

E = modulus of elasticity

δ_0 = allowable deflection under load P (*1 mm* for this example)

The moment of inertia for this thin-walled section is:

$$I = 2\left(\frac{rb}{2}\right)^2 bt + 2\frac{(rb)^3 t}{12} = tb^3 r^2\left(\frac{1}{2} + \frac{r}{6}\right) \tag{9.2}$$

where:

b = section width (120 mm for this example)

r = ratio of section height to width, r = height/width (r=1.5 for this example)

t = material thickness

Substituting Equation 9.2 into 9.1:

$$\delta_0 = \frac{PL^3}{48Etb^3 r^2\left(\frac{1}{2} + \frac{r}{6}\right)}$$

$$tb^3 = \frac{PL^3}{48Er^2\left(\frac{1}{2} + \frac{r}{6}\right)}\frac{1}{\delta_0} \tag{9.3}$$

Solving for thickness:

$$t = \frac{PL^3}{48Er^2\left(\frac{1}{2} + \frac{r}{6}\right)}\frac{1}{b^3}\frac{1}{\delta_0} \tag{9.4}$$

The mass of the beam is:

$$m = 2t(b + rb)L\rho = 2tbL\rho(1 + r) \tag{9.5}$$

where ρ = mass density

Substituting 9.4 into the above:

$$m = 2\rho(1+r)\frac{L}{b^2}\left[\frac{PL^3}{48Er^2\left(\frac{1}{2} + \frac{r}{6}\right)}\frac{1}{\delta_0}\right] \tag{9.6}$$

Now substituting values for density, ρ, and Young's modulus, E, for each material into Equation 9.6 allows us to compare materials. At the top of Figure 9.6 the material properties are shown, the bottom of Figure 9.6 shows the resulting beam mass and thickness for each material for this constrained section size example. It can be seen that for this constrained section size case, only CFRP offers a significant mass reduction compared to steel. Note that despite its low density, aluminum does not reduce mass compared to steel for this constrained case. This is because the thickness must be increased to meet the deflection requirement.

	Steel	Al	CFRP	GFRP	Mg	Ti
E (N/mm^2 x 10^3)=	200	70	160	25	45	120
density ρ (kg/mm^{3}x10^{-6})=	7.86	2.80	1.60	1.85	1.83	4.43
μ=	0.3	0.3	0.2	0.2	0.3	0.3
Example 1						
constrained b (mm)=	120	120	120	120	120	120
t (mm)=	1.95	5.58	2.44	15.62	8.68	3.25
constrained mass (kg)=	23.39	23.04	5.95	44.03	24.20	21.97

Figure 9.6 Example: Rocker sections for various materials.

Example 2: Material selection for rocker-unconstrained section size

In this example, we relax the vehicle package constraints and are free to vary rocker height and width as well as thickness to minimize mass for each material. In addition to the stiffness requirement used in Example 1, we add a strength requirement; the top plate of the section will just begin to buckle at load $P=3335\,N$, Figure 9.5 bottom.

From basic beam equations, the stress at the top plate of the section at mid span is:

$$\sigma_D = \frac{Mc}{I} = \frac{\left(\frac{PL}{4}\right)\left(\frac{rb}{2}\right)}{I} = \frac{PLrb}{8I} \tag{9.7}$$

where:
M = bending moment at the center of the span, $M=PL/4$

σ_D= design stress taken as the minimum of either yield, σ_y, or plate buckling, σ_{CR}:

$$\sigma_D = \min\left\{ \begin{array}{l} \sigma_Y \\[2ex] \sigma_{CR} = \dfrac{k\pi^2 E}{12\left(1-\mu^2\right)}\dfrac{t^2}{b^2}, \quad k=4 \end{array} \right\} \tag{9.8}$$

Assuming failure will be by plate buckling, we substitute Equations 9.8 and 9.2 into Equation 9.7 to arrive at the optimal thickness:

$$t^* = \left[\frac{3PL}{8r\left(\frac{1}{2}+\frac{r}{6}\right)} \frac{\left(1-\mu^2\right)}{\pi^2 E} \right]^{1/3} \tag{9.9}$$

Substituting this value for t into Equation 9.3, the stiffness constraint, gives the optimal section size:

$$b^* = \left[\frac{PL^3}{48Er^2\left(\frac{1}{2}+\frac{r}{6}\right)} \frac{1}{t} \frac{1}{\delta_0} \right]^{1/3} \tag{9.10}$$

with the resulting rocker mass:

$$m = 2\rho(1+r)L \left[\frac{PL^3}{48Er^2\left(\frac{1}{2}+\frac{r}{6}\right)} \frac{1}{\delta_0} \right]^{1/3} \left(\frac{3PL}{8r\left(\frac{1}{2}+\frac{r}{6}\right)} \frac{\left(1-\mu^2\right)}{\pi^2 E} \right)^{2/9} \tag{9.11}$$

Substituting material properties for each material into Equations 9.9, 9.10, and 9.11 gives the results tabulated at the bottom of Figure 9.6 and shown graphically in Figure 9.7b. Now we see considerable mass reduction relative to steel; for instance, aluminum at *64%* of steel mass and CFRP at *23%* of steel mass.

However, in addition to mass reduction we must consider other objectives. First, consider the objective of rocker packaging within the vehicle. As we optimized the rocker shape for alternative materials, we arrived at the optimum section size indicated by *b**. At the top of Figure 9.8, we see that all other materials will require more space (and step over-height) than steel. Therefore, to fully take advantage of alternative materials, the vehicle package will need to be designed to accommodate the larger sections early in the design process.

A second objective in material selection is to minimize material cost. At the bottom of Figure 9.8, the resulting rocker beam cost for various materials is shown (the material costs per unit mass used are those shown in Figure 9.4). Note that the cost axis is logarithmic. This graph shows the trade-off between material cost and mass. It can be seen that CFRP, magnesium, aluminum, and steel form the best set of alternatives. Depending on the value assigned to a unit mass reduction, one of these materials will be best: if the value is low, steel is best; if the value is great, CFRP is best; with magnesium and aluminum falling in between. Note also that both GFRP and titanium are always less desirable choices regardless of the value of mass reduction. For instance, aluminum is both lighter and less costly than either GFRP or titanium. Ultimately, the designer will make a material selection decision based on total cost, which includes material cost, part shaping cost, and assembly cost. Here we limit our current analysis to material cost alone, even though consideration of these other costs will influence material ranking.

		Steel	Al	CFRP	GFRP	Mg	Ti
Example 2	b^* *(mm)=*	*145.6*	*183.9*	*152.1*	*229.8*	*202.9*	*163.2*
	t^* *(mm)=*	*1.09*	*1.55*	*1.20*	*2.22*	*1.80*	*1.29*
unconstrained mass *(kg)=*		*15.87*	*10.13*	*3.70*	*12.00*	*8.46*	*11.88*
mass relative to steel=		*1.00*	*0.64*	*0.23*	*0.76*	*0.53*	*0.75*

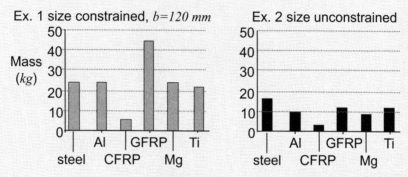

Figure 9.7 Example: Comparison of rocker mass for various materials.

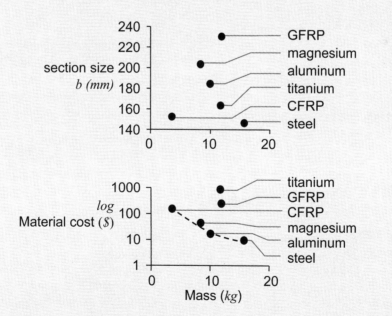

Figure 9.8 Example: Trade-offs for rocker sections of various materials.

The above examples show that material selection can have a large influence on mass, material cost, and size of structural elements. They also illustrate that mass performance of a material depends strongly on the nature of the constraints on section dimensions. In the next section, we develop a quick method to screen materials, accounting for structural requirements and dimensional constraints.

9.1.3 Automobile body material selection methodology

To develop a material selection methodology, consider the result of Example 1 above where we derived an expression, Equation 9.6, for beam mass meeting the deflection requirement, δ_0, under a load, P. Here we have rearranged the terms:

$$m = \left[\frac{(1+r)L^4}{12r^2\left(1+\dfrac{r}{3}\right)b^2} \right] \left[\frac{P}{\delta_0} \right] \left[\frac{\rho}{E} \right] \tag{9.12}$$

In this expression, the first bracketed term is related to geometry constraints, the second to structural requirements, and the third term to material properties. It can be seen that by minimizing the third term we are also minimizing mass. We call the reciprocal of the third term a material index for this particular function (stiff beam) and objective (minimize mass), and by maximizing the material index, we minimize mass [8, 9, 10].

$$\text{Material index for light, stiff beam:} \left[\frac{E}{\rho} \right] \tag{9.13}$$

For other functions such as strength, plastic moment, and panel stiffness, other forms of the material index will result. Figure 9.9 summarizes the material indices for thin-walled beam elements for several functions. Figure 9.10 summarizes

Structural element and loading	Constrained dimensions	Material index for minimum mass (for minimum cost replace $\rho \Rightarrow \rho C_P$)				
		function stiff	function strong		function stiff and strong	
			σ_D=yield stress	σ_D=plate buckling	σ_D=yield stress	σ_D=plate buckling
beam	b, r	E/ρ	σ_Y/ρ	$E^{1/3}/\rho$	—	—
	r	—	—	—	σ_Y/ρ	$\dfrac{E^{5/9}}{\rho}$
plastic hinge		adequate ductility is required to place full section at yield				
	b, r	—	σ_Y/ρ	—	—	—

Figure 9.9 Material indices for thin-walled beam elements.

Structural element and loading	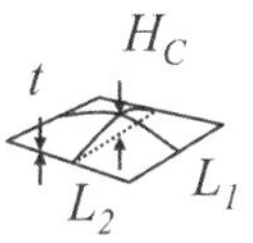 H_C t L_2 L_1 constrained dimensions	Material index for minimum mass *(for minimum cost replace $\rho \Rightarrow \rho C_P$)* Function: Stiff	Material index for maximum resonant frequency all dimensions constrained
Panel: highly crown 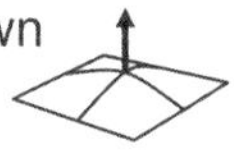	L_1, L_2, H_C	$\dfrac{\left(E/\sqrt{1-\mu^2}\right)^{1/2}}{\rho}$	$\dfrac{E/(1-\mu^2)}{\rho}$
Panel: flat 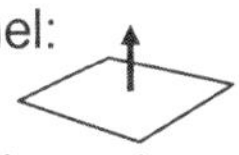 s.s. boundary	L_1, L_2	$\dfrac{\left[E/(1-\mu^2)\right]^{1/3}}{\rho}$	$\dfrac{E/(1-\mu^2)}{\rho}$

Figure 9.10 Material indices for normally loaded panels.

them for flat and crown panels under a central normal load. (No material index is included for energy absorption due to the high dependence between energy absorption and section shape. Forming processes, which vary between metals, allow shape optimization differences, which can overshadow material property differences. For example, aluminum can be extruded, producing sections with very efficient internal ribs of varying thickness. For fiber-reinforced composites, the energy absorption mechanism is by flaying rather than by plastic deformation as it is in metals, making comparisons by a simple material index impossible.)

Often the objective of material selection is to minimize material cost rather than mass. The same form of material index may be used with a simple substitution. In deriving the material index we have minimized part mass given by:

$$\text{Mass} = (\text{material volume}) \, (\text{mass/unit volume})$$

$$\textit{Mass} = (\textit{material volume}) \, (\rho)$$

Now compare the above to a similar expression for part cost:

$$\text{Cost} = (\text{material volume}) \, (\text{mass/unit volume})(\text{cost/unit mass})$$

$$\textit{Cost} = (\textit{material volume}) \, (\rho) \, (C_p)$$

where C_p = cost per unit mass

It can be seen that we can minimize cost by applying the corresponding material index but replacing $\rho \Rightarrow \rho C_p$. We can now apply the idea of material index to quickly rank materials for a specific function and the objective—minimizing material mass or cost to perform the function.

Example: Material selection for automotive hood panel

The shaded area in Figure 9.11 indicates a region of the outer hood panel supported around its perimeter by the inner hood structure. The dimensions, L_1, L_2, and H_C are constrained. The normal stiffness of the panel, K, must meet a minimum value, K_{REQ}. We are free to select the panel material and to set the thickness. Our objective in material selection for this example is to minimize the hood panel mass.

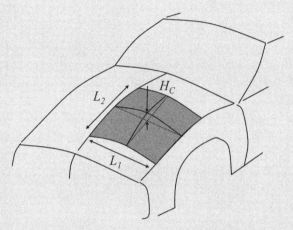

Figure 9.11 Example: Automobile hood material selection.

1. Primary function: meet normal stiffness requirement

2. Objective in material selection: minimize panel mass

3. Derive an appropriate material index. The normal stiffness of a panel, Equation 3.35 and 3.37b, is:

$$K = B \frac{4H_C}{L_1 L_2} \frac{Et^2}{\sqrt{1-\mu^2}}$$

Solving for thickness:

$$t = \left[\left(\frac{L_1 L_2}{4BH_C} \right) (K_{REQ}) \left(\frac{\sqrt{1-\mu^2}}{E} \right) \right]^{1/2}$$

The mass of the panel is then:

$$m = (\text{Area})tp = \left[L_1 L_2 \left(\frac{L_1 L_2}{4BH_C} \right) (K_{REQ}) \left(\frac{\sqrt{1-\mu^2}}{E} \right) \right]^{1/2} \rho$$

m=function of [given panel dimensions] [given structural requirement] [material properties]

Therefore, to minimize mass, m, we should maximize the material index:

$$\frac{\left(E / \sqrt{1-\mu^2} \right)^{1/2}}{\rho}$$

Here we have derived the material index; it is also included in Figure 9.10. Substituting values for material properties from Figure 9.4 into the above gives the following:

Table 9.2 Material index for stiff, light hood panel.

Material	E (N/mm²)	μ	ρ(kg/mm³x10⁻⁶)	$\dfrac{\left(E/\sqrt{1-\mu^2}\right)^{1/2}}{\rho}$
Aluminum	70,000	0.3	2.71	10.00×10^7
Steel	200,000	0.3	7.86	5.83×10^7
CFRP	160,000	0.2	1.60	25.26×10^7
GFRP	25,000	0.2	1.85	8.63×10^7
SMC	10,000	0.3	2.00	5.12×10^7

From Table 9.2, the material which maximizes the material index is CFRP and would therefore provide the minimum mass hood panel. Note that we can make this selection without knowing the specific values for panel dimensions or stiffness requirement. Hence, the material index is a very convenient method to rank material alternatives before spending extensive analysis time.

Example: Material selection for automotive hood panel-graphical solution

An alternate means to apply the material index is graphically using a cross plot of material properties [6]. Consider again the mass of the panel for this example:

$$m = (\text{Area})t\rho = \left[L_1 L_2\left(\frac{L_1 L_2}{4BH_C}\right)(K_{REQ})\left(\frac{\sqrt{1-\mu^2}}{E}\right)\right]^{1/2} \rho$$

This can be simplified to:

$$m = C''\left(\frac{\sqrt{1-\mu^2}}{E}\right)^{1/2} \rho$$

where the constant C'' depends on constrained dimensions, L_1, L_2, and H_C, and the given stiffness requirement. Taking logs of each side gives:

$$\log(m) = \log C'' + \frac{1}{2}\log\frac{\sqrt{1-\mu^2}}{E} + \log\rho$$

$$\log\left(\frac{E}{\sqrt{1-\mu^2}}\right) = 2(\log\rho) + (\log C'' - \log m)$$

which has the form of a straight line on a log-log plot:

$$y = (slope) \, x + intercept$$

where:

$$y \Rightarrow \log\left(\frac{E}{\sqrt{1-\mu^2}}\right)$$

$$x \Rightarrow (\log \rho)$$

slope = 2 for this case

intercept = (log *C″*–log *m*) which is constant for a given mass. As mass, *m*, decreases the intercept increases.

Figure 9.12 illustrates a log-log graph of $\left(\frac{E}{\sqrt{1-\mu^2}}\right)$ vs. ρ, on which are shown properties for the four

materials under consideration for the hood panel. A line with slope=2 is shown. All materials falling on this line will result in the same hood mass. Materials falling on a line upward to the left (a greater intercept) will result in a lower hood mass. It can be seen that steel and SMC will result in a similar

mass while CFRP will be lighter. In a similar way, Figure 9.13 illustrates a log-log graph of $\left(\frac{E}{\sqrt{1-\mu^2}}\right)$

vs. (ρC_p). A line of slope=2 now represents materials resulting in an equal material cost. Materials falling on a line upward to the left will result in a lower-material-cost hood.

By combining the results of Figure 9.12 for mass, and Figure 9.13 for cost, we can investigate the material mass / material cost trade-off, Figure 9.14. Steel, aluminum, and CFRP are efficient alternatives, depending on the value assigned to mass reduction. Note that the performance of a plywood hood has been included to illustrate the potential of a nonconventional hood material.

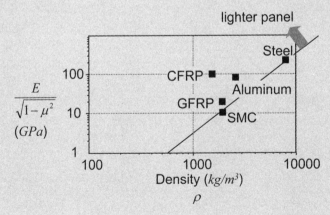

Figure 9.12 Example: Material index for light, stiff hood.

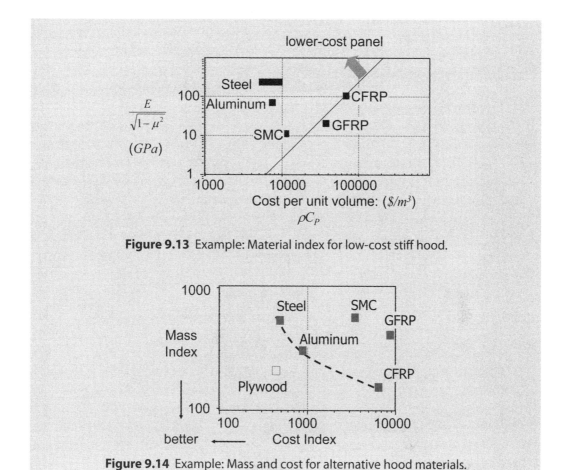

Figure 9.13 Example: Material index for low-cost stiff hood.

Figure 9.14 Example: Mass and cost for alternative hood materials.

Material property graphs for frequently considered body structure materials, Figures 9.15 to 9.18, are included to facilitate this graphical approach to ranking materials using material indices.

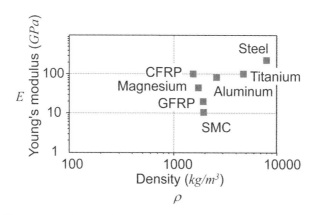

Figure 9.15 Material properties: Young's modulus vs. density.

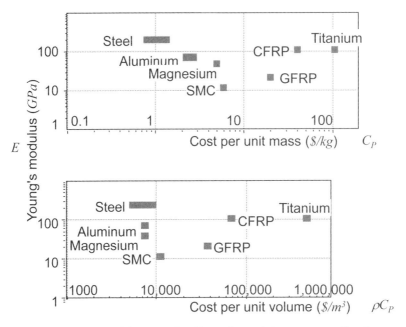

Figure 9.16 Material properties: Young's modulus vs. normalized cost.

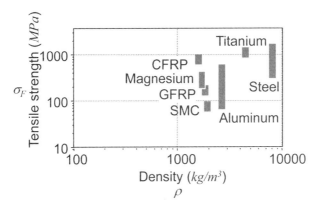

Figure 9.17 Material properties: Tensile strength vs. density.

9.1.4 Summary: Material selection

In the examples above, a method was demonstrated for preliminary selection of materials. This method was derived from reference [6] to which the reader is referred for a more in-depth discussion. The following are the suggested steps for ranking material alternatives.

1. Determine the primary function of the structural element: stiffness, strength, and vibration performance for panels.

2. Determine the objective of material selection: minimize mass or minimize material cost.

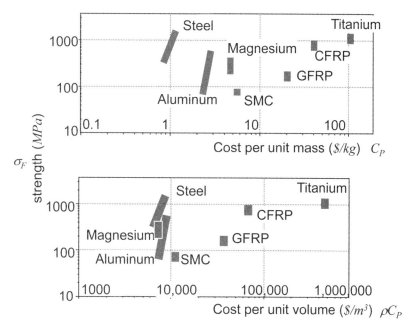

Figure 9.18 Material properties: Tensile strength vs. normalized cost.

3. Select or derive the material index corresponding to the type of structure, the function, the objective, and constrained dimensions. Figures 9.9 and 9.10 summarize common indices.

4. Rank materials by descending material index, either using a tabular form or graphically on material property cross-plots, Figures 9.15 to 9.18. The alternative which maximizes material index is the most preferred.

5. Make the final material selection after gathering additional information. Manufacturability, part integration opportunities, performance under secondary requirements, compatibility with adjoining materials, and fastening methods will vary greatly with material selection. Frequently, cost reductions in these other areas will allow a material with a higher material cost to be competitive on a total cost basis.

9.2 Preliminary Mass Analysis

A careful review of structural requirements will show that most of the applied loads are inertial loads from the mass of vehicle subsystems. For example, refer back to Figure 4.1 showing the bending moments applied by subsystem weight. Therefore, to determine structure requirements we need estimates of subsystem mass. Unfortunately when we begin a new vehicle design, we do not have existing subsystems to weight, so a means to generate preliminary mass estimates to begin the design process is critical. In this section, we discuss a means to generate these initial mass estimates.

9.2.1 Vehicle design stages

Mass estimation occurs throughout the vehicle development process, Figure 9.19, but with different objectives. During the product planning stage, or pre-configuration stage, the purpose of mass estimation is to determine general feasibility of a program, to set initial mass goals for vehicle subsystems, and to develop structural requirements based on mass. At this stage, only a broad vehicle mission is known; for example, vehicle type and number of passengers. Because of the sparse data, mass estimation is often based on first-order statistical models using data from contemporary vehicles.

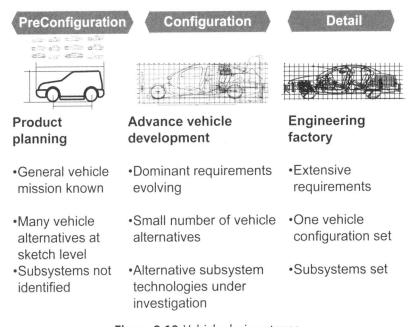

PreConfiguration	Configuration	Detail
Product planning	**Advance vehicle development**	**Engineering factory**
•General vehicle mission known	•Dominant requirements evolving	•Extensive requirements
•Many vehicle alternatives at sketch level	•Small number of vehicle alternatives	•One vehicle configuration set
•Subsystems not identified	•Alternative subsystem technologies under investigation	•Subsystems set

Figure 9.19 Vehicle design stages.

In the next stage of design, the configuration stage, the purpose of mass estimation is to aid in decision making about which mass-reducing technologies to include in the program, and also to add additional precision to the subsystem mass estimates. In this stage, mass estimation is based on both semi-empirical models and statistical data.

During the detail-design, post-configuration stage, the goal of mass estimation is to monitor mass growth to meet targets. In this stage, mass estimation is based on actual subsystem designs and prototypes.

Here we focus on the preconfiguration and configuration stages, where we must generate an initial mass estimate from very sparse data. Before discussing a means to estimate mass in these stages, some of the frequently used conditions for vehicle mass are defined below:

Curb mass- The mass, MCURB, of the vehicle with fluids.
Test mass- The mass of vehicle with two passengers, 140 kg (300 lb) and high penetration options. This condition is used to determine the test weight class, TWC, for fuel-economy testing.

Gross vehicle mass- The maximum mass, MGVM, of the vehicle including all passengers, cargo, and options. The gross mass is significant, as many subsystems requirements depend on this condition.

9.2.2 Regression-based mass estimation

The objective of mass estimation in the pre-configuration phase is to arrive at an estimate for the curb and gross vehicle mass and for the mass for each of the vehicle subsystems. Because the vehicle definition is very sparse at this stage, often limited to vehicle length and width, a regression-based approach is useful [11, 12, 13, 14]. Figure 9.20 shows curb mass plotted against vehicle plan view area for 72 sedans with transverse front-wheel-drive configuration and steel body frame integral construction [15, 16]. From this graph, we can infer a convenient relationship for preliminary estimation of curb mass:

$$M_{CURB} = \beta_0 + \beta_1(Area) \tag{9.14}$$

where:

M_{CURB} = estimated curb mass of typical vehicle

β_0, β_1 = coefficients estimated by regression

$Area$ = plan view area (overall length x overall width)

An estimate for the coefficients of this equation is found at the top of Figure 9.20 for the data shown. Further, the mass of each subsystem may be estimated as a fraction of the curb mass:

$$m_i = \alpha_i M_{CURB} \tag{9.15}$$

where:

m_i = estimated mass for subsystem i of typical vehicle

α_i = mass of subsystem i per unit of curb mass estimated from mass data. Figure 9.21 provides estimates for α for the same 72 vehicles.

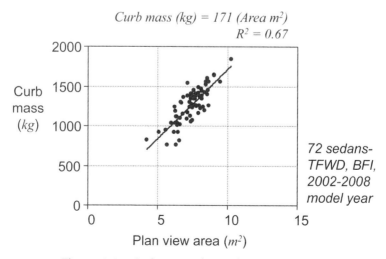

Figure 9.20 Curb mass relationship to vehicle size.

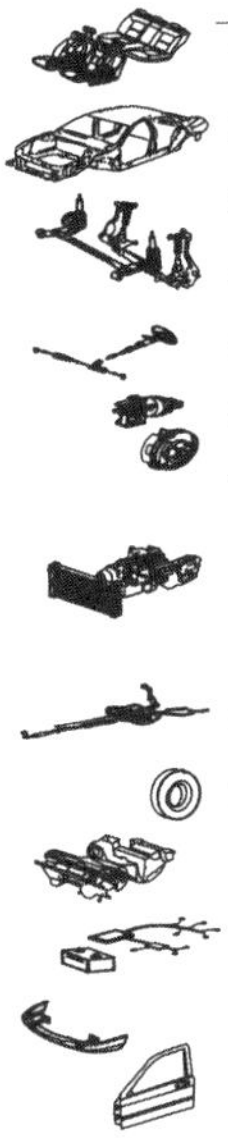

Functional subsystem	Mass as fraction of curb mass, α_i	
1 Body non-structure	0.215	
2 Body structure	0.243	*72 Sedans-*
3 Front suspension	0.036	*TFWD, BFI,*
4 Rear suspension	0.032	*2002-2008 model year*
5 Steering	0.016	
6 Braking	0.040	
7 Powertrain	0.139	
engine	0.094	
transmission	0.045	
8 Fuel & exhaust	0.065	
9 Tires & wheels	0.066	
10 Air conditioning	0.027	
11 Electrical	0.049	
12 Bumpers	0.027	
13 Closures	0.045	

Figure 9.21 Functional subsystem mass fractions.

Example: Initial mass estimation in the preliminary design stage

A new vehicle is in the planning stage (pre-configuration), Figure 9.22. It is targeted at 5 passengers with a *120-kg* cargo capacity. From human accommodation and dimensional benchmarking, the vehicle length is estimated at *4.7 m* and width at *1.8 m*. It is anticipated that it will be a transverse front drive configuration and body frame integral construction. The target test weight for fuel economy evaluation is *3250 lb* resulting in a *2950 lb (1341 kg)* curb mass. Our objective in mass estimation at this stage of design is to determine the curb mass for an average vehicle of this size and compare it to the target test weight class of *3250 lb*, and to determine mass estimates for each subsystem.

A new vehicle in the planning (*pre-configuration*) stage.

Target values
number of passengers: 5
cargo capacity = *120 kg*
vehicle width = *1800 mm*
vehicle length = *4700 mm*
fuel economy:
   Test Weight = *3250 Lb*
    curb mass = *2950 Lb (1341 kg)*

Figure 9.22 Mass estimation example: Vehicle specifications.

The curb mass can be estimated, given the plan view area, using Equation 9.14 with the coefficients shown in Figure 9.20. This calculation is shown in Figure 9.23a. Once curb mass has been estimated, subsystems' mass may be estimated using Equation 9.15 and the coefficients shown in Figure 9.21. An example of this calculation is shown in Figure 9.23b for the body non-structure subsystem. The resulting subsystem mass estimates are shown in Figure 9.23c. These mass estimates are for an average or typical vehicle of the size and type we are designing. Even though this is only a first-order estimate, we have determined that the curb mass estimate, *1446.6 kg*, exceeds the target curb mass, *1341 kg*, by *105.6 kg*, and that some action must be taken to reduce mass. In the next section, we develop a model to analyze such mass reductions.

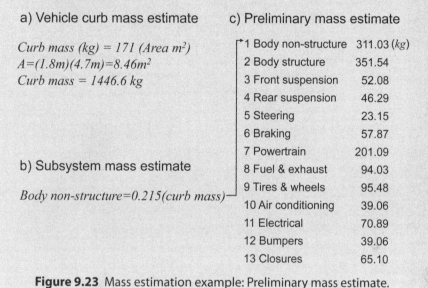

a) Vehicle curb mass estimate

Curb mass (kg) = 171 (Area m²)
A=(1.8m)(4.7m)=8.46m²
Curb mass = 1446.6 kg

b) Subsystem mass estimate

Body non-structure=0.215(curb mass)

c) Preliminary mass estimate

1 Body non-structure	311.03 (*kg*)
2 Body structure	351.54
3 Front suspension	52.08
4 Rear suspension	46.29
5 Steering	23.15
6 Braking	57.87
7 Powertrain	201.09
8 Fuel & exhaust	94.03
9 Tires & wheels	95.48
10 Air conditioning	39.06
11 Electrical	70.89
12 Bumpers	39.06
13 Closures	65.10

Figure 9.23 Mass estimation example: Preliminary mass estimate.

9.2.3 Mass-compounding model

Vehicle design engineers intuitively know that an unplanned mass increase in a component during vehicle design has a ripple effect throughout the vehicle; other components need to be resized, increasing vehicle mass even more. The phrase *mass begets mass* describes this phenomenon. A more encouraging view of this behavior is that a reduction in the mass of a component enabled by a new technology will result in an even greater mass reduction—a secondary mass reduction—for the overall vehicle. These secondary mass changes can be considerable.

A means to quantify the secondary mass change is using a mass-compounding model [15]. In this model, each subsystem (denoted by i) is assigned a mass influence coefficient, γ_i. The influence coefficient is the change in the subsystem mass when gross vehicle mass undergoes a unit change, Figure 9.24. The physical interpretation of the influence coefficient is each subsystem is sized to some degree by the mass of the vehicle, and as the vehicle mass changes the subsystem may also be resized.

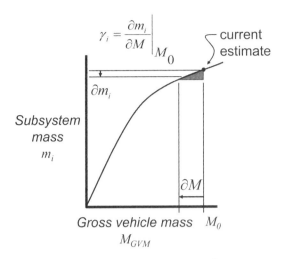

Figure 9.24 Mass influence coefficient.

The total change in mass for the vehicle per unit reduction in gross mass is then given by:

$$\gamma_V = \sum \gamma_i \tag{9.16}$$

where:

γ_V = the vehicle mass influence coefficient

γ_i = the influence coefficient for subsystem i

Example: Mass compounding

The typical question we are interested in answering is: Given a balanced vehicle under design, a primary mass change now occurs before the design is finalized. What is the final vehicle mass after resizing subsystems? To visualize this situation consider a vehicle with M_{GVM} = 1000 kg. All subsystems have been sized based on this gross vehicle mass. Now assume a 10-kg reduction is made in a subsystem; for example, a reduction in seat frame mass. Further assume that for this vehicle, the vehicle mass influence coefficient is $\gamma_v = 0.4$.

Figure 9.25 illustrates the changes in gross vehicle mass as this vehicle is redesigned in response to the initial 10-kg mass reduction. The first change is the primary reduction of 10 kg resulting in new M_{GVM} =990 kg. All subsystems were based on a 1000-kg vehicle mass, and subsystems can now be resized to this lower gross vehicle mass. This resizing results in a further reduction of (0.4)(10kg) =4 kg, and a new M_{GVM}=990−4=986 kg. Now the components are sized for 990 kg, but the vehicle mass is 986 kg so another resizing can occur, as shown in Figure 9.25. The resizing repeats for an infinite number of iterations, but does converge to a final value of M_{GVM} =983.3 kg. This final vehicle mass represents a primary reduction of 10 kg, and a secondary reduction due to this resizing of 6.7 kg. Note that the secondary reduction is the sum of the mass reductions for each of the infinite series of resizings. Written in the form below, the secondary reduction is the sum of a geometric series which can be expressed by the well-known expression:

$$(0.4)(10) + (0.4)\left[(0.4)(10)\right] + (0.4)\left[(0.4)\left[(0.4)(10)\right]\right] + \ldots$$

$$= \sum_{n=1}^{\infty} (0.4)^n (10)$$

$$= \frac{0.4}{1 - 0.4}(10)$$

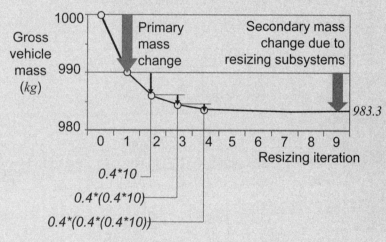

•*1000 kg* GVM
•*10 kg* of initial mass change (primary change)
•vehicle mass influence coefficient = *0.4*

Figure 9.25 Mass-compounding example.

This resizing in response to a primary change shown in the previous example is captured by a mass-compounding model:

$$M_V^C = M_V + \Delta + \Delta S_V \tag{9.17}$$

where:

$$S_V = \left[\frac{\gamma_V}{(1 - \gamma_V)}\right] = \text{the Secondary mass coefficient for the vehicle} \tag{9.18}$$

M_V = the initial vehicle mass for which the subsystems are sized

Δ = the initial (primary) mass change for the vehicle. This is the sum of the primary changes of each subsystem, $\Delta = \Sigma \Delta_i$

Δ_i = primary change in subsystem i

M_V^C = the resized (compounded) vehicle mass

ΔS_V = the additional (secondary) mass change due to resizing subsystems

γ_V = the mass influence coefficient for the vehicle given by $\gamma_V = \sum_{i=1}^{n} \gamma_i$

γ_i = the mass influence coefficient for subsystem i

The resulting mass for subsystem i due to an initial increase of Δ is:

$$m_i^C = m_i + \Delta_i + \Delta S_i \tag{9.19}$$

where the Secondary Mass Coefficient for Subsystem i is:

$$S_i = \left[\frac{\gamma_i}{(1-\gamma_V)}\right] \tag{9.20}$$

where:

   m_i = initial subsystem mass

   m_i^C = the resized (compounded) subsystem mass

   Δ_i = the primary mass change in subsystem i

   S_i = the additional secondary change

   γ_V, γ_i, and Δ are given above

To apply these equations, estimates of subsystem influence coefficients, γ_i, are required. This may be accomplished using regression, as illustrated in Figure 9.26, where a function is fit relating the mass of each subsystem i to gross vehicle mass. The influence coefficient is the slope of the fit curve. Figure 9.27 is a summary of subsystem influence coefficients found by regression of 72 sedans with transverse front-wheel-drive configuration and steel body frame integral construction. Only those subsystems whose sizing depends upon vehicle mass will have a non-zero influence coefficient. So, for example, the body nonstructural subsystem (interior trim) has a zero influence coefficient.

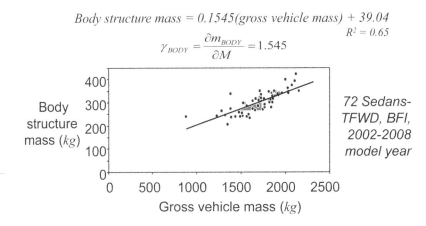

$$\textit{Body structure mass} = 0.1545\textit{(gross vehicle mass)} + 39.04$$

$$R^2 = 0.65$$

$$\gamma_{BODY} = \frac{\partial m_{BODY}}{\partial M} = 1.545$$

Body structure mass (kg)

72 Sedans-
TFWD, BFI,
2002-2008
model year

Gross vehicle mass (kg)

Figure 9.26 Estimating a mass influence coefficient.

Functional subsystem	Influence coefficient, γ Change in subsystem mass per unit change in GVM	
Body structure	0.1545	*72 Sedans-*
Front suspension	0.0291	*TFWD, BFI,*
Rear suspension	0.0251	*2002-2008*
Steering	0.0085	*model year*
Braking	0.0224	*Note: Body non-*
Powertrain	0.1284	*structure, electrical,*
Fuel & exhaust	0.0646	*air conditioning, and*
Tires & wheels	0.0505	*closures subsystems*
Bumpers	0.0360	*assumed to have no*
Total	0.5191	*dependence on gross vehicle mass*

Figure 9.27 Mass influence coefficients.

Note that for the data where the vehicle influence coefficient $\gamma_V = 0.5191$, the secondary mass reduction per unit primary mass change

is $\left[\dfrac{\gamma_V}{(1-\gamma_V)} \right] = \left[\dfrac{0.5191}{1-0.5191} \right] = 1.079 kg / kg$, a substantial reduction.

An alternative means to estimate the mass influence coefficient is by the ratio method [11]. In this method only a single reference vehicle is needed. The ratio method is therefore recommended when no statistical data on influence coefficients are available. In this method, the influence coefficient is given by Equation 9.20.

$$\gamma_i = \frac{m_i}{M_{GVM}} \qquad (9.21)$$

where:

m_i = subsystem i mass of reference vehicle

M_{GVM} = gross vehicle mass of reference vehicle

Example: Mass adjustment using compounding

Continuing with the previous example of Figure 9.22 where we estimated the vehicle mass to be *105.6 kg* above the target value; now we will analyze the effect of mass reduction ideas using the mass compounding model. We begin by building a list of mass-reducing technologies. Figure 9.28 shows a list of six hypothetical technologies. The mass reductions are based on sizing the subsystem for the average vehicle (*1446.6 kg*), as estimated in the previous example. Along with the mass reduction, the unit cost to achieve each reduction is included, and for each technology, a marginal cost (cost to implement / mass reduction) is shown. The technologies have been sorted by increasing marginal cost. In this example, we have assumed the value of mass to be $6/kg, and we exclude all technologies with marginal cost above this value. From this list, we have identified a primary mass reduction for the remaining five technologies of $\Delta = -40$ kg.

subsystem	mass reduction technology	mass saving kg	unit cost $\$$	marginal cost $\$/kg$	cumulative mass saving	cumulative cost
body non-structure	sound treatment optimization	5	-5	-1	5	-5
body non-structure	change trim	10	0	0	15	-5
body structure	body joint optimization	15	15	1	30	10
front suspension	lower control arm change	5	10	2	35	20
closures	material change: hood	5	30	6	40	50
rear suspension	material change: control arms	10	100	10	50	150
		excluded due to high marginal cost				

Figure 9.28 Mass estimation example: Mass reduction technologies.

If we accept these technologies without resizing, the new vehicle mass is *1446.6–40=1400.6 kg.* However, if we have the ability to resize components, we may apply the mass-compounding model of Equations 9.17and 9.18:

$$M_V^\varsigma = M_V + \Delta + \Delta S_V$$
$$M_V^\varsigma = 1446.6 - 40 - 40(1.079)$$
$$M_V^\varsigma = 1363.48 \text{kg}$$

where:

$$S_V = \left[\frac{\gamma_V}{(1-\gamma_V)}\right] = \left[\frac{0.5191}{1-0.5191}\right] = 1.079 kg/kg \text{ using the influence coefficients of Figure 9.27.}$$

We can further estimate the mass for each subsystem. For example, the body structure which we designate as subsystem 2, *i=2*:

The original body structure mass is m_2=*351.54 kg* from Figure 9.23 as sized for a "typical car" of curb mass (*1446.6 kg*). The primary mass change made directly to the body structure by the application of the technologies of Figure 9.28 is Δ_2= *–15 kg*. The body influence coefficient from Figure 9.27 is γ_2=*0.1545*. The resized body mass is then given by Equations 9.19 and 9.20:

$$m_2^\varsigma = m_2 + \Delta_2 + \Delta S_2$$
$$m_2^\varsigma = 351.54 - 15 - 40(0.3213) = -323.69 \text{ kg}$$

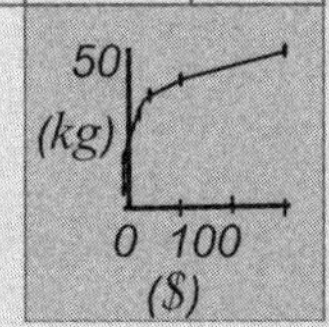

where $S_2 = \dfrac{\gamma_2}{1-\gamma_V} = \dfrac{0.1545}{1-0.5191} = 0.3213$

The mass changes for this example are summarized in Figure 9.29. Note that in this example we have assumed that all subsystems were free to be resized. Often this is not the case. For example, in many new vehicle programs, an existing powertrain is used and will not be resized. For this situation, an influence coefficient of zero would be applied to the powertrain subsystem.

	subsystem mass for average vehicle (kg)	primary mass change from technology (kg)	secondary mass change from compounding (kg)	Final mass (kg)
1 Body non-structure	311.03	-(15+5)	-0.00	296.03
2 Body structure	351.54	-15	-12.85	323.69
3 Front suspension	52.08	-5	-2.42	44.66
4 Rear suspension	46.29	-	-2.09	44.21
5 Steering	23.15	-	-0.71	22.44
6 Braking	57.87	-	-1.86	56.00
7 Powertrain	201.09	-	-10.68	190.41
8 Fuel & exhaust	94.03	-	-5.37	88.66
9 Tires & wheels	95.48	-	-4.20	91.28
10 Air conditioning	39.06	-	-0.00	39.06
11 Electrical	70.89	-	-0.00	70.89
12 Bumpers	39.06	-	-2.99	36.07
13 Closures	65.10	-5	-0.00	60.10
	1446.66	-40	-43.18	1363.48

Figure 9.29 Mass estimation example: Compounded mass.

9.2.4 Summary: Preliminary mass estimation

A method for mass estimation has been demonstrated in which estimates can be made using only the sparse vehicle definition typically available at the beginning of a new vehicle program. The steps of this methodology are:

1. *Estimate the mass of an average or typical vehicle of the size and type under consideration:* Using the regression-based models, apply Equation 9.14 to estimate curb mass, followed by Equation 9.15 to determine the mass for each subsystem. These estimates may be used to decide if the vehicle mission is consistent with vehicle type and size.

2. *Identify mass reduction ideas:* If mass reductions must be made to meet the vehicle mission, potential mass-reduction technologies may be identified for each subsystem. Each technology should be developed consistent with the vehicle mass identified in Step 1.

3. *Select mass-reduction technologies to apply*: For each technology, identify the mass reduction and unit cost. The marginal cost is the unit cost per unit mass reduction to provide the technology. For all identified mass-reduction technologies, sort by increasing marginal cost. Filter technologies using marginal cost for inclusion in the vehicle configuration. The threshold depends on the inferred value of a unit mass reduction. This value would include, most significantly, the reduction in fuel consumption over the life of the vehicle. An approximate sensitivity of change in fuel consumption per unit mass reduction is based on a regression of vehicles in the U. S. Environmental Protection Agency database resulting in a value of *0.5 L/100 km/100 kg* for internal combustion engines [17, 18]. For those technologies passing the marginal cost threshold, sum the mass reductions to arrive at the primary mass reduction, Δ.

4. *Estimate vehicle mass with mass-reduction technologies using mass compounding*: The mass-reduction opportunities adopted in Step 3 are *primary* mass reductions. They are sized for the vehicle mass found in Step 1. Now as each technology is applied to the vehicle, the vehicle mass is reduced. Due to this reduction, the subsystems may be resized in an iterative fashion. This resizing results in a *secondary* mass reduction given by Equations 9.17 and 9.18 for the vehicle, and Equations 9.19 and 9.20 for each subsystem.

This methodology is a rational procedure to estimate a vehicle's mass based on current practice, and then to estimate the influence of mass-reduction technologies using a mass-compounding model. Once the preliminary mass is established, structural requirements may be set and structure design started.

References

Materials in Body Design

1. Pafumi, M., "Advanced High Strength Steel Technology in the 2006 Honda Civic, American Iron and Steel Institute Great Designs in Steel Conference," April, 2007

2. J.R. Shaw, B. K. Zuidema, "New High Strength Steels Help Automakers Reach Future Goals for Safety, Affordability, Fuel Efficiency, and Environmental Responsibility," SAE Paper 2001-01-3041.

3. *Automotive Steel Design Manual*, American Iron and Steel Institute, Southfield MI, August 2002, p. 2.4–4.

4. Adams, D. G., et al, "High Strength Materials and Vehicle Weight Reduction Analysis," SAE 750221.

5. Tsai, S. W., *Composites Design-1986*, Think Composites, Dayton, Ohio, 1986 (also www.thinkcomposites.com).

6. MatWeb, Material Properties Data, www.matweb.com.

7. Baumester, T., Editor, *Marks' Standard Handbook for Mechanical Engineers*, McGraw-Hill, NY, 1978.

8. Ashby, M. F., *Materials Selection in Mechanical Design*, Elsevier Science, Burlington MA 1992.

9. Ashby, M. F., *Materials Engineering, Science, Processing and Design*, Elsevier Science, Burlington, MA 2007.

10. Chang, D. and Justusson, J. W., "Structural Requirements in Material Substitution for Car-Weight Reduction," Society of Automotive Engineers Paper 760023.

Mass Estimation

11. Tornbeek, E., *Synthesis of Subsonic Airplane Design*, Delft University Press, Boston, 1988.

12. *Introduction to Aircraft Weight Engineering*, Society of Allied Weight Engineers, Los Angeles CA, 1996.

13. de Weck, O., "A Systems Approach to Mass Budget Management," Paper AIAA 2006-7055, 11th AIAA/ISSMO Multidisciplinary Analysis and Optimization Conference, September 2006.

14. Papila, M., "Accuracy of Response Surface Approximations for Weight Equations Based on Structural Optimization," PhD Dissertation, University Of Florida 2001, pp. 1–4.

15. Malen, D. & Reddy, K., *Preliminary Vehicle Mass Estimation Using Empirical Subsystem Influence Coefficients*, American Iron and Steel Institute, April 2007.

16. A2mac1 Automotive Benchmarking, www.a2mac1.com, 3901 Bestech road, Ypsilanti, MI .

17. Wohlecker, R., et al, "Determination of Weight Elasticity of Fuel Economy for ICE, Hybrid and Fuel Cell Vehicles," SAE 2007-01-0343, Society of Automotive Engineers, PA, 2007.

18. Anderson, M., Lecture notes: Powertrain Design and Integration, Integrated Vehicle System Design, AUTO 501, University of Michigan, October 28, 2009.

Appendix A
Exercises

1. Strength of Materials Review

1.1 Free bodies and moment diagrams

Two *1.8-m*-diameter steel rolls are carried on a flat bed trailer, each has a mass of *18200 kg*. The trailer is supported at the front by the semi and at the rear by the axle. Each acts as a simple support. The trailer consists of two beams; one on each side. Assume the trailer weight is small compared to the steel rolls.

a) Compute the loads at the front trailer pivot and at the rear axle of the trailer.

b) Compute the shear loads on the two side beams and plot shear force versus length.

c) Compute the bending moment on the two side beams and plot bending moment versus length.

d) What is the maximum bending moment and location?

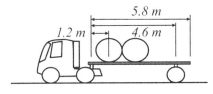

Exercise 1.1

1.2 Moment of inertia of thin-walled sections

Body structure typically consists of beams with thin-walled sections in which the width-to-thickness ratio, b/t, is large ($b/t > 60$).

a) Write an expression for the exact area moment of inertia about the x axis, Ixx.

b) When t is very small compared to the other section dimensions, we can approximate the exact Ixx with an expression linear in t. Using either a Taylor expansion of part a or by elimination of terms of t of second and higher order, write the linear approximation for Ixx.

c) For $b=50$ mm, $h=100$ mm, and $5 < b/t < 150$, plot Ixx versus b/t for the two expressions, part a and b, on the same graph.

d) For what range of b/t is the linear expression for Ixx accurate if we desire to be within ±5% of the exact value?

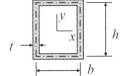

Exercise 1.2

1.3 Beam bending stress

For the trailer of Exercise 1.1 each frame beam is a hollow rectangular section with $b=100$ mm, $h=255$ mm, and $t=6.4$ mm. At the point of maximum bending moment from Exercise 1.1d:

a) Compute the bending stress at the top of the right frame beam.

b) Plot the bending stress versus vertical position on the section.

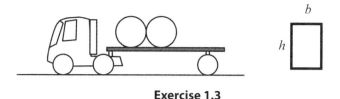

Exercise 1.3

1.4 Strain energy

The steel coils of Exercise 1.1 are unloaded by a crane with a steel cable which has an equivalent cross section area A and length L. The weight of the steel coil is P.

a) Write the expression for the direct stress, σ, in the cable.

b) The strain energy, e, stored in the cable is $e = \dfrac{1}{2} \displaystyle\int \sigma \varepsilon dV$, where V is the volume of the cable, and ε is the strain in the cable. Show that this can be rewritten

as $e = \displaystyle\int \dfrac{\sigma^2}{2E} dV$ where E is Young's Modulus.

c) When the coil weight, P, is applied to the cable, it stretches an amount Δ and the force, P, does work through the distance Δ. The work done by the external force is $1/2P\Delta$ and this must equal the total strain energy in the stretched cable. Solve for the deflection, Δ, using this equality and the relations from parts a and b.

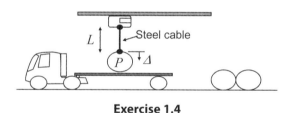

Exercise 1.4

1.5 Mechanical vibration

While the steel coil is suspended by the cable, its motion in the vertical direction can be described as a single-degree-of-freedom vibratory system. The cable can be represented as a spring, held fixed at the top and stiffness $K=P/\Delta$, and the coil as a rigid mass free to move up and down. Use the result from Exercise 1.4 for Δ.

What is the natural frequency of this system for vertical motion (not pendulum motion)? Take $L=9.1\ m$ and area of cable A $=645\ mm^2$, and mass of coil as in Exercise 1.1.

2. Developing structure requirements

2.1 Brake mount requirement

The cowl structure is a closed section extending from side to side. The brake pedal loads must be reacted by the cowl structure without yielding. Also there can be no more than $1\ mm$ deflection at the master cylinder when the maximum pedal load, $P=900\ N$, is applied. Assume that the centroid, point A, of the cowl structure does not move and the brake master cylinder and pedal rotate as a rigid body around point A. The cowl structure strength requirement is expressed as the maximum moment, M, to be reacted, and the stiffness requirement is expressed as the minimum stiffness, $K=M/\phi$. What are the values for M and K which meet the load and deflection requirements?

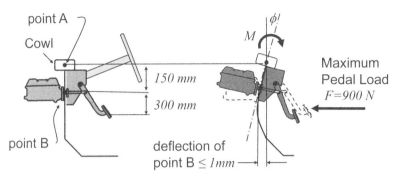

Exercise 2.1

2.2 Seat belt anchor requirement

Shown in the figure are the torso loads in a severe crash.

a) In illustration a, the seat belt is attached at the top and bottom to the B pillar. Determine the fore-aft and vertical strength requirements at the top and bottom B pillar for this configuration. Assume uniform tension in the seat belt and neglect the weight of the occupant.

b) In illustration b, the seat belt is attached to the seat frame. Determine the strength requirements at the seat mounts for the belts-to-seat configuration. Assume the fore-aft loads are reacted equally by the front and rear seat attachments, and neglect the weight of the occupant.

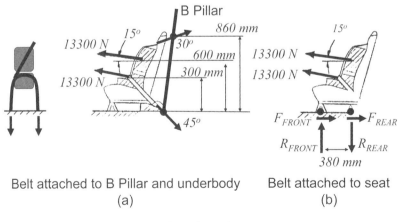

Belt attached to B Pillar and underbody
(a)

Belt attached to seat
(b)

Exercise 2.2

2.3 Rear suspension requirement

For the twist axle rear suspension, compute loads at the body attachments, S, B, Q, *and* V for each of the modes of use listed below. Assume suspension is a rigid body and that each bushing shares equally in the lateral load reaction; i.e., $Q_{RIGHT} = Q_{LEFT} = 1/2 L$. Also assume the lateral load at the spring/shock attachment is zero.

a) Vertical bump: Vertical upward load, R, at each tire patch.

b) Braking: Rearward load, F, at each tire patch.

c) Severe cornering: Lateral load, L, at outside tire patch only.

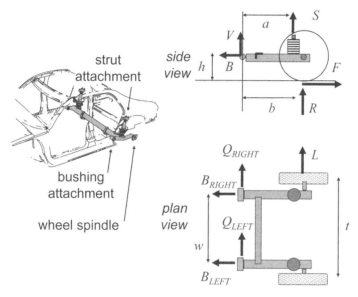

Exercise 2.3

2.4 Powertrain mount requirement

For front-wheel-drive powertrains, the engine and transmission are an integral unit. To establish strength requirements for the powertrain mounting brackets, we wish to estimate the most severe load case. We choose to use the loads generated when the engine stalls. Let T_0 be the maximum stall torque of the engine. To stall the engine, a sufficient torque must be applied at the axles of the vehicle, Ta. The transmission gear ratio, n, is such that $Ta = nT_0$.

Four powertrain mounts are spaced at a fore-aft distance apart, a. Neglect the weight of the engine and transmission. Write an expression for the powertrain mount bracket force, R, required to balance the effects of engine stall torque.

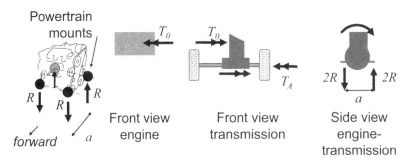

Exercise 2.4

2.5 Short and long arm suspension: Vertical bump requirement

For the short-and-long-arm front suspension shown above, the applied load is a maximum bump load, R, applied at the tire patch. Derive the expressions for the strength requirements at the structure interfaces:

a) strut tower: S_V

b) upper control arm: A_U

c) lower control arm: A_L, A_V

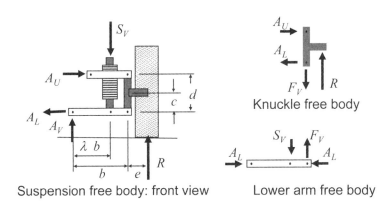

Suspension free body: front view   Lower arm free body

Exercise 2.5

3. Structural Element Behavior

3.1 Seat mount cross member

a) Find maximum stress and deflection at the load point for simply supported end conditions.

b) Find maximum stress and deflection at the load point for fixed end conditions.

c) For the simply supported case with requirements, $k=100$ N/mm, $P_Y=2000$ N, determine the required thickness; consider only yielding behavior.

d) Which requirement dominates? (That is, which requires the greater thickness?)

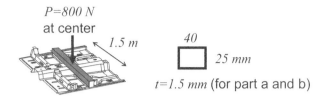

Exercise 3.1

3.2 Motor compartment rail

The motor compartment rail must meet the package constraints imposed by the suspension and powertrain. These constraints often force nonsymmetrical loading and section shape. For the illustrations shown in *a, b,and c*, determine the deflection at the tip of the beam and maximum direct stress and location for a *2000-N* load.

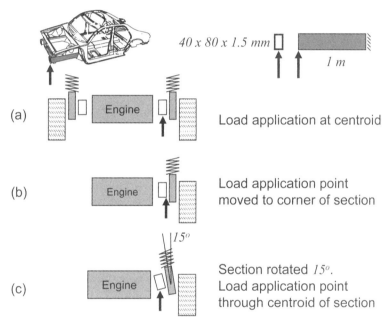

Exercise 3.2

3.3 Local compliance: Tie-down point

A shipping tie down is located at the center of the rocker: a rectangular section *100 mm × 80 mm*, *1.0 mm* thick, *1 m* long, and effectively simply supported at the ends, illustration a. The initial design is unreinforced, illustration b. The total deflection along the beam is the deflection of an ideal beam plus the local deflection. The local distortion is a half sine wave along the length of the beam.

a) Compute the beam deflection under a unit load assuming there is no local distortion of the section.

b) Compute the local deflection of the section under a unit load.

c) Compare the local deflection at the point of load application with that of an ideal beam.

d) What is the total deflection of the beam under a unit load including both beam deflection and local distortion.

e) Repeat parts a, b, and c when a reinforcement is added for the length of the rocker. Assume the reinforcement is a uniform plate *0.6 mm* thick.

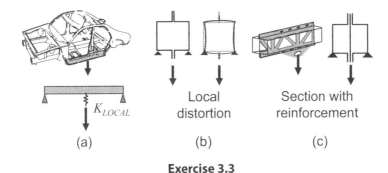

(a) K_{LOCAL}

Local distortion (b)

Section with reinforcement (c)

Exercise 3.3

3.4 Frame cross member

The rear powertrain mount is at the midspan of a cross member. The maximum vertical load is *1000 N*. The end condition for the cross member at the frame rails is simply supported. Calculate the deflection at the load point, and maximum direct stress and its location for:

a) A closed rectangular cross section with load applied at centroid.

b) A rear-facing C section with load applied at the web with warping unconstrained.

c) Same as part b but with warping constrained at each end.

Use $C_W = \dfrac{h^2 b^3 t(2h+3b)}{12(h+6b)}$, where h= height, b= flange length, t= thickness.

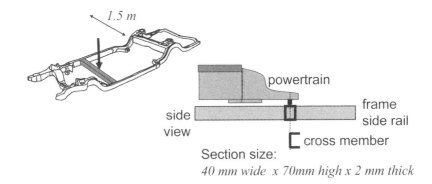

1.5 m

powertrain

frame side rail

side view

cross member

Section size:
40 mm wide x 70mm high x 2 mm thick

Exercise 3.4

3.5 Steering column mount

The steel steering column mounting beam extends from side to side with end conditions unconstrained for warping. A torque is applied to the beam at the steering column location.

a) Determine rotation at the point of load application under *0.5 kNm* torque with a closed section.

b) Determine the rotation at the point of load application under *0.5 kNm* torque when the section contains narrow slit (unrestrained warping).

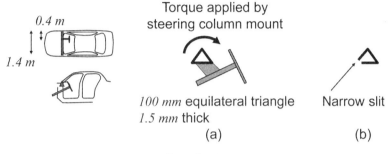

100 mm equilateral triangle
1.5 mm thick

Narrow slit

(a)

(b)

Exercise 3.5

3.6 Steering column mount

The steel steering column mounting beam extends from side to side with end conditions unconstrained for warping. A torque is applied to the beam at the steering column location, now at the midspan of an open C section. The warping constant for the section is $Cw=4.3837x10^7 \ mm^6$

a) Determine angular deflection when both end conditions do not constrain warping.

b) Determine angular deflection when both end conditions do fully constrain warping.

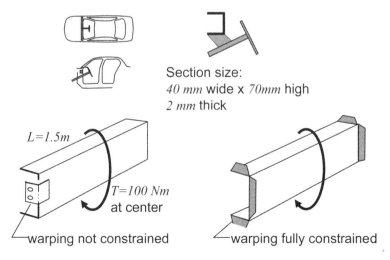

Section size:
40 mm wide x *70mm* high
2 mm thick

$L=1.5m$

$T=100 \ Nm$ at center

warping not constrained

warping fully constrained

Exercise 3.6

3.7 Spot weld effect on beam torsion stiffness

A long steel beam is made of two hat sections spot welded together. The beam is loaded by a torque, T. The beam twists through an angle, θ under the action of T. The ends are unrestrained and free to warp.

a) Assume the welded flange makes the beam act as a closed section. For the values *L=1000 mm, a=75 mm, t=0.9 mm, and w=12 mm* calculate the torsional stiffness: $(T/\theta)_{IDEAL}$.

b) Now assume the welded flange is flexible. Write an expression to describe the angle of twist, θ. Hint: equate the work by the external torque, T, to the energy stored in the ideal section plus the energy stored in the distorted flanges.

c) Plot the normalized beam stiffness, $(T/\theta)_{WELDED}/(T/\theta)_{IDEAL}$, versus the weld pitch, p, for the range {*12 mm < p < 75 mm*}.

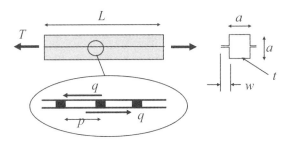

Exercise 3.7

3.8 Geometrical analysis of sections

(This exercise requires CARS section analysis software available from the American Iron and Steel Institute.)

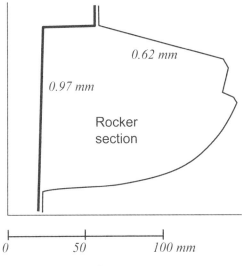

Exercise 3.8a

a) By scaling the drawings, calculate the *nominal section* properties. Use the vertical and horizontal orientation in the drawing for the axis system. Sections: Rocker, A pillar, Roof rail, Upper B pillar, Upper C pillar.

b) Calculate the *effective section* properties at yield under a uniform axial compressive load. Use yield stress $\sigma_y = 269\ MPa$.

c) Compare the *effective* with *nominal* moments of inertia about the horizontal axis by taking the ratio $I_{EFFECTIVE}/I_{NOMINAL}$ for each section.

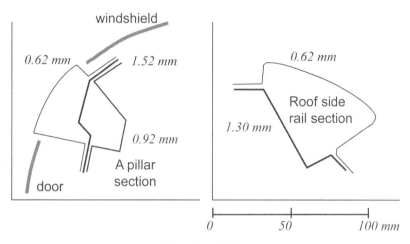

Exercise 3.8b

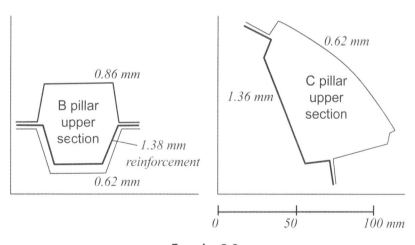

Exercise 3.8c

3.9 Vision Obscuration versus A pillar size

A common design trade-off involves A pillar structure and vision obscuration. Consider the rectangular A pillar shown, which is the reference design. The moment of inertia around the xx axis, *Ixx*, currently meets requirements. There is a desire to reduce the vision obscuration. To do this, hold rear vision line and decrease the rectangular section width along windshield. Go from base section shown to a two-degree reduction in obscuration angle, adjusting thickness to continue to meet the *Ixx* requirement. Plot the cross section area vs. obscuration angle as shown.

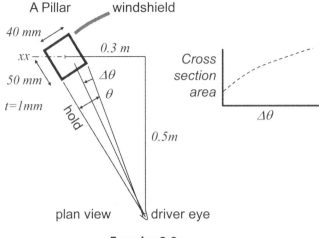

Exercise 3.9

3.10 Step-Over Height versus Rocker Size

A common design trade-off involves rocker structure and step-over height. Consider the rectangular rocker shown, which is the reference design. The moment of inertia around the horizontal axis, *Ixx*, currently meets requirements when the thickness is *1 mm*. There is a desire to reduce the step-over height, h_0. To do this, hold the bottom of the rocker at ground clearance, and lower the top surface by up to $\Delta h = 20$ mm. Adjust thickness to achieve the required moment of inertia. Plot the rocker cross section area vs. the reduction in step-over height, Δh.

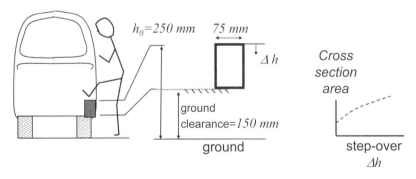

Exercise 3.10

3.11 General buckling width-to-thickness ratio

For a flat plate with simply supported edge constraints (*k=4*):

a) Write an expression for *b/t* ratio such that σ_{yield} and σ_{cr} will be equal.

b) For mild steel $\sigma_Y = 207$ N/mm², what is the numerical *b/t* ratio at which $\sigma_Y = \sigma_{CR}$?

c) Repeat part b for an aluminum alloy where $\sigma_Y = 345$ N/mm² and $E = 69000$ N/mm²

3.12 General buckling coefficient

a) For a flat plate of dimensions a and b, show that when the normal deflections, w, are given by:

$$w = A_{mn} \sin\left(\frac{m\pi x}{a}\right)\sin\left(\frac{n\pi y}{b}\right)$$

that the edge constraints are $Mx=0$ and $My=0$ (that is, simply supported). Refer to Equation (3.21)

b) Show that when w is given as above, the plate equation (Equation 3.21) yields

$$f_{CR} = \frac{D\pi^2}{tb^2}\left[m\left(\frac{b}{a}\right) + \frac{n^2}{m}\left(\frac{a}{b}\right)\right]^2$$

where the terms are defined at Equation 3.23.

c) Plot $\left[m\left(\frac{b}{a}\right) + \frac{n^2}{m}\left(\frac{a}{b}\right)\right]^2$ vs. (a/b) for $n=1$, $m=1,2,3$ and for $n=2$, $m=1,2,3$.

Also show that the bracket term has a lower limit of 4 for these n and m values.

3.13 Buckling stress for a section element-1

Consider the top of a rocker as a long, horizontal flat plate of thickness $t=0.62$ mm. Treat the edge conditions as simply supported.

a) Compute the stress at which it will buckle using hand calculations. (Scale the illustration for needed dimensions.)

b) At what bending moment does the stress in part a occur? (The moment of inertia of the section is I and distance from the centroid to the buckling plate is c.)

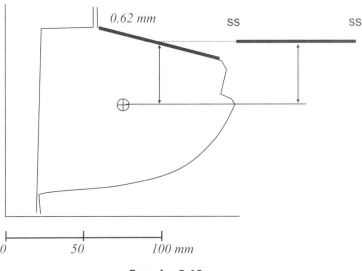

0.62 mm   SS   SS

0   50   100 mm

Exercise 3.13

3.14 Buckling stress for a section element-2

Consider the top flange of a rocker. Treat each flange as an independent long, flat plate (ignore spot welds). Treat the edge conditions as shown.

a) Compute the stress at which the flange on the right side (t=0.62 mm) will buckle using hand calculations.

b) At what bending moment does the stress in part a occur? (The moment of inertia of the section is I and distance from the centroid to the buckling plate is c.)

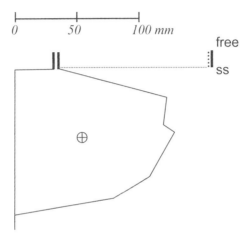

Exercise 3.14

3.15 General buckling effective width

The illustration shows the stress across a buckled plate on the left, and the stress across an effective plate.

a) Assume the stress is distributed as a cosine function with the maximum stress, σ_S, and minimum stress, σ_{CR}, as shown. Derive the below expression for the effective width by assuming for the effective plate the maximum stress acts uniformly over the width, w, and that both the real and effective plates react the same force, P.

$$w = \frac{b}{2}\left(1 + \frac{\sigma_{CR}}{\sigma_S}\right)$$

b) Plot the effective width w versus the maximum stress-to-critical stress ratio σ_S/σ_{CR}.

c) For a flat steel plate with simply supported edges, where b=100 mm and t=0.86 mm, plot the effective width versus applied compressive stress.

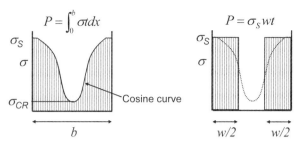

Exercise 3.15

3.16 Effective width

Using hand calculations determine the following:

a) At what bending moment, M_{CR}, will the top cap just buckle?

b) What is the effective width of the top cap when $\sigma_s = 1.1 \, \sigma_{CR}$, $1.5 \, \sigma_{CR}$, and $2.0 \, \sigma_{CR}$?

c) What is the effective moment of inertia Ixx when $\sigma_s = 2.0 \, \sigma_{CR}$?

d) What is the applied moment when the maximum stress in the top cap is $\sigma_s = 2.0 \, \sigma_{CR}$?

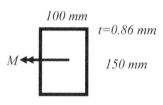

Exercise 3.16

3.17 Effective properties after buckling

Determine the bending moment vs. tip deflection for the beam shown. The result should look similar to the graph shown. Consider only plate buckling behavior of the upper cap of the section. Plot the range $0 < M < 5 \, M_{CR}$.

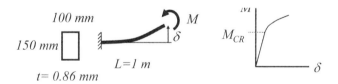

Exercise 3.17

3.18 Between-weld flange buckling

Spot-weld spacing, p, affects the stress at which flange buckling occurs. The flange design stress is the lower of yield or buckling. For a mild steel flange of thickness, t, and height $17 \, mm$, plot the flange design stress vs. weld pitch, p, for the range $24 \, mm < p < 75 \, mm$, with separate curves for $t = 0.5 \, mm$, $t = 1 \, mm$, and $t = 1.5 \, mm$.

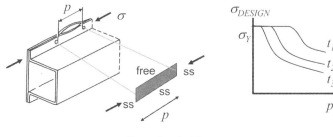

Exercise 3.18

3.19 Rear rail with bumper loading

a) For the rectangular section shown, what is the lowest bumper load at which one of the plate elements will buckle?

b) All four flat sides of the section are now replaced with curved elements of $R=500$ *mm* which go through the corners of the original square section. What is the lowest bumper load at which one of the curved plate elements will buckle?

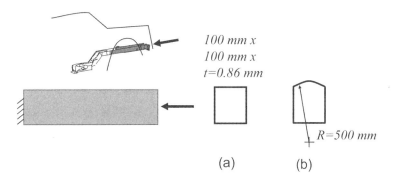

100 mm x
100 mm x
t=0.86 mm

R=500 mm

(a)   (b)

Exercise 3.19

3.20 Panel behavior

Consider the pickup truck roof panel shown as a flat plate for simplicity in analysis. The thickness of the roof is $t=0.8$ *mm*, and the width is *1200 mm*, while the total length is *800 mm*. The material is steel with Young's modulus $E=200,000$ *N/mm²* and Poisson's ratio $\mu=0.3$.

a) Determine the stiffness of the roof when it is subjected to a *100 N* normal point load applied at the center of the panel. The stiffness is defined as the ratio of the force applied and the maximum deflection of the roof. For a rectangular flat plate with sides a and b, loaded at the center by a load P, the deflection, y_{CENTER}, is [Chapter 3, reference 6]:

$$y_{CENTER} = \frac{\alpha \, P b^2}{E t^3}$$

a/b	1.0	1.5	2	∞
α	0.1267	0.1668	0.1805	0.1851

b) Repeat the question in part a, but now the roof has a crown of *50 mm*.

c) Make a comparison of the stiffness of the roof for the flat and crown conditions.

d) The flat roof is now subjected to a compressive, uniformly distributed, in-plane edge loading in the length direction (fore-aft); find the critical stress and also the resulting critical load on the front or rear edge. The boundary conditions are now simply supported.

e) The simply supported flat roof is now subjected to an in-plane shear loading along the edges of the plate. Find the critical stress and also the resulting critical load on the right or left edge of the roof. The boundary conditions are now simply supported. Note, for a plate under shear, the buckling constant is given by $K = 5.34 + 4(b/a)^2$, where b is the length of the short edge and a is the length of the long edge; *(b<a)* [Reference 10, Chapter 3].

f) A *0.10 kg* mass (a hail stone) hits the center of the flat roof with a speed of *10 km/hr*. Find the maximum deflection of the roof at the center, y_{CENTER}, and the equivalent static load, P, which would cause the same deflection. Hint: use energy balance. P.105

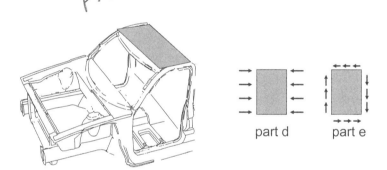

part d   part e

Exercise 3.20

3.21 Panel shaping: Front strut tower

The applied loads at a front strut tower are primarily down the axis of the shock absorber. This axial load is reacted by a shaped panel which is rigidly restrained at its boundary. The idealization in the right illustration shows the axial force being applied to a rigid circular plate. The tower panel is restrained by a circular boundary with the dimensions shown.

a) Write an equation which describes the generating curve for the membrane panel. For corrosion purposes the minimum gage for the panel is *1 mm*.

b) What is the resulting membrane stress under the *22000-N* load?

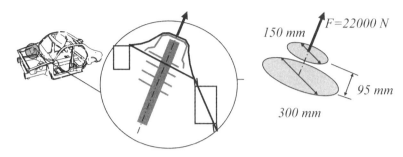

Exercise 3.21

3.22 Distortion of a thin-walled section

Equation 3.6 describes the stiffness of a slice of a rectangular section one unit long under a point load; see illustration. Derive this expression by considering the left side of the section as three beams. This is a redundant structure with an unknown moment, M. Nature will minimize the strain energy caused by this moment, and the value of M may be determined by setting the derivative of the strain energy in the beams to zero as captured in the expression:

$$\frac{\partial}{\partial s}\int\frac{y''M}{2}\,ds = \frac{\partial}{\partial s}\int\frac{M^2}{2EI}\,ds = 0$$

where:

s = the perimeter coordinate along the beams

y'' = curvature at location s

M = bending moment at location s

E = Young's Modulus

I = the moment of inertia of a unit width of the plate (take as $t^3/12$)

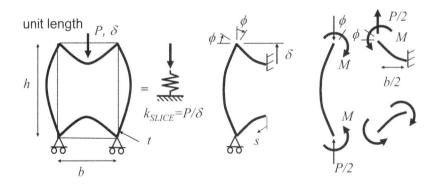

Exercise 3.22

4. Design for Bending

4.1 Bending moment requirements

a) For the base vehicle, sketch the bending moment diagram for the loading shown with supports at the suspensions, and with supports during front towing. Note the following accurately: The maximum bending moment (absolute value); the position where maximum moment occurs; and the basic shape of the bending moment diagram.

b) Complete part a for the stretch limousine made by slicing the base vehicle in half and adding a *1200-mm*-long section welded to the two halves. The same weight per unit length occurs over this center portion.

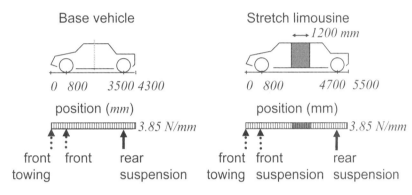

Exercise 4.1

4.2 Body bending: Backbone structure

One topology for resisting bending loads is a backbone beam down the center of the vehicle. The back-bone can be idealized as a uniform beam supported at the axle position. The test load condition is shown. The material is steel with $E=207,000$ N/mm^2 and $\sigma_{DESIGN} = 175$ N/mm^2. Use the basic beam equations.

a) What is the minimum thickness to meet the strength requirement of supporting a *6670-N* load?

b) What is the minimum thickness if the deflection at center span is to be at most *1mm*?

c) Given the thickness computed in part a, what is the deflection in that case?

d) Given the thickness computed in part b, what is the maximum stress in that case?

e) If we desire to minimize mass (i.e. minimum *t*), what is the dominant requirement—strength or stiffness?

f) How short does the wheel base, *L*, need to be before the strength and stiffness requirements result in the same thickness?

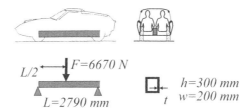

Exercise 4.2

4.3 Load paths for bending

A three-wheel-car concept is shown. The structure is made of shear panels which can only react loads within their plane. We are interested in the bending performance of the structure. The car body is supported at three points as shown and loaded with a bending load, F, which is applied in the plane of the shear panel going down the center of the car as shown. Using free body diagrams for each shear panel and bar element (labeled A through G), determine numerical values for the loads on each structural element. Take $F=6000$ N.

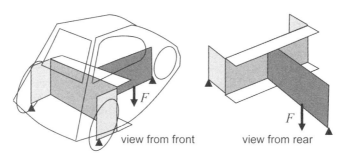

view from front   view from rear

Exercise 4.3a

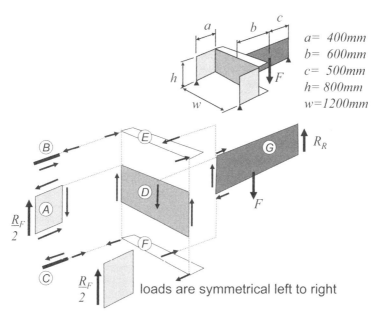

$a= 400mm$
$b= 600mm$
$c= 500mm$
$h= 800mm$
$w=1200mm$

loads are symmetrical left to right

Exercise 4.3b

4.4 Rocker sizing: Bending load for convertible

The desired maximum deflection for the convertible is *1mm* elastic under a load of *6670 N*, or the stiffness requirement is:

$$K \geq (6670\ N)/(1\ mm) = 6670\ N/mm$$

Also the rocker fails at a minimum load of *6670 N* in yield or buckling. Determine *a* and *t* to minimize rocker mass.

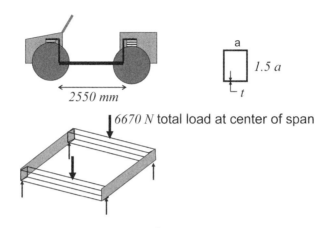

Exercise 4.4

4.5 B pillar: Door sag-1

A downward load is applied to the back edge of the convertible door, as shown. This results in a horizontal load, *P*, at the upper hinge attachment on the B pillar. The B pillar is a flexible cantilever beam attached through a flexible joint to a rigid rocker, as shown in the illustration.

a) What is the rearward deflection at the load for the values shown?

b) Determine the strain energy in the beam and joint.

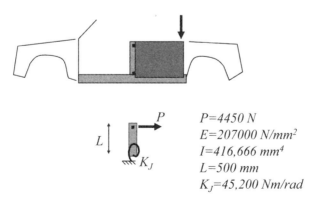

$P=4450\ N$
$E=207000\ N/mm^2$
$I=416,666\ mm^4$
$L=500\ mm$
$K_J=45,200\ Nm/rad$

Exercise 4.5

c) If the deflection from part a is too excessive, which member—B pillar or joint—should be improved? Use strain energy to decide.

d) What is the joint efficiency?

4.6 B pillar: Door sag-2

A more representative model for the condition described in Exercise 4.5 is shown in the illustration. Now the rocker is also viewed as a flexible beam.

a) What is the rearward deflection at the load for the values shown?

b) Determine the strain energy in the B pillar, the rocker, and joint.

c) If the deflection from part a is too excessive, which member—B pillar, rocker, or joint—should be improved? Use strain energy to decide.

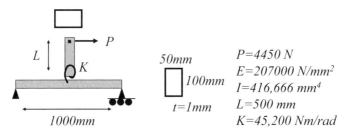

$P=4450\ N$
$E=207000\ N/mm^2$
$I=416,666\ mm^4$
$L=500\ mm$
$K=45,200\ Nm/rad$

50mm
100mm
t=1mm
1000mm

Exercise 4.6

4.7 Side frame deformation

(Requires access to finite element analysis)

Data are provided for a side frame structure. The stiffness requirement is *7000 N/mm* for the vehicle, or *3500 N/mm* for each side frame.

a) Compute the initial vehicle bending deflection, Δ, with rigid joints (a joint stiffness of $Kj=1x10^{10}\ Nmm/rad$ may be considered as rigid). Does the resulting stiffness meet the requirement?

To improve bending stiffness, any of the sections may be increased in size (w, h) by up to 200% of the initial dimensions, except the rocker which is restricted to an increase in size up to *125%* due to entry constraints. Thickness on all sections may be increased up to *3 mm*. (Do not reduce the size of any beams from the given initial size.)

b) Continuing with rigid joints, adjust the side frame beams to meet the stiffness requirement in the most mass efficient way. Do at least two iterations of resizing. Which beams did you adjust; why did you choose them; what are the final beam sizes; and what is the final stiffness?

c) After doing part b, enter the joint stiffness values shown and determine the bending stiffness with flexible joints. What is the new bending stiffness? What is the fraction of stiffness with flexible joints to stiffness with rigid joints?

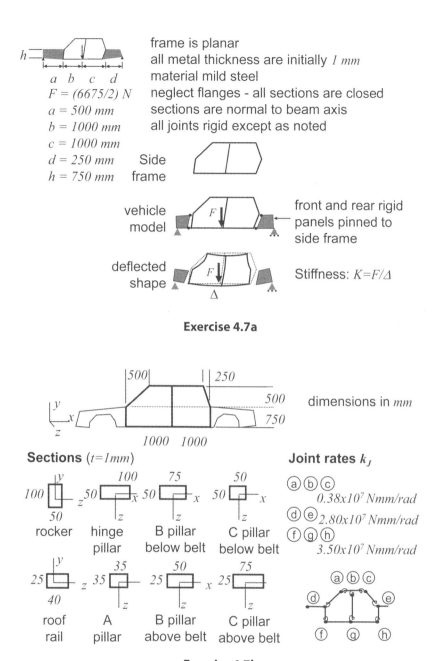

frame is planar
all metal thickness are initially *1 mm*
material mild steel
neglect flanges - all sections are closed
sections are normal to beam axis
all joints rigid except as noted

$F = (6675/2)$ N
$a = 500$ mm
$b = 1000$ mm
$c = 1000$ mm
$d = 250$ mm   Side
$h = 750$ mm   frame

vehicle model

front and rear rigid panels pinned to side frame

deflected shape

Stiffness: $K = F/\Delta$

Exercise 4.7a

dimensions in *mm*

Sections ($t=1mm$)

rocker
hinge pillar
B pillar below belt
C pillar below belt
roof rail
A pillar
B pillar above belt
C pillar above belt

Joint rates k_J

ⓐ ⓑ ⓒ 0.38×10^7 Nmm/rad
ⓓ ⓔ 2.80×10^7 Nmm/rad
ⓕ ⓖ ⓗ 3.50×10^7 Nmm/rad

Exercise 4.7b

4.8 Bending stiffness sensitivity

(Requires access to finite element analysis)

Often trade-offs must be made during design. Styling may desire an increased windshield angle or an increased tumble home (the dimension from the belt line to the roof side rail), or manufacturing constraints may make achieving a stiff joint construction difficult. For this exercise use the model of Exercise 4.7

a) Investigate the effect on body bending stiffness of A pillar angle. For A pillar angle from $45°<\theta<60°$ plot the resulting bending stiffness against windshield angle.

b) Investigate the effect on body bending stiffness of B pillar tumble home. For tumble home from $0<T_B<200\ mm$, plot the resulting bending stiffness against tumble home.

c) Investigate the effect of hinge pillar/rocker joint stiffness on body bending stiffness. Take the nominal value of hinge pillar joint stiffness to be $K_0=3.5x10^7$ Nmm/rad. For joint stiffness from $0.1K_0< Kj <10K_0$ Nm/rad, plot the resulting bending stiffness against joint stiffness.

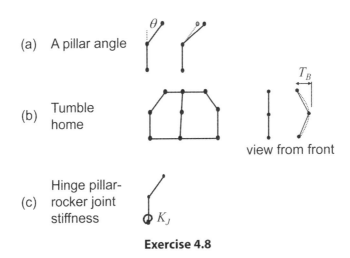

(a)   A pillar angle

(b)   Tumble home

view from front

(c)   Hinge pillar-rocker joint stiffness

Exercise 4.8

4.9 Internal loads on a joint

The thin-walled T joint shown is loaded by a force F along side edges. The reactions to ground are limited to R and Q_2.

a) Compute all the internal forces for the first-order shear panel/bar model shown. (Note: shear panels, representing panels, can only react in-plane shear forces; bars, representing corners, can only react axial forces.)

b) Dimensions are: $b=100\ mm$, $h=200\ mm$, $a=100\ mm$, $w=75\ mm$, $L=500\ mm$, $l=200$ mm. The thickness of panel D is $1\ mm$; thickness of panels A, B, and C is $0.89\ mm$. The material is steel: $E=207000\ N/mm^2$, $\sigma_Y=207\ N/mm^2$. Compute the loads, F, at which shear buckling occurs in panels A, B, C, and D (use buckling constant $k=5$ in the buckling equation).

c) Which panel dominates the strength of the joint (i.e. which will fail at the lowest applied load)?

d) Using the fact that external work equals internal strain energy, what is the deflection in the direction of load F?

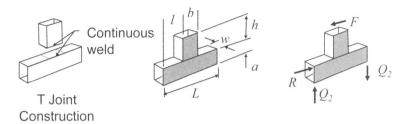

T Joint
Construction

Exercise 4.9a

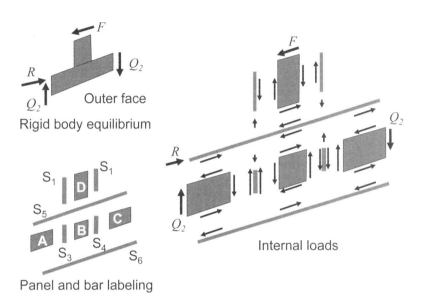

Outer face

Rigid body equilibrium

Panel and bar labeling

Internal loads

Exercise 4.9b

4.10 Effective joint in curved beams

Under bumper loading the rectangular section at the toe pan is subject to a moment $M=Fc$. The member is curved with a constant radius, R, which is greater than the section height, h. Assume that the moment is reacted solely by corner loads, P, as shown such that $(Fc)=(Ph/2)$.

a) Derive an expression for the compressive load per unit length, w, to which the side walls are subjected under corner load, P.

b) Consider an element of unit length on the side of the section as shown in the illustration on the right. Consider this element a unit width beam; write an expression for the Euler buckling load of this beam.

c) Using the results of parts a and b, write an expression for the minimum side wall thickness to just buckle when load F is applied at the bumper.

d) Using the result of part c, what is the effect on minimum thickness if the radius R is doubled?

e) For a steel section and $F=40000$ N, $c=250$ mm, $h=50$ mm, and $R=150$ mm, what is the required thickness to just buckle at load F?

411

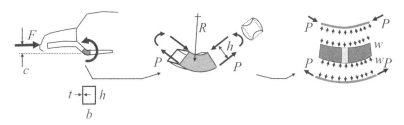

Exercise 4.10

4.11 Joint rate: Tube joint construction

(Requires access to finite element analysis)

Consider the A-pillar/hinge pillar/upper rail joint shown. The upper rail and hinge pillar are square tubes *40 mm × 40 mm × 0.9 mm*. The A pillar is a square tube, *40 mm × 40 mm × 0.9 mm*. For first-order modeling, the attachment of each beam to another is taken to transmit zero moment, as shown in the center illustration (pin joint). Determine the equivalent joint rate, K_J, in *Nm/Rad*.

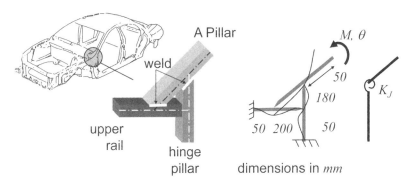

Exercise 4.11

5. Design for Torsion

5.1 Torsion strength requirements: Twist ditch

Determine the maximum twist ditch torque requirement for:

a) the sedan case

b) the limousine

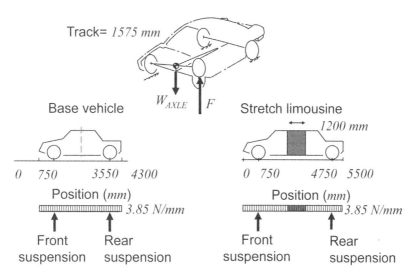

Track= *1575 mm*

Base vehicle

W_{AXLE} | F

Stretch limousine

1200 mm

0   *750*   *3550   4300*

Position (*mm*)

3.85 N/mm

Front   Rear
suspension   suspension

0   *750*   *4750   5500*

Position (*mm*)

3.85 N/mm

Front   Rear
suspension   suspension

Exercise 5.1

5.2 Rocker sizing: Torsion load for convertible

The torsional requirement for this convertible is *680 KNm/rad*.

Determine the rocker size, *a, t,* to meet the stiffness requirement and the requirement that the section will just begin to buckle under a maximum torque of *8.0 KNm*. Do this for two different assumptions regarding the bulkheads.

a) The bulkheads shown are very flexible, i.e., they apply no bending moments to the rocker.

b) The bulkheads are very rigid and constrain the rocker to have zero slope at either end.

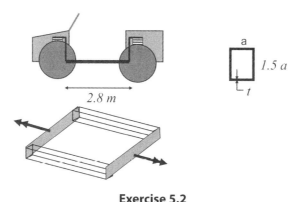

2.8 m

a

1.5 a

t

Exercise 5.2

5.3 Body torsion: Backbone structure

For the steel back-bone structure of Exercise 4.2, the twist ditch torque is *6780 Nm* with an allowable shear stress of τ_{DESIGN} = 86 N/mm². Also, the stiffness requirement for torsion is *12,000 Nm/º* as measured between the axles (*L=2790 mm*).

a) Compute the required thickness to meet the twist ditch strength requirement.

b) Compute the required thickness to meet the torsional stiffness requirement.

c) Which is the dominant requirement?

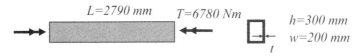

$L=2790\ mm$ $T=6780\ Nm$ $h=300\ mm$
$w=200\ mm$
t

Exercise 5.3

5.4 Van box model for torsion

a) Determine the shear loads in each panel when $F=8000\ N$.

b) What is the torsional stiffness if all panels are perfectly flat steel panels *1 mm* thick?

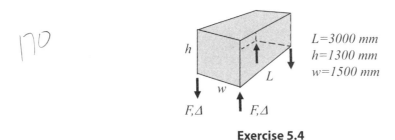

$L=3000\ mm$
$h=1300\ mm$
$w=1500\ mm$

Exercise 5.4

5.5 Torsional stiffness of van with crown roof panel

The van in Exercise 5.4 now has a roof with *20-mm* crown height. Determine the van torsional stiffness with this crown panel.

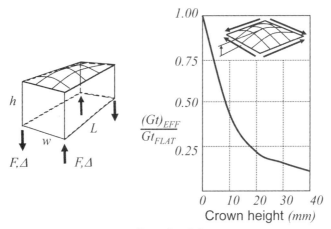

$\dfrac{(Gt)_{EFF}}{Gt_{FLAT}}$

Crown height *(mm)*

Exercise 5.5

5.6 Torsional stiffness of van with adhesive-bonded windshield

The windshield of the van in Exercise 5.4 consists of the entire front panel. The glass is bonded with an adhesive having shear modulus $G_a = 0.69 \ N/mm^2$. The adhesive thickness is $t_a = 6 \ mm$, and the width $w_a = 12 \ mm$. Assume the glass is rigid compared to the flexibility of the adhesive bond.

a) What is the effective shear rigidity $(Gt)_{EFF}$ for the windshield/adhesive system?

b) Determine the van torsional stiffness with the windshield (the roof is again a flat panel).

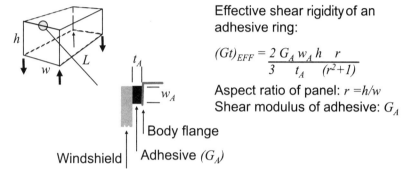

Effective shear rigidity of an adhesive ring:

$$(Gt)_{EFF} = \frac{2}{3} \frac{G_A w_A}{t_A} h \frac{r}{(r^2+1)}$$

Aspect ratio of panel: $r = h/w$
Shear modulus of adhesive: G_A

Body flange

Windshield | Adhesive (G_A)

Exercise 5.6

5.7 Torsional stiffness of van with rear hatch frame stiffness

A test is done on the rear hatch opening of the van in Exercise 5.4. The test results are shown in the illustration. The hatch door does not contribute to torsional stiffness and can be neglected.

a) What is the effective shear rigidity $(Gt)_{EFF}$ for the hatch frame?

b) Determine the van torsional stiffness with the flexible hatch opening and no hatch door (all panels are again flat panel steel panels).

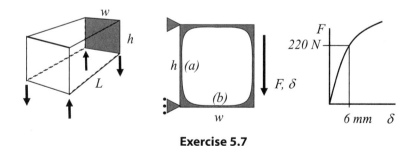

Exercise 5.7

5.8 Torsional stiffness of van with side frame

The right and left side frames of the van now consist of four steel pillars, $40 \ mm \times 40 \ mm \times 1.8 \ mm$ thick, and rigid roof and rocker beams. All other panels are as in Exercise 5.4. The pillars form three door openings. The doors in this case do not contribute to torsional stiffness and can be neglected. Consider the side frame to be isolated, restrained at the bottom, and loaded by a shear load, Q, at the top. The resulting horizontal deflection is Δ.

Because the roof beam and rocker are rigid, each pillar acts as two cantilever beams of length $h/2$, as shown. Each cantilever reacts a load, $Q/4$, and deflects by an amount $\Delta/2$.

a) What is the shear stiffness, Q/Δ, of each side frame?

b) What is the effective shear rigidity, $(Gt)_{EFF}$ of each side frame?

c) Determine the van torsional stiffness with a right and left side frame replacing the flat steel sides of Exercise 5.4 (all other panels are flat panels).

d) Write a formula for the maximum bending stress in each pillar as a function of Q, the applied shear force. (Use $\sigma_{MAX}=cM_{MAX}/I$ on the cantilever illustrated.)

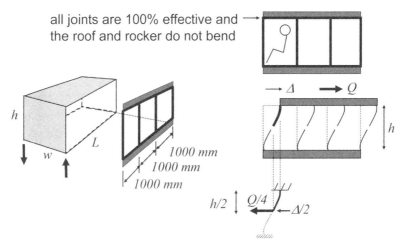

all joints are 100% effective and the roof and rocker do not bend

Exercise 5.8

5.9 Torsional stiffness of van with side frame

The side frame of the van in Exercise 5.8 is modified from the original design to the alternative design shown.

a) Using beam equations, what is the effective shear rigidity, $(Gt)_{EFF}$ of the modified frame (all joints are *100%* effective)?

b) Determine the van torsional stiffness with the alternative side frame (all other panels are again flat panels).

c) Compare the results of this exercise with those of Exercise 5.8.

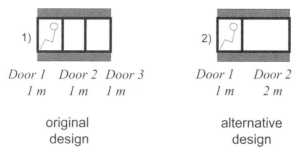

1)					2)		
Door 1	Door 2	Door 3			Door 1		Door 2
1 m	1 m	1 m			1 m		2 m

original
design

alternative
design

Exercise 5.9

5.10 Side frame deformation with torsional load

(Requires access to finite element analysis)

For the side frame of Exercise 4.7:

a) Compute $(Gt)_{EFF}$.

b) Say the value for $(Gt)_{EFF}$ computed in part a is too low compared to the requirement, which beam would you alter first and why?

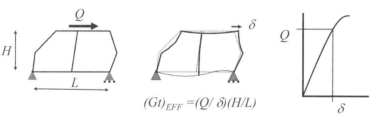

$$(Gt)_{EFF} = (Q/\delta)(H/L)$$

Exercise 5.10

5.11 Torsion stiffness sensitivity

(Requires access to finite element analysis)

Investigate the effect on body torsion stiffness of A pillar angle, tumble home, and joint rate as in Exercise 4.8.

5.12 Frame as a torsional member

A cross frame will be used for the main torsional load bearing member. The maximum section size for the members shown is *50 mm* wide and *100 mm* high. The torsional requirement is *680 KNm/rad*. Determine the frame rail thickness, t, to meet the stiffness requirement and the requirement that the section will just begin to buckle under a maximum torque of *8.0 KNm*. (The front and rear cross members shown are very flexible: they apply no bending moments to the frame rails.)

417

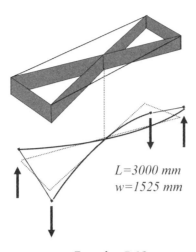

$L=3000\ mm$
$w=1525\ mm$

Exercise 5.12

5.13 Torsion of sedan

The sedan cab shown is loaded by a torque acting on the dash panel, and a reaction torque acting on the rear panel, as shown. For the dimensions in mm given:

a) Use moment equilibrium for each panel to relate unknown edge loads. This set of seven equations can be expressed in matrix form as **AQ=T,** where **A** is a square matrix of coefficients, **Q** is a column matrix of the edge loads, and **T** is a column matrix of applied torques.

b) For $T=12.7x10^6\ Nmm$, solve the equations from part a for the internal loads, Qi, $i=1\ to\ 7$.

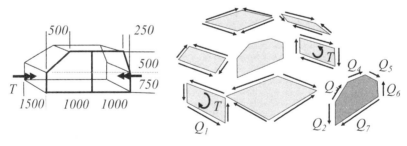

Exercise 5.13

5.14 Flow of torsion strength loads

A structure consists of a central thin-walled backbone beam and shear panels, which can only react loads within their plane. We are interested in the torsion performance of the structure. The car body is supported at the two rear suspension points and loaded with a twist ditch torque applied by a couple (Fw) to the front suspension points.

a. Solve for the reaction forces, F, based on applied twist ditch torque.

418

b. Using free-body diagrams for each shear panel and bar elements (labeled A through G), determine numerical values for the loads in the illustration (only the internal loads shown are generated by the structure).

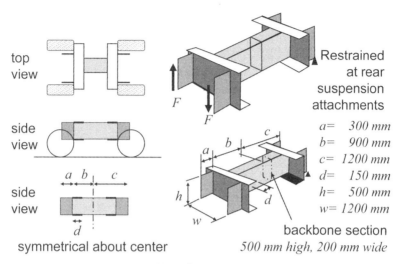

Exercise 5.14a

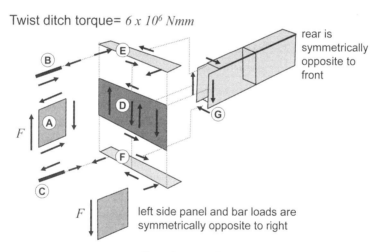

Exercise 5.14b

5.15 Shear loads on side frame

Under a torque applied to the dash panel of the sedan cabin, each shear-resistant surface of the cabin must be in static equilibrium. Equation 5.8 is a matrix equation which summarizes these equilibrium equations, the last being static equilibrium for the side frame. The illustration shows a side frame. Derive the expression for static equilibrium for the side frame under the shear loads shown. It is helpful to sum moments about point 0, shown.

$$(h_o - h_1)Q_4 + \frac{\left[L_F(h_o - h_2) + L_2(h_2 - h_1)\right]}{L_B} Q_5 + -L_F Q_6 + h_1 Q_7 = 0$$

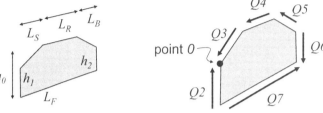

side frame dimensions   side frame shear loads

Exercise 5.15

5.16 Shear rigidity of an adhesively bonded windshield

Derive an expression for the shear rigidity, $(Gt)_{EFF}$, of an adhesively bonded windshield. Parameters are shown in the illustration.

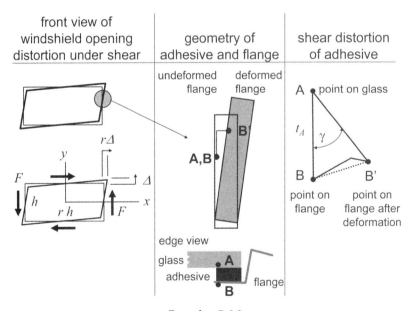

Exercise 5.16

6. Design for Crashworthiness

6.1 Front barrier requirements

A *1000-Kg* car impacts a rigid barrier at V_0=48 km/h. It is desired that the maximum deceleration level be a_{MAX}=20 g. The anticipated crush efficiency is η=0.80. Assume fully plastic behavior.

a) What is the required crushable space, Δ

b) What is the mean crush force which must be generated by the vehicle, F_{AVG}?

initial conditions   after impact

Exercise 6.1

6.2 Sizing the front mid rail for crumpling

For the vehicle in Exercise 6.1, assume that the two midrails will provide 50% of the required average crumpling force, F_{AVG}. The body will be tested statically for this force level. The material is steel with yield $\sigma_Y=207\ N/mm^2$.

a) Determine the size of the square section required to generate $0.25\ F_{AVG}$ if the width to thickness ratio $b/t=65$. What is the mass of the two rails?

b) A hexagonal section is being considered with the same b/t ratio. What is the resulting mass?

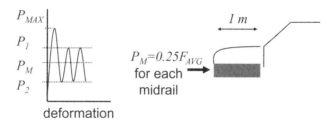

Exercise 6.2

6.3 Sizing the rear mid rail to react crumpling loads

For the vehicle in Exercise 6.1, the two midrails provide 50% of the average force required. The body will be tested statically for this force level.

a) Determine the maximum load for the square section P_{MAX} from Exercise 6.2a.

b) What moment is needed at the toe pan to react the load in part a?

c) Determine the required section thickness at the toe pan so the load of part a can be reacted if $\sigma_Y=220\ N/mm^2$. The section is rectangular with height=40 mm and width=70 mm.

d) Crush initiators are now added to the front part of the crumple zone, which lower the maximum load from P_{MAX} to $P_{1'}$ the accordion buckling load. Recalculate your answer to parts b and c with this lower load.

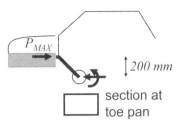

P_{MAX}

200 mm

section at
toe pan

Exercise 6.3

6.4 Reducing pitch upon impact

It is desired to avoid pitching upon impact. Let F_L be the average midrail force calculated in Exercise 6.2a. Assume that the upper rail force, F_{UP}, is the only other force to balance the pitching of the vehicle (i.e. all other forces have a net zero moment about the center of gravity (CG).

a) For $h=500\ mm$, $h_L=400\ mm$, $h_{UP}=900\ mm$, and F_L is the value from Exercise 6.2a, what is the required average crush load for each upper rail, F_{UP}?

b) What is the required upper rail square section size to meet the crush load from part a if $b/t = 80$ and $\sigma_{MAX}=207\ N/mm^2$

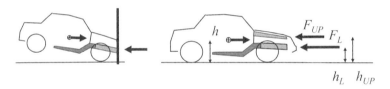

Exercise 6.4

6.5 Load capacity of a beam with plastic hinges

The instrument panel beam must react impact loads in a frontal collision. The torso must be decelerated by the air bag, and inertial loads are applied in the forward direction, as shown. Find an expression for peak force, F, using a limit analysis model with three plastic hinges and two rigid links, as shown. The plastic moment for all plastic hinges is given as M_p. Use small angle assumptions.

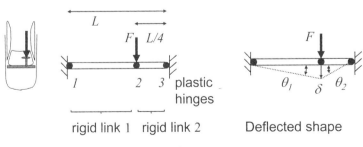

Exercise 6.5

6.6 Plastic moment of a section

The thin-walled section for the beam of Exercise 6.5 is shown. For limit analysis, the effective section and the stress distribution are shown. The compression cap has been fully buckled and has been eliminated; the remainder of the section is at yield. Take $w=120\ mm$, $t=2\ mm$, also $\sigma_Y=207\ N/mm^2$.

a) Find the neutral axis position, X_{NA}, of the effective section so that $A_U = A_L$. (use $t<<w$ assumptions).

b) Find the center of gravity position of the upper area, X_U, and lower area X_L.

c) Find the plastic moment: $M_p = (\sigma_Y(A_U X_U + A_L X_L)$.

d) Use this value of M_p to find the limit load in Exercise 6.5 using the results above and the beam length, $L=1500\ mm$.

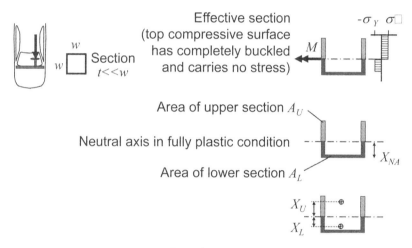

Exercise 6.6

6.7 Limit load analysis for upper load path in barrier

The upper rail load is reacted by the side frame members as shown. Using limit analysis, write an expression for F_p. The dimensions are shown and the plastic joint capacities are M_{p1}, M_{p2} . . . .

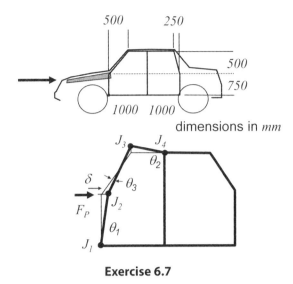

Exercise 6.7

6.8 Limit load analysis for upper load path in barrier

Using the sections given, and assuming fully plastic behavior with the compression cap for each section eliminated, compute the limit load, F_p, for Exercise 6.7. Use mild steel properties.

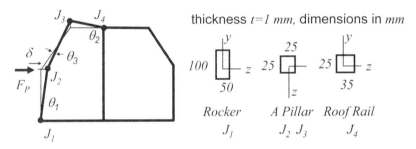

Exercise 6.8

6.9 Limit load analysis: Roof header

The illustration shows the result of an impact between a vehicle traveling to the left at freeway speeds with a wheel that had become detached from a vehicle traveling in the opposite direction. The originally straight roof header is modeled as three rigid links and four plastic hinges with $M_p=20x10^5$ Nmm for all joints. The deformed geometry is shown. Note that the link between joints 2 and 3 remains horizontal, and all deformation occurs in the plane illustrated. Determine the peak load, F_p, using limit load analysis. Use small deflection and angle assumptions.

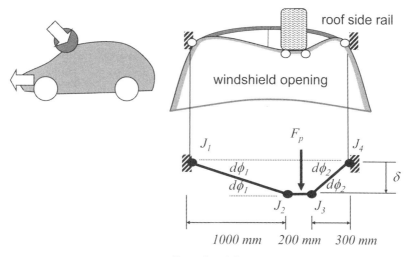

Exercise 6.9

6.10 Side impact

A change to the underbody structure is proposed to reduce injury in the standard side impact test shown in Figure 6.44. The current underbody crush capacity is F_2=150,000 N. It is proposed to reinforce this structure to achieve a capacity of F_2=200,000 N.

a) For both design proposals, use Equations 6.8 to 6.14 to construct velocity-time histories for the struck vehicle, the moving barrier, and the occupant similar to that shown in Figure 6.56. The parameters are:

M_1 = Barrier mass (*1365 kg*)

M_2 = Vehicle mass (*1590 kg*)

V_0 = Lateral impact speed (*13.25 m/sec*)

Δ = Crushable space within door (*175 mm*)

Δ_0 = Space between door inner panel and occupant shoulder (*125 mm*)

b) Assuming injury is directly proportional to the change in occupant velocity during the impact, V_{TF}, compute the percent reduction in injury with this change.

c) Assuming injury is directly proportional to the occupant acceleration during the impact, a_{OCC}, compute the percent reduction in injury with this change.

7. Design for Vibration

7.1 Acoustic cavity modes in the automobile

A new sedan is being designed with cabin length of 2.5 m, as shown in the illustration. All boundaries of the air cavity are rigid except the flat panel at the rear. When the panel resonance coincides with the cavity resonance, structure-borne noise can become loud. The rear close out dimensions are *a=0.6 m, b=1 m, and t=0.9 mm*. The panel is steel with E=207,000 N/mm²; density, ρ=7.83 x10⁻⁶ kg/mm³.

a) What are the first two resonances of the cavity, assuming rigid walls on all sides?

b) Calculate the flat panel resonances in the range $0 < f < 60\ Hz$.

c) Create a mode map of the panel and cavity resonances. Include zones to be avoided around each cavity resonance $\pm 10\ Hz$.

d) The panel can be crown up to $25\ mm$, but at least $12\ mm$ crown is required for manufacturability. What is the minimum crown, δ, in the range $12\ mm < \delta < 25\ mm$ which decouples the panel frequencies and two acoustic cavities' resonance zones?

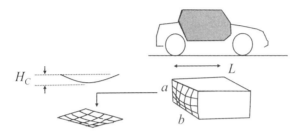

Exercise 7.1

7.2 Steering column shake

Often the steering column is supported by an instrument panel beam extending across the car from hinge pillar to hinge pillar. The supports at the beam ends can be taken as simply supported. We are interested in the natural frequency of the vertical vibration of the steering column when mounted on the beam. The steering column behaves as a rigid mass, m. The system is modeled as a massless thin-walled steel beam with a point mass attached (i.e. no twisting). Find the section size, b, t, so that $f_n \geq 35\ Hz$ and $(b/t) \geq 60$, where $m=10\ Kg$ and $L=1.5\ m$.

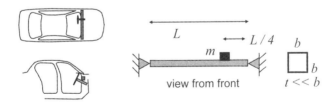

Exercise 7.2

7.3 Modal model

A trimmed car body of mass $m=500\ kg$ is vibration tested as shown. A sinusoidal force of amplitude, F_1, is applied at the engine mount, and acceleration amplitude is measured at the input, Y_1, and also at the steering column mount, Y_2. The log-log plots of the frequency response at each point are shown (acceleration amplitude / input force amplitude).

a) Create a modal model of body of the form:

$$\frac{Y_i}{F_1} = \frac{1}{m} \frac{\varphi_i \varphi_1}{\left(1 - \dfrac{\omega_N^2}{\omega^2}\right) + j\dfrac{2\zeta\omega_N}{\omega}} + \frac{1}{M_i}, \; i = 1,2$$

Where:

ω_n = natural frequency for the mode we are fitting

m = body mass

ζ = damping factor for mode 1 (ζ is found from the half power points, where $\zeta = 2\,\Delta\omega/\omega$)

M_1 = effective rigid body mass for point 1 (engine mount location)

M_2 = effective rigid body mass for point 2 (steering column mount location)

ϕ_1 = modal influence coefficient for point 1 (engine mount location)

ϕ_2 = modal influence coefficient for point 2 (steering column mount location)

$j = \sqrt{-1}$

unknowns are ζ, ϕ_1, ϕ_2, m, M_1, M_2. Take m= actual mass of body, $\zeta=0.03$ a typical value, and the remaining four variables may be determined from the frequency responses. These four are the amplitude at resonance and the amplitude as frequency approaches zero for each point.

b) Verify the answer to part a by plotting Y_1/F_1 and Y_2/F_1 versus frequency on both log-log and linear axes; compare to the given data plots.

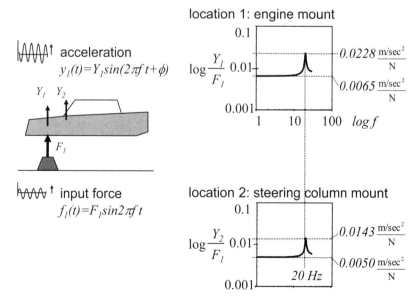

Exercise 7.3

7.4 Effect of added mass

a) An instrument panel is added to the body in Exercise 7.3. It does not add stiffness to the body but does add a mass of *45 kg*. The instrument panel is attached rigidly at the steering column mount. Provide an estimate of the new natural frequency of the body.

b) Two locations for a *22-kg* battery are being considered: next to the engine mount, or at the cowl (near the steering column mount). Using the modal model from Exercise 7.3 (and without the instrument panel mass) compute the new natural frequencies for each position. If maximizing the natural frequency is preferable, which position is best?

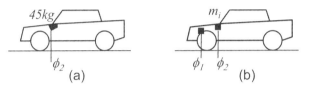

Exercise 7.4

7.5 Powertrain mounting

A powertrain is installed in the body of Exercise 7.3. To save cost, the powertrain is mounted directly to the body (no compliant engine mounts). Consider only the vertical motions of the powertrain and assume the powertrain has negligible mass compared to the body.

The four-cylinder engine generates a vertical unbalance force caused by the reciprocating pistons equal to:

$$f(t) = 4mr\omega^2 \frac{r}{L}\cos(2\omega t)$$

where:
  m=1-kg reciprocating mass

  r=50-mm crank radius

  L=120-mm connecting rod length

  ω=engine speed (*rad/sec*)

a) Plot the excitation force versus engine speed on log-log axes for *500 RPM<ω<900 RPM*.

b) Plot the steering column mount acceleration amplitude, Y_2, vs. engine RPM.

c) For the range *500 RPM<ω<900 RPM*, are the vibration levels calculated in part b acceptable? Use the human response to vertical vibration shown.

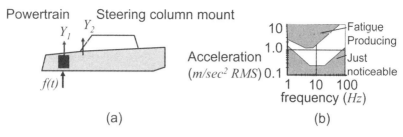

Powertrain   Steering column mount

Y_1 Y_2

Acceleration
(m/sec^2 RMS)

10
1.0
0.1

1   10   100
frequency (Hz)

Fatigue
Producing
Just
noticeable

$f(t)$

(a)   (b)

Exercise 7.5

7.6 Vehicle system vibration analysis

The bending stiffness requirement for the new vehicle needs to be determined. This stiffness requirement will be based on the desired first bending frequency of the vehicle. Using a mode map of the most important vibration sources and paths, we will place the body resonance away from major vibration sources and subsystem resonances. Data for this vehicle are shown below:

Vehicle	Engine: One-cylinder engine	
Overall length= *370 mm*	Maximum engine speed= *900 RPM*	
Wheelbase= *2700 mm*	Engine mount vertical spring rate= *1210 N/mm*	
Effective cabin length= *3000 mm*	Powertrain mass= *100 kg*	
Mass of stiffly mounted components= 250 kg	Crank shaft offset= *70 mm*	
	Connecting rod length= *100 mm*	
	Reciprocating mass= *1.5 kg*	
Front suspension	**Steering column**	**Floor panel size (steel)**
Maximum vehicle speed= *100 km/hr*	Column mount stiffness= *250 N/mm*	Density ρ=7.85×10³ kg/m³)
Tire spring rate= *170 N/mm*	Effective column mass= 5 kg	Width= *400 mm*
Ride rate= *18 N/mm*		Length= *300 mm*
Unsprung mass= *48.75 kg*		Thickness= *1.0 mm*
Tire rolling radius= *300 mm*		
Wheel radius= *225 mm*		
Unbalance mass at wheel radius= *0.03 kg*		

In answering parts a, b, and c, consider only single-degree-of-freedom models for the vibration systems. Note: watch units in these calculations; in general use *kg, m, N, ω (rad/sec)*, then convert to *Hz (cycle/sec)*.

a) Suspension: Calculate the vertical wheel unbalance force as a function of vehicle speed. Calculate the wheel hop resonance frequency. Sketch the force into the body through the ride spring under the wheel unbalance forcing function as a function of frequency.

b) Engine: Calculate the vertical engine unbalance force as a function of engine speed. Calculate the engine bounce frequency. Sketch the force into the body through the engine mounts under the engine unbalance forcing function as a function of frequency.

c) Mounted steering column: Calculate the steering column resonance frequency.

d) Panel vibration: Calculate the floor panel resonance frequencies below 150 Hz. Calculate the acoustic cavity resonances below 150 Hz.

e) Plot the data from part a, b, c, and d on the mode map shown. Show natural frequencies with a ±2½ Hz box.

f) Based on the mode map, identify a target bending frequency, fn, for the vehicle which avoids major vibration inputs (the answer should be a whole number value between *25 Hz and 35 Hz*).

g) Based on your answer to part f, identify the bending stiffness requirement for the body.

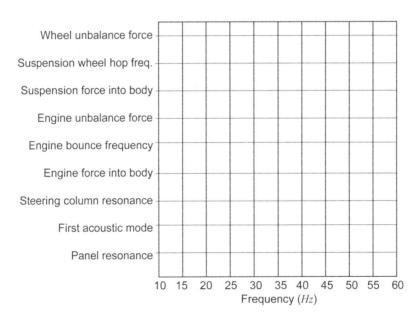

Exercise 7.6

8. Body Layout Exercise

A preliminary layout for a new vehicle is shown. The initial parameters are:

Powertrain mass (P)	*180 kg*
Fuel tank and fuel mass (F)	*45 kg*
Passengers mass	*each 70 kg*
Radiator mass (R)	*10 kg*
Suspensions mass (4)	*total 90 kg*
Cargo mass (C)	*45 kg*
Battery mass (B)	*20 kg*
Mass uniformly distributed along vehicle length	*total 600 kg*
Gross vehicle mass	*1200 kg*
Center of gravity height (*mm*)	750
Suspension travel: Up (*mm*)	125
Suspension travel: Down (*mm*)	125
Turn angle	±35°

Design a structure which meets the constraints shown in the illustrations.

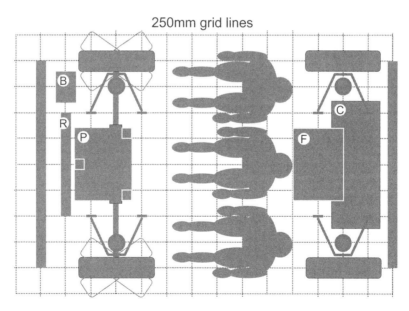

250mm grid lines

Exercise 8a

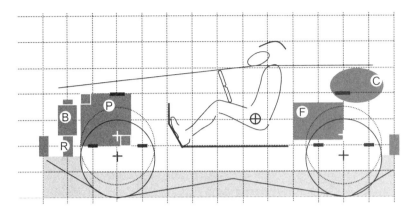

Exercise 8b

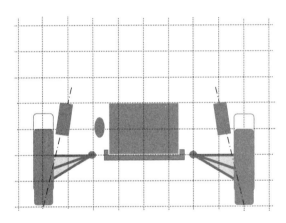

Exercise 8c

8.1 Bending and torsion strength requirements

a) Compute bending moment diagram for *1 g, 2 g* loads. Supports are at the front and rear suspensions.

b) Compute bending moment diagram for front towing, and rear towing with the appropriate restraints.

c) Determine the value for a single load, applied at the H point which is an envelope to the 2g diagram from part a.

d) Determine the twist ditch torque.

8.2 Section sizing for overall bending

e) Say a *50-mm*-wide × *100-mm*-high steel frame section can be packaged on either side of the vehicle, and that this frame must bear all load. Using the bending load from part c, what is the required section thickness considering only yield stress? What is buckling stress for the top cap?

f) Using the results for section size and thickness from part e, will this section meet the twist ditch torque from part d? Assume the cross members are very flexible and the frame rails are subject only to a torque (no differential bending).

8.3 Frame layout

g) Lay out a frame topology using the section from parts e and f which meets the package constraints illustrated and is positioned to react fore-aft bumper loads and suspension lower control arm loads. Show the topology in both the side and plan view.

8.4 Crashworthiness requirements

h) Determine the required crush space, Δ. The structure will be 80% efficient, and the allowable maximum deceleration is 20 g. The impact speed is 48 km/h.

i) Is the space available to package the required crush space from part h? Consider the powertrain and bumper to be rigid. The radiator can collapse to 50% of its initial width. The end of the crush space is the driver's front of foot.

j) Compute the average total crush force in a 48-km/h frontal impact with a rigid barrier.

k) Each motor compartment side rail must generate 25% of the total force from part j. A 100-mm-square section size is to be used. What is the required thickness for the steel section?

9. Material Selection and Mass Estimation

9.1 A pillar material

An A pillar performs several structural functions: 1) contributes to body stiffness as a stiff beam, 2) reacts roof crush loads as a strong beam with failure by plate buckling, 3) reacts front barrier loads by providing a high plastic hinge capacity. The size of the section is constrained by the package environment (only thickness is free to choose).

a) If the objective is minimum material mass, use the appropriate material index to rank the materials shown in Figure 9.4 for each of the three cases above.

b) Repeat part a if the objective is to minimize material cost.

9.2 Derivation of material index

Derive the material index for a light, strong beam under moment loading when the size of the section is constrained by the package environment (only thickness is free to choose), and:

a) Failure is by yield.

b) Failure is by plate buckling of the compression cap.

9.3 B pillar steel grade substitution

The B pillar shown in Figure 9.3 is 90 mm wide and 50 mm deep, with thickness 1.5 mm. The material is steel with σ_y=200 N/mm². The section is loaded by a required moment, M_0, and when loaded by M_0 has been sized to just reach plate buckling at the yield stress. Consider only buckling of the compression cap, not the side walls, and consider the section to be a rectangle (no flanges).

a) Calculate M_0 for this case.

b) To reduce mass, a high-strength steel, $\sigma_Y = 600 \, N/mm^2$, will be substituted. Your manager reasons (erroneously) that because the yield stress has increased to three times the original, the thickness can be reduced to one-third of the original. This material is directly substituted, and $t = 0.5 \, mm$. What moment will yield the effective section?

c) What thickness of the high-strength steel section is required to meet the M_0 found in part a?

d) Add features to the section which inhibit buckling and recalculate the thickness to meet M_0.

9.4 Mass estimation

You must determine if each of the two vehicle proposals below will likely meet its fuel economy and acceleration performance targets. A new family of four-cylinder high-performance engines will be used, having the performance chart illustrated.

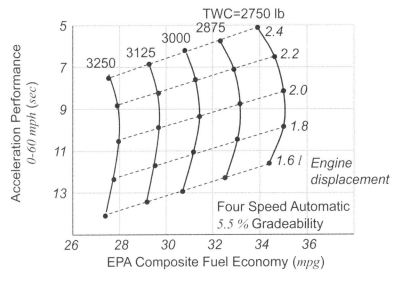

Source: Course notes, U of MI, AUTO501, Vehicle System Integration, M. Anderson

Exercise 9.4

	(A)	(B)
Target market	Generation X urban males	Young married professionals with no children
vehicle type	Sedan	Sedan
OAL (*mm*)	3710	4100
OAW (*mm*)	1810	1805

number of passengers	5	4
cargo (*kg*)	*100*	*100*
Target fuel economy (*mpg*)	28	35
Target 0–60 *mph* acceleration performance (*sec*)	6	8

a) The performance target and fuel economy target for each vehicle must be met or exceeded. Using the engine performance chart, what is the TWC target for each vehicle?

b) Estimate the mass for each vehicle using Equations 9.14 and 15 with the data provided in Chapter 9.

c) Based on the estimated curb mass, calculate the estimated TWC.

d) Refer back to the engine performance chart. Is each car feasible at the current market specifications? Also, recommend the engine displacement for each vehicle which best meets target fuel economy and performance.

e) An increase in trim of *30 kg* is being planned for vehicle B. Use mass compounding to determine how this will impact the curb mass and test weight class for this vehicle. All vehicle subsystems may be resized. Use the influence coefficients provided in the text.

9.5 Ratio method of mass compounding

A single-passenger, one-of-a-kind advanced vehicle concept is now moving to the engineering factory, and a mass budget must be provided. The subsystem masses in *kg* for the experimental vehicle are shown below:

Body Non-structure	*17.00*
Body Structure	*102.00*
Front Suspension	*15.00*
Rear Suspension	*22.00*
Steering	*20.00*
Braking	*18.00*
Powertrain	*96.00*
Fuel & Exhaust	*30.00*
Tires & Wheels	*41.00*
Air conditioning	*14.00*
Electrical	*21.00*
Bumpers	*9.00*
Closures	*30.00*
Cargo	*50*
Passenger	*70*

Testing has shown that improvements to the body structure are needed to meet future side-impact performance requirements. Structural analysis shows that these side impact improvements will increase body structure mass by *15 kg*.

a) Use the ratio method to determine subsystem influence coefficients. All subsystems may be resized with the exception of body-nonstructural, air conditioning, electrical, and closures.

b) Use mass compounding to determine the final curb mass after resizing for the *15-kg* mass increase.

Appendix B
Nomenclature

A	cross section area
a	section dimension, acceleration, plate length
b	plate width
c	largest distance from neutral axis, speed of sound
C_w	warping constant
D	plate stiffness
e	efficiency, strain energy
E	modulus of elasticity, energy
f	joint efficiency, frequency (Hz)
f_x	compressive stress
F	force
g	acceleration due to gravity
G	shear modulus
$(Gt)_{EFF}$	effective shear rigidity of a shear-resistant member
h, H	height
i	current
I	section moment of inertia
j	$\sqrt{-1}$
J, J_{EFF}	torsion constant
k	buckling constant, stiffness, factor in warping equations
K	stiffness
K	torsional stiffness
l, L	length
m	mass, integer 1, 2, 3 . . .
M	bending moment, mass
n	integer 1, 2, 3 . . .
p	weld pitch
P	force, power, perimeter
q	shear flow, unit normal load
Q	shear force
r	aspect ratio, radius
R	reaction force, radius, resistance
s	perimeter coordinate
S	perimeter, developed length, force
t	thickness, vehicle track width, time
T	torque
u	principle coordinate
U	stored energy

v	principle coordinate
V	shear force, velocity
w	plate effective width, plate out-of-plane deflection
W	weight, width, work
x, y, z	coordinate values
Y, Z	vibration amplitude
z	distance from neutral axis
α	angle
β	angle of twist per unit length
δ	linear deflection
Δ	crush space, linear deflection
ε	direct strain
ϕ	angle, modal model influence coefficient
γ	shear strain
η	structural damping coefficient, crush efficiency, lateral deflection, width-to-thickness ratio (b/t)
λ	acoustic standing wave length
μ	Poisson's ratio
θ	angle of twist
ρ	radius of curvature
σ	direct stress
σ_{CR}	critical plate buckling stress
σ_Y	yield stress
σ_S	maximum stress in plate
τ	shear stress
ω, Ω	frequency (rad/sec)
ω_n	natural frequency
ζ	viscous damping coefficient
ψ	general factor grouping

Appendix C
English & Metric Units & Typical Values for Key Parameters

Metric-English Units

Length	*25.4 mm = 1 in*
Mass	*1 kg = 2.2 lb*
Force	*4.448 N = 1 lb*
Volume	*1 L = 0.2642 gal*
Velocity	*88 ft/sec = 60 mph, 1.609 km/hr = 1 mph,*
	1 km/hr = 0.278 m/sec
Acceleration	*0.3048 m/sec^2 = 1 ft/sec^2*
Acceleration of gravity (g)	*9.8 m /sec^2= 32.2 ft /sec^2*
Stress, Modulus of elasticity	*145 psi = 1 MPa*
Stiffness	*0.1751 N/mm = 1 lb/in*
Density	*2.768 x10^4 kg/m^3 = 1 lb/in^3*
Fuel Economy	*0.4251 km/L = 1 mpg*
Fuel Consumption	*2.352 L/km=1 gal/mi*

Fundamental Metric Units

Length	*1000 mm = 1 m, 1 km = 1000 m*
Force	*1 N = 1 kg (m/sec^2)*
Volume	*1000 L = 1 m^3*
Stress, Modulus of elasticity	*Pascal = 1 N/m^2, 1 MPa = 1 N/mm^2*

Values for steel used in text examples

Modulus of Elasticity (E)	*207 x10^3 MPa (30 x 10^6 psi)*
Yield stress: mild steel (σ_Y)	*207 MPa (30 x 10^3 psi)*
Shear Modulus	*78 x 10^3 MPa (11.3 x 10^6 psi)*

$$G = \frac{E}{2(1+\mu)}$$

Density (ρ)	*7.83 x 10^{-6} kg/mm^3 (0.283 lb /in^3)*
Poisson's Ratio (μ)	*0.3*

Values for aluminum used in text examples

Modulus of Elasticity (E)	*1/3 value for steel*
Yield stress (σ_Y)	*same as steel*
Density (ρ)	*1/3 value for steel*
Poisson's Ratio (μ)	*same as steel*

Index

About the Author

Dr. Donald E. Malen is an adjunct faculty member at the University of Michigan where he teaches graduate level courses in *Automobile Body Structure* and *Product Design*. Prior to this, he was an engineer with General Motors Corporation for 35 years. His background at GM was in automotive body structure design and analysis, and systems engineering. While at GM, he worked on many new vehicle programs, and he has brought this experience to his teaching and writing.

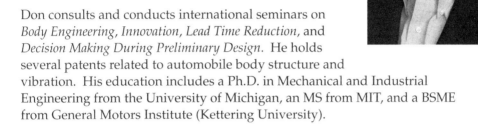

Don consults and conducts international seminars on *Body Engineering*, *Innovation*, *Lead Time Reduction*, and *Decision Making During Preliminary Design*. He holds several patents related to automobile body structure and vibration. His education includes a Ph.D. in Mechanical and Industrial Engineering from the University of Michigan, an MS from MIT, and a BSME from General Motors Institute (Kettering University).

Visit www.donaldemalen.com

Supplementary Resources
for

Fundamentals of Automobile Body Structure Design

CARS software (*Computerized Application and Reference System*) is available at no cost to qualifying students and engineers through the *Steel Market Development Institute*, 2000 Town Center, Suite 320, Southfield, MI 48075. (248.945.4777)

Contact Jody Hall, jhall@steel.org, to obtain a copy.

The software contains *Geometrical Analysis of Sections (GAS)* which estimates buckling and post buckling behavior of automotive thin walled sections. (See Exercises 3.8-3.10.)

Instructor resources are available for teachers who adapt this book as a text for their course. These resources include a sample syllabus, a course calendar with readings and homework assignments, a solutions manual, and sample exams.

Contact Donald E Malen, dmalen@umich.edu, for more information.

Validation is required of instructors, and there is a nominal cost for the instructor packet.